Ciarcia's Circuit Cellar Volume V

Volume V

Ciarcia's Circuit Cellar

by Steve Ciarcia

McGRAW-HILL BOOK COMPANY

New York St. Louis San Francisco Auckland Bogotá Hamburg
Johannesburg London Madrid Mexico Montreal New Delhi Panama
Paris São Paulo Singapore Sydney Tokyo Toronto

Library of Congress Cataloging-in-Publication Data
(Revised for volume 5)

Ciarcia, Steve.
  Ciarcia's Circuit cellar.

  Articles written by the author for a Byte magazine
column, Ciarcia's Circuit cellar, which began in Nov.
1977.
  Vol.   has imprint: New York: McGraw-Hill.
  Includes bibliographies and indexes.
  1. Microcomputers — Collected works. I. Circuit cellar.
II. Title.
TK7888.3 C58   621.3819'5   78-20920
ISBN 0-931-71807-4 (v. 1)

ISBN 0-07-010963-X (v. 2)

ISBN 0-07-010965-6 (v. 3)

ISBN 0-07-010966-4 (v. 4)

ISBN 0-07-010967-2 (v. 5)

Copyright © 1986 by McGraw-Hill, Inc. All rights reserved.
Printed in the United States of America.
Except as permitted under the United States
Copyright Act of 1976, no part of this publication may
be reproduced or distributed in any form or by any
means, or stored in a data base or retrieval system,
without the prior written permission of the publishers.

  34567890   SEM/SEM   89

ISBN 0-07-010967-2

The editors for this book were Stephen G. Guty and Georgia Kornbluth,
and the production supervisor was Sally Fliess.

Printed and bound by Semline, Inc.

*The author of the programs and hardware schematics provided with
this book has carefully reviewed them to ensure their performance in
accordance with the specifications described in the book. Neither the
author nor BYTE Publications Inc., however, make any warranties
whatever concerning the programs or hardware schematics. They assume
no responsibility or liability of any kind for errors in the programs or
hardware schematics or for the consequences of any such errors. The
programs and hardware schematics are the sole property of the author
and have been registered with the United States Copyright Office.*

This book is dedicated to the people on the Circuit Cellar *and* ASK BYTE *staff whose support and help have contributed to making many of these projects possible.*

Contents

Preface	ix
1 Build the RTC-4 Real-Time Controller (July 1983)	1
2 Build a Power-Line Carrier-Current Modem (August 1983)	12
3 Build the Micro D-Cam Solid-State Video Camera, Part 1: The IS32 Optic RAM and the Micro D-Cam Hardware (September 1983)	19
4 Build the Micro D-Cam Solid-State Video Camera, Part 2: Computer Interfaces and Control Software (October 1983)	31
5 Build the H-Com Handicapped Communicator (November 1983)	56
6 Keep Power-Line Pollution Out of Your Computer (December 1983)	71
7 Build the Circuit Cellar Term-Mite ST Smart Terminal, Part 1: Hardware (January 1984)	80
8 Build the Circuit Cellar Term-Mite ST Smart Terminal, Part 2: Programming and Use (February 1984)	92
9 Build a Third-Generation Phonetic Speech Synthesizer (March 1984)	107
10 Build a Scrolling Alphanumeric LED Display (April 1984)	121
11 Trump Card, Part 1: Hardware (May 1984)	138
12 Trump Card, Part 2: Software (June 1984)	153
13 A Musical Telephone Bell (July 1984)	161
14 Build the AC Power Monitor (September 1984)	169
15 An Ultrasonic Ranging System (October 1984)	180
16 The Lis'ner 1000 (November 1984)	192
17 Build the Power I/O System (December 1984)	207
Index	221

Preface

Writing *Circuit Cellar* articles is very different from other literary endeavors. Unlike reclusive fiction writers who take refuge on a mountain top or white-water-raft down the Colorado for inspiration, I hide away in the Circuit Cellar surrounded by electronic test equipment and trade magazines. Novelists who have an inspiration can merely sit at a typewriter and document the ramblings of their minds. In my case, the need to document hardware assembly eliminates cerebral speculation and elicits cold functional performance.

I've often mused about what it would be like on the other side. Because of my engineering background and interests, the kind of fiction that I would most enjoy writing would be science fiction. I've always been fascinated by the concept of time travel and the relationship of mass and time at light speed. Considering the fast cars I like to drive, I may find the answers sooner than I'd like.

Unfortunately, every time I start fantasizing about warp drives, thought-activated control, and other science fiction plots, I get frustrated because they are not real and must still be invented. But as a practical thinker, I do not delude myself trying to invent that for which science has not yet created the raw materials. Newton surely could have accomplished a bit more if he had had a personal computer. All things considered, however, he did pretty well with a regular apple.

Ultimately, I am motivated to accomplish something worthwhile with the tools at hand. We now have the personal computer and many other electronic marvels but still no warp drives or teleportation mechanisms. Rather than fret about what isn't, I prefer to emphasize and demonstrate what is.

Circuit Cellar projects are designed to be built, not just read. They are not speculation. They are reality presented in a how-to format accompanied by an explanation of previous techniques and present benefits. Do you have a fantasy about a robot servant? The *Circuit Cellar* projects describing motor controls, voice synthesis, voice recognition, remote control, and video imaging will help you design what you dream about.

Over the past 8 years I have presented about 80 *Circuit Cellar* projects. There is a continuing evolution of ideas, and the subject matter of the projects has generally followed the trends in the personal computer industry. Early articles, circa 1977, were primarily oriented toward explaining basic computer concepts and input/output configurations. As the industry matured and fewer readers needed "hand-holding," the projects became more sophisticated and higher-performance. Even as the column matures, however, I always remember that there are new people entering the field of computers who need supporting explanation to accompany the projects. I try to keep my designs conservative and avoid esoteric components where possible. As a result, projects that I presented in 1978 are still viable and cost-effective today.

Ciarcia's Circuit Cellar Volume V documents 15 more projects in the continuing evolution of my practical approach to high-performance computing. The projects in this volume fall primarily into catagories of "control" and "real world I/O." Scattered among them are tutorials on subjects as varied as the intricacies of the telephone system, moving display sign technology, and the effects of radio frequency interference.

Within the subject of computer control, there are five projects: a single-chip time-activated programmable controller, a complete AC/DC power I/O control system suitable for industrial or home environmental control applications, an AC power line communication system, a single-board intelligent terminal, and a coprocessor expansion board for the IBM PC. This last project, the Trump Card, is supported by a 5-worker-year software effort and adds 16-bit Z8000 processing capability to the IBM PC.

No control system would be adequate without the ability to acquire data about its surroundings and detect the effects of its control decisions. Suitable data acquisition projects include an AC watt-hour power monitor, an ultrasonic distance-measuring system, a solid-state video image sensor, a voice synthesizer, and a voice recognition system.

Finally, since *Circuit Cellar* projects are meant to be built and are guaranteed to work, companies have put their money behind my soldering iron. All projects are fully supported by me, and many have also been put into commercial production and are available to readers in various configurations. Concern about the functional reality of these projects evaporates with the knowledge that thousands are in use.

Ultimately, your perception may be that these monthly musings come from a frustrated engineer who would rather be writing science fiction. While I may someday surrender to temptation and give Jerry Pournelle a run for his money, today I am satisfied with converting a little of that fiction into science fact a month at a time.

Steve Ciarcia

Ciarcia's Circuit Cellar Volume V

Build the RTC-4 Real-Time Controller

A 4-bit single-chip microcomputer from Texas Instruments comes preprogrammed for timed automatic control.

Bee-beep . . . bee-beep . . . bee-beep. I fumbled for the alarm button on my digital watch and briefly wondered what I had set it for.

"Oh yeah. Time to turn on the recorder."

The audio-cassette recorder in question was set up in one corner of the Circuit Cellar and connected to an FM radio. I had set it up to record a local news program, which is broadcast daily at 5:30 p.m. in my area. Because 5:30 is not usually a convenient time for me to listen to the radio, I often record the program so that I can listen to it later in the evening or the next day while I'm driving someplace.

So it was 5:28. Time to turn on the recorder. The trouble was I was outside working on my satellite dish antenna. As I dashed into the house and down the stairs, I thought, "There's got to be a better way to do this."

Once the recording was safely under way, I began to consider various automatic methods. I could use a mechanical timer, a simple digital alarm clock (with appropriate Circuit Cellar modifications), a BSR X-10 Home Control System timer, or my existing home-security and control system. None of these seemed completely satisfactory. Instead, I decided to look for a more universal solution.

The problem of timed activation pointed out the need for a generalized, cost-effective timer/controller that might, while solving my particular problem, have additional capabilities. As I was evaluating various circuits for real-time controllers, I came across some preprogrammed TMS1000-series 4-bit microcomputers-on-a-chip from Texas Instruments. One of these chips became the essential element in this month's project, the Circuit Cellar RTC-4 real-time controller.

The RTC-4 is a four-channel time-activated device or appliance controller that can be built for less than $100, complete with keypad, display, power supply, and relay outputs. It can be used in the home or laboratory for general time-dependent applications or, as in my case, to solve one particularly nagging problem.

Before jumping into how to build the RTC-4, let's take a look at the TMS1000 family tree and the particular branch of use in our project.

The TMS1000 Family

Texas Instruments (TI) makes a large family of single-chip MOS/LSI (metal-oxide semiconductor/large-scale integration) components called the TMS1000 series, which all contain 4-bit microprocessors. While the approximately 50 members of the family share a common subset of about 40 instructions, they differ in the varying amounts of read-only memory (ROM) and random-access read/write memory (RAM), and varying numbers of I/O- (input/output) control circuits. The family's members also differ in packaging and power requirements, but, in the 28-pin dual-inline package, the basic TMS1000 and TMS1100 are virtually the same with the exception of internal memory capacity. TI intends that the ROM in TMS1000-series components be mask-programmed for specific computing tasks; the devices are not general-purpose microprocessors, as are the familiar Z80, 6502, and 8088.

One major member of the product line is the TMS1100. This chip has 2048 bytes of internal ROM, 128 4-bit words (nybbles) of RAM, 4 input lines, and 19 output lines. (The basic TMS1000 contains less memory.) Figure 1 is a block diagram of its internal structure.

TMS1121C UTC

Somewhere on the 1100 branch of the TMS1000 family tree is an offshoot called the TMS1121C Universal

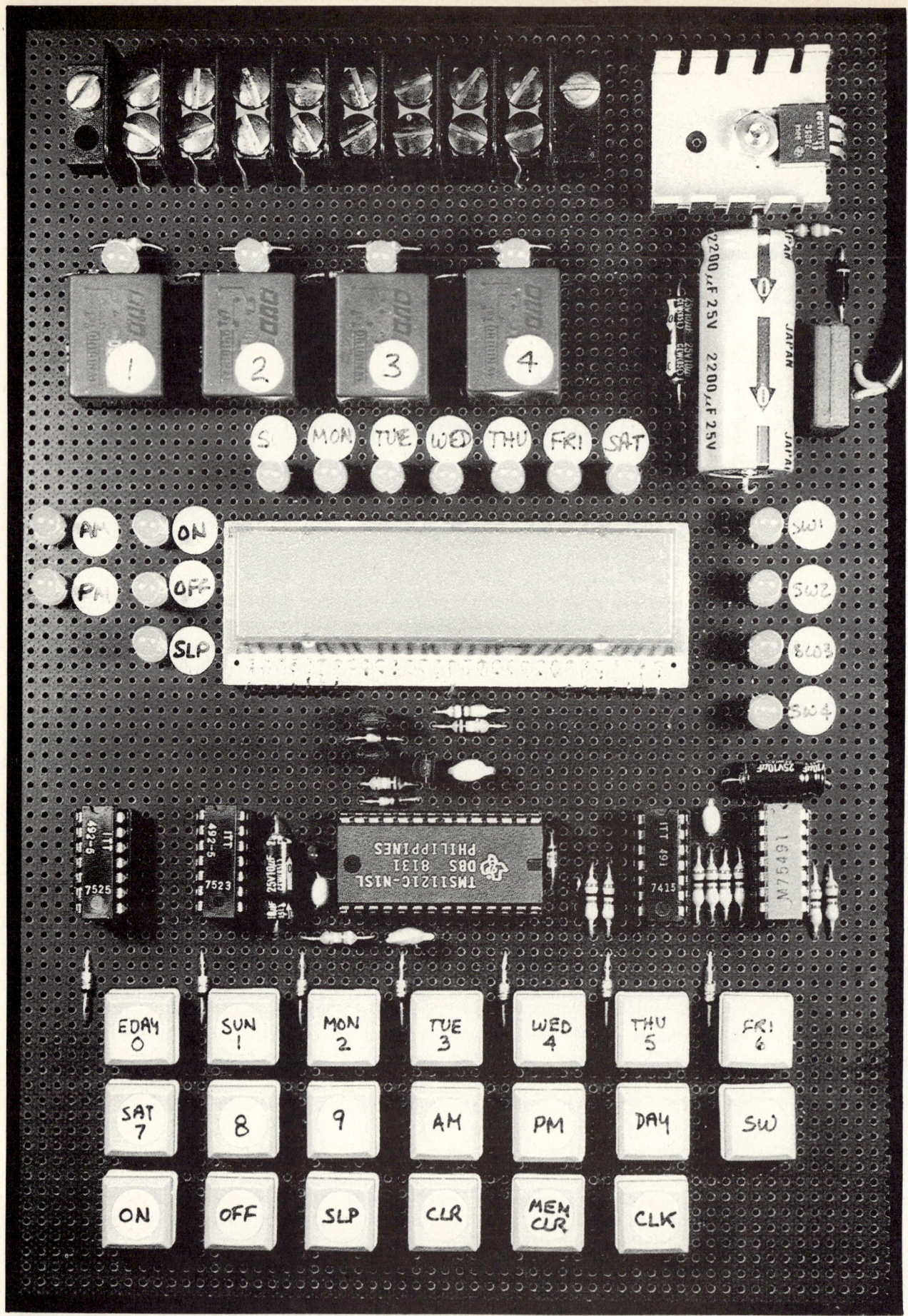

Photo 1: *Prototype of the Circuit Cellar RTC-4 real-time controller, a unit suitable for switching low-current external devices on and off according to immediate commands or stored time-lapse event settings.*

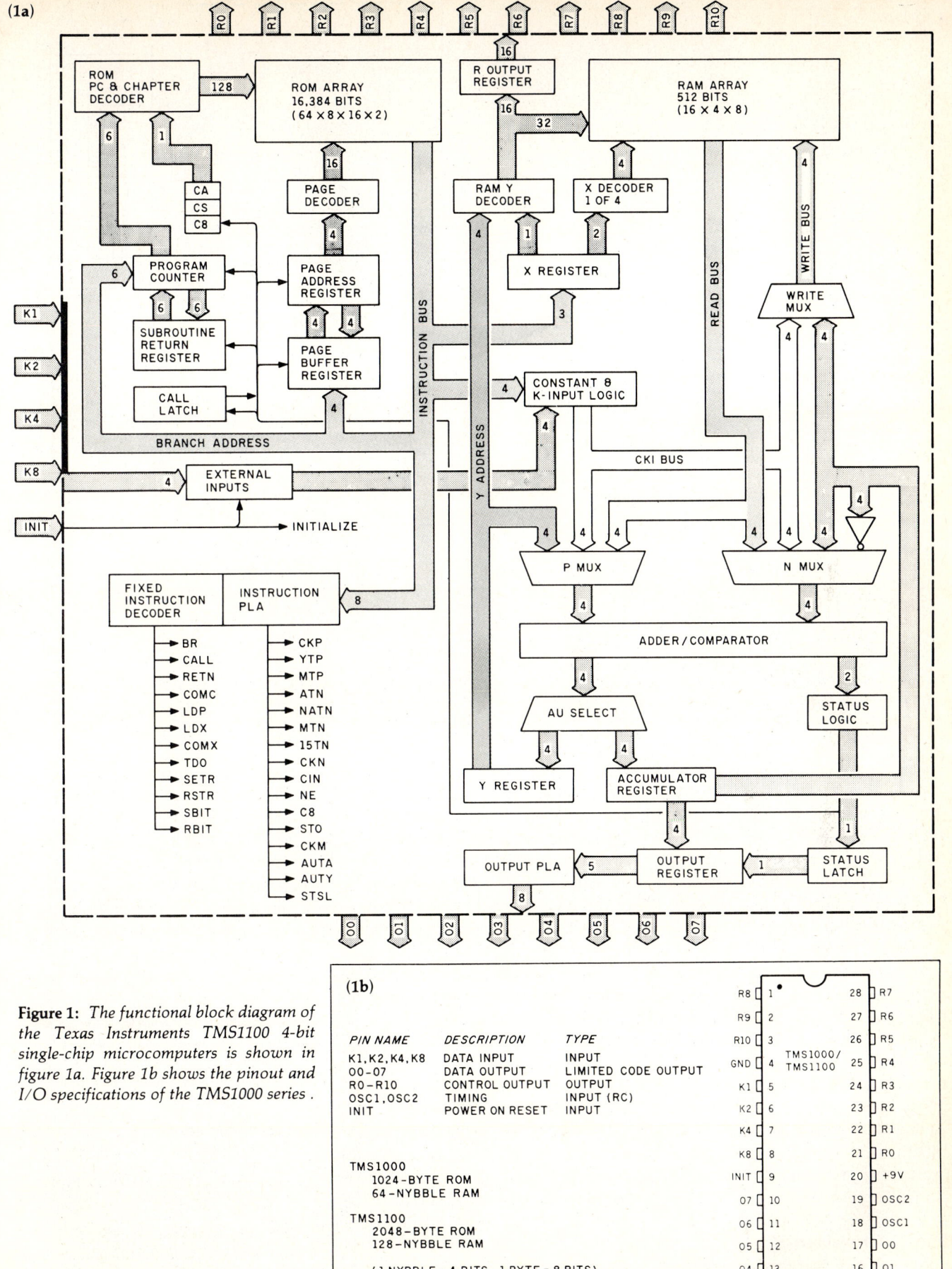

Figure 1: *The functional block diagram of the Texas Instruments TMS1100 4-bit single-chip microcomputers is shown in figure 1a. Figure 1b shows the pinout and I/O specifications of the TMS1000 series.*

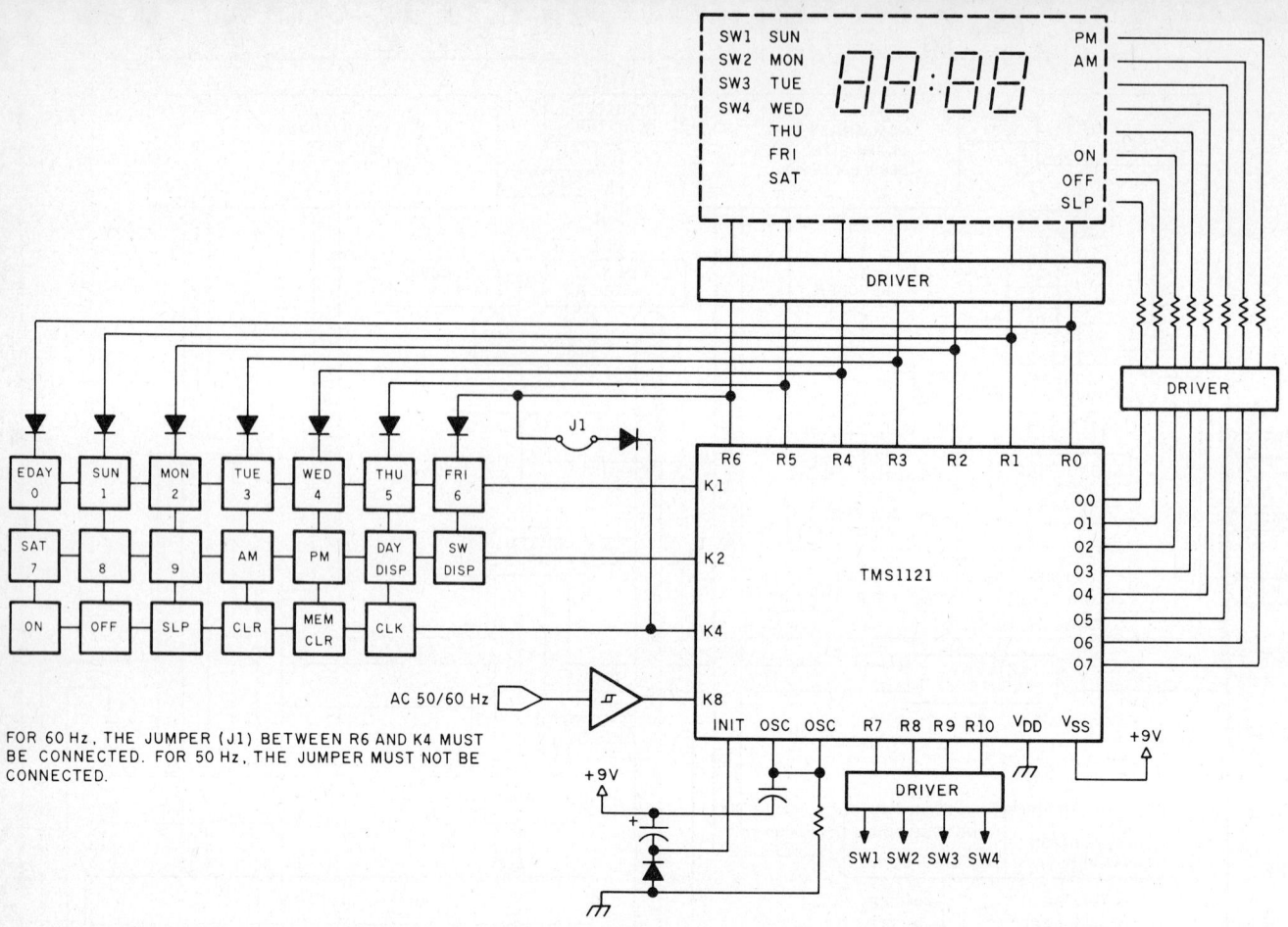

Figure 2: *Block diagram of the Circuit Cellar RTC-4 real-time controller, showing the TMS1121C 4-bit microcomputer and its associated user-interface components.*

Timer Controller. When I found out about the TMS1121C UTC, I realized that it was the perfect basis for the Circuit Cellar RTC-4 real-time controller. It is a variant of the 1100, designed and mask-programmed by Texas Instruments primarily to be a demonstration device for the TMS1000 product line. The 1121 is intended to provide a simple and inexpensive means for a prospective customer to become familiar with the workings of the 1000 series in an easily understood application that does not require costly custom programming. Fortunately for us, someone at TI did a little thinking and produced a rather neat (even considering its limitations) real-time programmable controller on a single chip.

RTC-4 Characteristics

The RTC-4 features a variety of control possibilities and operating aids: various status displays are available, many events can be programmed at once, and several options can be set up. Part of the RTC-4's task is merely to operate as a digital clock, displaying the time of day and the day of the week. The capabilities of the RTC-4 are essentially those of the TMS1121. A list of the unit's features is shown in table 1.

The RTC-4 contains additional circuitry necessary to interface the chip to the real world. In addition to the TMS1121, the unit includes an LED (light-emitting diode) display and associated drivers, a 20-key keypad, and four output relay switches.

The 20-key keypad enables you to enter instructions and data into the TMS1121. The instructions may be in the form of a stored program telling the RTC-4 what to do in the future, or they may be intended for immediate action. Using the pad, you can turn any of the output relays on or off directly without storing the commands in memory. Also, using the pad you can originate, change, inspect, or delete any timer programs.

The RTC-4 is capable of retaining up to 18 timer programs (on/off settings), which are entered through the keypad. Each of these programs can control one of the four output relay switches. The onboard output relays are small, low-current devices, so if

1. Four independent switch outputs with buffer
2. Display of time and day and status of its own switches
3. As many as 18 daily or weekly programmable setpoints
4. Memory display of programmed setpoints
5. Key entry for clock set and timer set
6. 50-Hz or 60-Hz operation

Table 1: *Capabilities of the RTC-4, which are essentially identical to those of the TMS1121C Universal Timer Controller.*

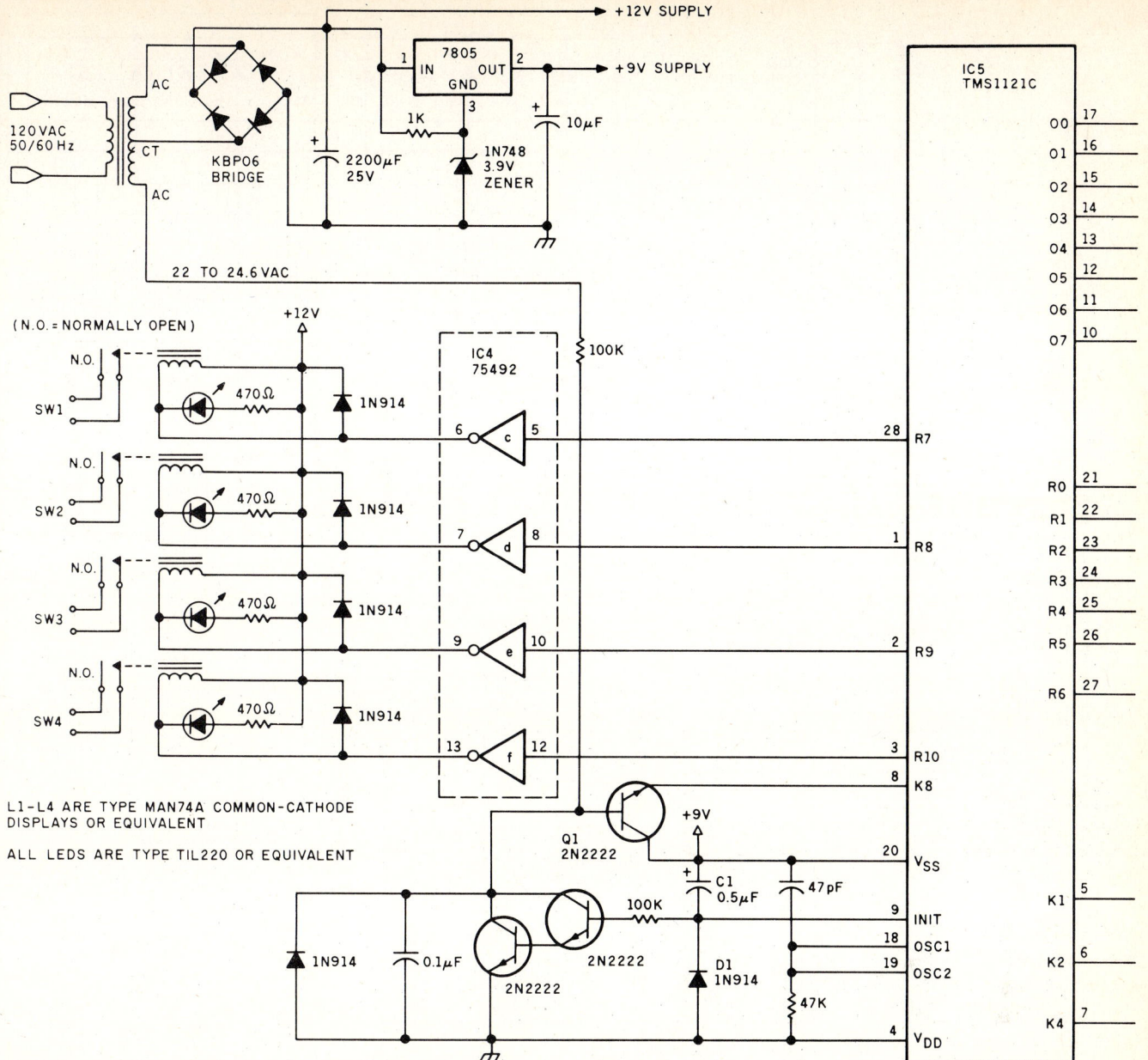

Figure 3: *Schematic diagram of the RTC-4. A type-MAN74A common-cathode LED display module is specified, a type somewhat different from the junk-box unit used in the prototype (see text box, page 9). Please note the position of the MSD (leftmost digit) and the LSD (rightmost digit) when wiring the unit.* (Figure continued)

you want to control a toaster, you'll need to let the onboard relay control a larger outboard relay, which in turn operates the high-current device.

There are two kinds of timer programs: fixed programs and interval programs. Fixed-time programs toggle an output switch at a specific time of day, while interval programs toggle an output switch when a certain interval of time has elapsed since the previous toggling of the switch. Fixed-time programs are retained in memory and repeatedly executed. Interval programs are automatically deleted after execution.

Each program setting toggles only one switch, but a special function exists to combine on/off operations into one program sequence: the SLP (sleep) function is used to turn a switch on and then off 1 hour later. I'll tell you more about programming the RTC-4 later.

RTC-4 Hardware

Figure 2 is a simplified block diagram of the RTC-4, and figure 3 is the schematic diagram. Most of the circuitry of the RTC-4 is associated with the keypad and display and is multiplexed.

As with the other members of the TMS1100 group, the input lines on the TMS1121 are designated K1, K2, K4, and K8 (after their binary significance), and the two sets of output lines are labeled O0 through O7 and R0 through R10. The eight O output lines are configured during the mask-programming of the chip such that only certain combinations of their output values are attainable. Rather than a full range of 256 values, these eight lines can be set only to a

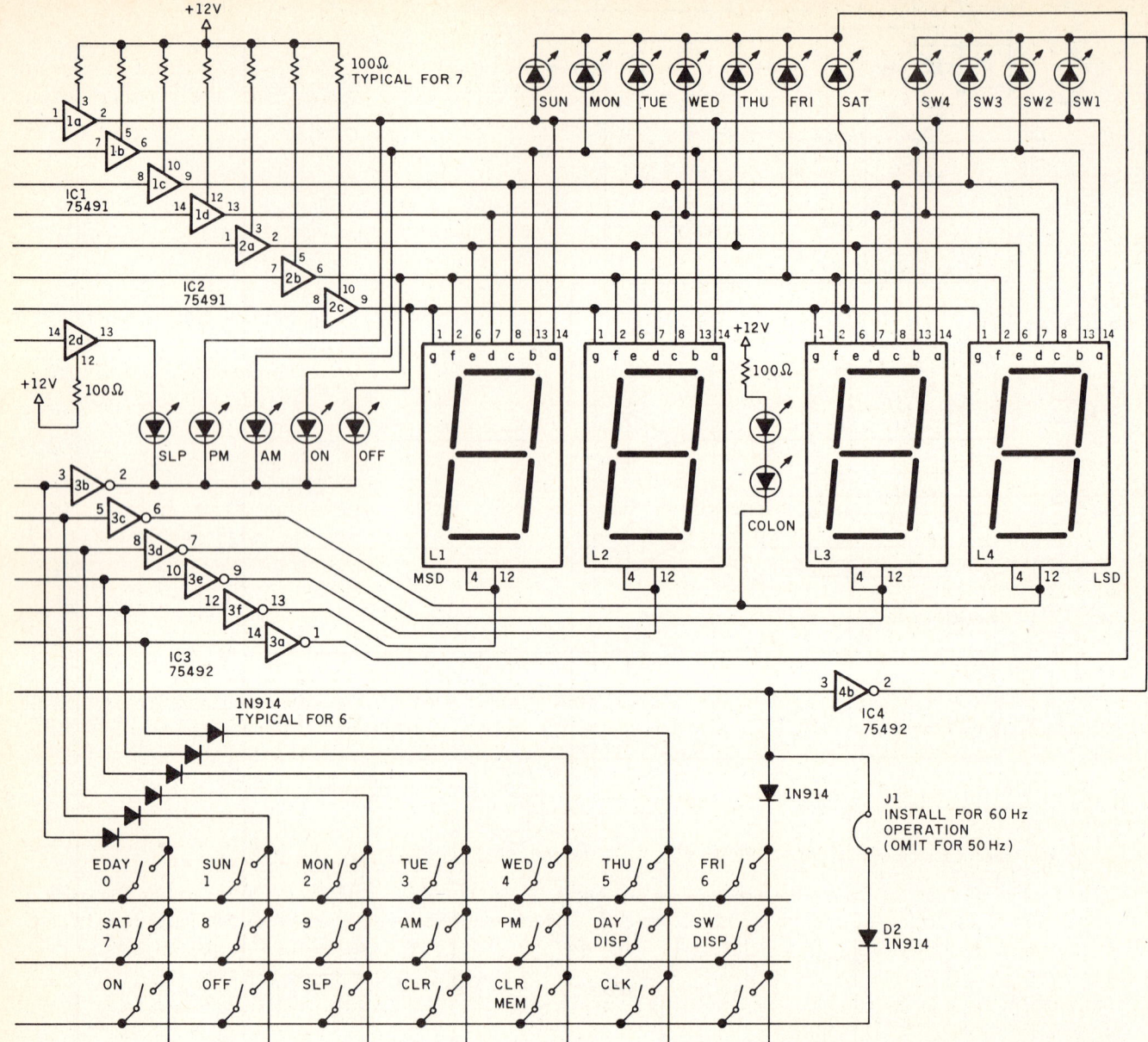

Figure 3 continued

Number	Type	+5V	GND
IC1	75491	11	4
IC2	75491	11	4
IC3	75492	11	4
IC4	75492	11	4
IC5	TMS1121C	20	4

specific subset of combinations defined by the accumulator and status flags. Most frequently the O lines are used as display drivers or status indicators. In this application, very few combinations of the eight O lines are necessary. The 11 R-series output lines, on the other hand, are treated separately; each R line can be set or cleared individually.

The O0 through O7 lines are data-output lines used here to convey the mode, day, switch, and seven-segment codes (for the display) along a time-multiplexed bus. Each O line is buffered through a type-75491 MOS-to-LED-segment driver. Line O0 is segment A, O1 is segment B, and so on through O6, which is segment G. Line O7 is called the decimal-point segment even though no decimal points are used in this application; O7 is attached to the LED that indicates operation of the SLP (sleep) command.

Output lines R0 through R6 are the digit-select lines. Each of the four numeric digits in the LED display module is composed of a cluster of seven segments; the O lines control the segments, while each of four of the R lines activates a digit cluster. Each of the digit-select lines is buffered through a type-75492 signal-inverting MOS-to-LED digit driver. Only one digit-select line is active and one display group illuminated at a time, but by sequencing (multiplexing) through all the digits rapidly, the system can make all the LEDs appear to be illuminated simultaneously.

The two MOS-to-LED drivers, the 75491 and 75492, are designed to

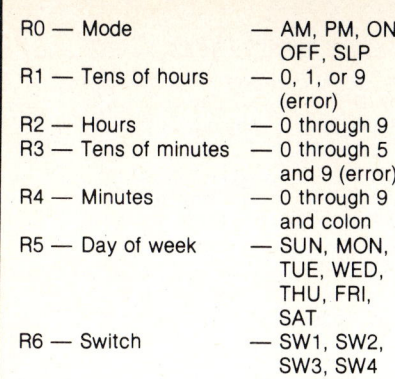

R0 — Mode	— AM, PM, ON, OFF, SLP
R1 — Tens of hours	— 0, 1, or 9 (error)
R2 — Hours	— 0 through 9
R3 — Tens of minutes	— 0 through 5 and 9 (error)
R4 — Minutes	— 0 through 9 and colon
R5 — Day of week	— SUN, MON, TUE, WED, THU, FRI, SAT
R6 — Switch	— SW1, SW2, SW3, SW4

Table 2: *Groups of LED status and numeric displays in the RTC-4, along with their corresponding select lines.*

drive common-cathode LED displays. The 4-digit numeric module is wired for common-cathode operation, and I set up the discrete LEDs in the three other display groups as common-cathode also. By impressing a logic high level (binary 1) on the O lines and logic low level (0) on one of the digit-select lines, the LEDs of each of the display groups can be lit. The seven display groups and their individual select lines are shown in table 2.

In addition to functioning as digit-select lines, the R0 through R6 outputs are used in combination with the K-group input lines in scanning the keypad matrix for user input. As each display group or digit cluster is selected in its turn by an R line, the same R line applies a logic 1 to a column of three keys (or two keys on the R6 line). If one of these three keys is pressed during the application of this scanning signal (thereby closing the circuit), the signal will flow into one of the three K input lines. The program in ROM determines which key has been pressed by reading the K lines in sequence and comparing the combined R and K addresses. This procedure includes use of a 10-ms (millisecond) software contact-debouncing subroutine.

The remaining four R output lines (R7 through R10) are the timer outputs. Buffered through some more sections of a 75492, these signals, in the case of the RTC-4, drive electromechanical relays; you could use solid-state relays also to control external equipment.

The rest of the RTC-4 consists of the power supply and timing sections.

The TMS1121 uses a resistor and a capacitor in a tuned circuit, rather than a crystal, to set its internal clock frequency. This is adequate because the real-time-clock function is synchronized to the 60- or 50-Hz AC power line, and the actual processor clock speed is not critical.

With the 47-picofarad/47k-ohm resistor/capacitor combination shown in the schematic, the clock frequency should be about 300 kHz. The 60- or 50-Hz timing signal is derived from one side of the power-supply transformer. This clock signal is made

> **The RTC-4 keyboard has 10 dual-function and 10 single-function keys; programming is relatively straightforward.**

more square by the transistor Q1 and applied to the K8 input line. If the frequency source is 60 Hz, then you should install jumper J1 (it is omitted for 50-Hz operation). The INIT input pin, connected to diode D1 and capacitor C1, functions as a power-on reset line.

The power supply requires a center-tapped (CT) step-down transformer that has an output between 20 and 24 V (volts). I used a 300-mA (milliampere), 22-V CT unit (Micromint PITB-109), but a 450-mA, 24-V center-tapped transformer from Radio Shack (catalog number 273-1366) should also work. The RTC-4 cannot be battery powered because the power line's frequency is used for timing. The rest of the power-supply circuit consists of a standard type-7805 three-terminal voltage regulator configured for a +9-V output with a zener diode to raise its ground reference potential.

If you don't have any 4- or 5-V zener diodes for this kind of circuit, I recommend that you use an LM317 regulator instead. The power for the displays and relays need not be well filtered or regulated; it can be derived directly from the rectifier output when using a 22-V transformer. In the case of higher-voltage transformers, you might need to add a 7812 regulator to keep the LED drivers from dissipating too much power.

Keypad Programming

Programming the RTC-4 is relatively straightforward. The keyboard has 10 dual-function and 10 single-function keys, as shown in table 3.

When the RTC-4 is turned on, the display will automatically read Sunday at 12:00 p.m. if the unit is configured for 60-Hz operation (with jumper J1 installed). If the RTC-4 is set for 50 Hz, then you must press the CLK key to start it. Obviously, after you turn it on in this fashion, the first thing to do is set the correct time on the clock.

For instance, entering the following keypress commands:

SUN, DAY, 1, 1, 2, 4, PM, CLK

sets the clock to 11:24 p.m. Sunday evening.

The pattern for setting the clock is always the same. The day of the week is registered by typing, in this case, the SUN and DAY keys. The SUN key could be interpreted as meaning "1", but the entry of the DAY key immediately afterward leaves no ambiguity, and the software can determine your intent. (Texas Instruments uses the name WEEK for this key, instead of DAY, but since all entries involve a day setting, I felt that DAY was better than WEEK as the key legend.)

After the day of the week has been set, you tell the computer the hour and minutes, and whether these are antemeridian or postmeridian (a.m. or p.m.). Once the proper day and time have been selected, you press the CLK key to start the real-time clock from that setting.

Direct Switch Control

The RTC-4 controls external devices and appliances through four relay-switch outputs designated SW1,

7

SW2, SW3, and SW4. These switches can be individually turned on or off at specific times under program control, or they may be directly controlled from the keypad. To directly turn switch 2 on and then off, the commands would be:

2, SW, ON, . . . 2, SW, OFF

An alternative to separate on/off command sequences is the SLP (sleep) command. The sequence:

2, SW, SLP

will turn switch 2 on immediately and then automatically off 1 hour later. Any of the three basic functions, SLP, ON, or OFF, may be specified for any of the four switches. But the direct control sequences are not stored in RAM.

Fixed-Time Programs

Fixed-time programs change the state of the switch when the clock reaches a preset time. A typical sequence for entering a fixed-time program would be:

3, SW, MON, DAY, 9, 0, 0, AM, ON

This series would turn on switch 3 on Monday morning at 9:00 a.m. The first two keys, 3 and SW, indicate a switching function for the specific output channel 3. Next, you enter the day and time in the same manner as if you were setting the clock. Finally, you designate the action desired, ON.

As the key sequence is entered, the digital readout and LED indicators display the program settings. The day, time, and program function are automatically stored but will continue to be displayed until another sequence is initiated. To return the display to the current-time digital-clock mode, press CLK.

If the switch being controlled or day of the week differs in the next sequence from the preceding setting, the preceding key sequence must be repeated in its entirety with the new parameters. If, however, the switch number and day of the week are the same, a shortened sequence can be used:

1, 1, 4, 5, AM, OFF

If both of the above sequences are entered, the combined result would be to activate switch 3 on Monday at 9:00 a.m. and deactivate it at 11:45 a.m. the same day. In the case of my tape recorder, I want the action to take place every day, so I use the EDAY (every day) command to turn the recorder on every day at 5:30 p.m. and off at 7:00 p.m. as follows:

1, SW, EDAY, DAY, 5, 3, 0, PM, ON
7, 0, 0, PM, OFF

Exceptions to this can be added as separate program lines. Because the radio program I record runs for only 30 minutes on weekends, instead of

You are able to display stored timer settings by the day or by the switch channel. Status LEDs show pertinent information.

the usual 90 minutes during the week, I can shut the recorder off 60 minutes sooner on Saturday and Sunday:

1, SW, SAT, DAY, 6, 0, 0, PM, OFF
1, SW, SUN, DAY, 6, 0, 0, PM, OFF

Interval Programs

In an interval program, the function is performed after the specified time interval has passed. For example, in the sequence

4, SW, 2, 3, 0, ON

switch 4 would be turned on 2½ hours after the last key in the sequence is pressed. ON, OFF, and SLP commands can be used in interval programs. If SLP were substituted for ON in the above entry, switch 4 would have been turned on after 2½ hours and off again after 3½ hours from the time of programming. The maximum time length for any interval is 11 hours and 59 minutes. Inter-

Double-Function Keys	
EDAY/0	- Everyday or 0
SUN/1	- Sunday or 1
MON/2	- Monday or 2
TUE/3	- Tuesday or 3
WED/4	- Wednesday or 4
THU/5	- Thursday or 5
FRI/6	- Friday or 6
SAT/7	- Saturday or 7
SW/DISP	- Switch or Display switch-program memory
DAY/DISP	- Day or Display daily-program memory

Single-Function Keys	
8	- Numeric 8
9	- Numeric 9
AM	- AM time setting
PM	- PM time setting
ON	- Switch on
OFF	- Switch off
SLP	- Sleep — switch output on for 1 hour then off
CLR	- Clear entry or Error
MEM CLR	- Clear Program Memory
CLK	- Set or display clock

Table 3: *Functions of keys on the RTC-4's keypad; some have two functions, others only one.*

val programs can also be combined in the same manner as the fixed-time programs:

4, SW, 3, 0, ON
3, 1, OFF

These commands cause switch 4 to be activated after 30 minutes and deactivated after 31 minutes. The result is a 1-minute "on" time that starts 30 minutes after command entry.

Program Display and Errors

Stored timer settings can be displayed by the day or the switch channel. For example:

2, SW, SW, . . . SW, SW

would display all programs affecting switch 2. As each step is displayed, the status LEDs and numeric display show pertinent information such as the time, day, and what relays are set on or off. Entering

SAT, DAY, DAY, . . . DAY, DAY

Prototype Construction Techniques

Photo 2: Back side of the RTC-4 prototype.

I receive at least half a dozen letters each month asking me how I put my project prototypes together and wire them. This month I'll reveal a few of the techniques I exercised in building the RTC-4.

To begin with, I don't recommend my techniques for everyone. I try to keep the circuit boards very small and lay the components out in an aesthetically pleasing arrangement. The results are not particularly easy to troubleshoot or modify.

I use standard integrated-circuit (IC) sockets (I prefer the Amp brand), and I hard-wire the power-supply and ground bus lines around the perforated project board using 22-gauge tinned bus wire. These soldered power connections fasten the IC sockets to the board.

Next, I insert the discrete components into the board and if possible directly route their leads to the appropriate pins on the IC sockets. When I absolutely must cross some leads, I use Teflon-insulated wire or sleeving. I finish by using 28- or 32-gauge wire-wrap wire to finish the connections, but I point-to-point solder them.

This may seem tedious, but I end up with a low-profile package, shown in photo 2, that looks very much like a commercially produced unit. It took me about 14 hours to assemble the RTC-4 prototype. Some of my other projects take much longer.

One detour I traveled during the RTC-4 project was the result of wanting large LED-display digits. I had some 0.3-inch type-MAN74A common-cathode single-digit display components, but I decided to adapt a 4-digit 0.7-inch display component I found in my junk box. Unfortunately, units of that type are not intended to be driven as multiplexed displays, and the digits are not wired individually. I had to disassemble and rewire it so that it functioned as 4 separate digits. Photos 3a through 3d show the process. It was so much trouble that I would just use the MAN74As if I were doing it over.

---

instead displays the programs for a particular day. You can also display those for every day (use EDAY).

Errors in command entry are indicated by the appearance of all nines ("99:99") in the display, which can be cleared by the entry-clear key, CLR, or the program-memory-clear key, MEM CLR. It is possible to selectively clear program segments. For example, entering

1, SW, MEM CLR

would erase all programs pertaining to switch 1. Similarly, entering

FRI, DAY, MEM CLR

erases any Friday programs.

Applications

Obviously, one of the more practical uses for the RTC-4 is to enhance your home's security by giving your house that lived-in look. If, for example, you have three lights and a stereo system plugged into the RTC-4, you could go away for weeks at a time and leave everything blinking on and off and sounding away. Because the RTC-4 can be programmed for sequences of one week, most normal home activities can be simulated. It is certainly more cost-effective and reliable than other timer-controllers I've seen.

A Previous Project's Woes

Some of you might have wondered why I didn't use the BSR X-10 Home Control System timer unit, which performs many of the same functions. Besides the obvious damper of losing a chance to experiment with some interesting components from TI, eliminating a good article topic, I've had less than consistent results with the X-10 devices over the years.

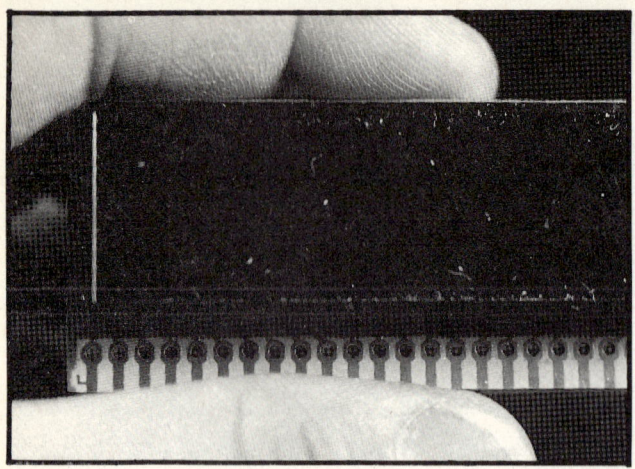

Photo 3a: *This typical National Semiconductor 4-digit LED display (which I believe to be one of the NSB7400 series) is most frequently used in digital clocks. It consists of three parts: a plastic circuit board with LEDs bonded to the circuit traces, a plastic reflective digit form, and a red lens cover.*

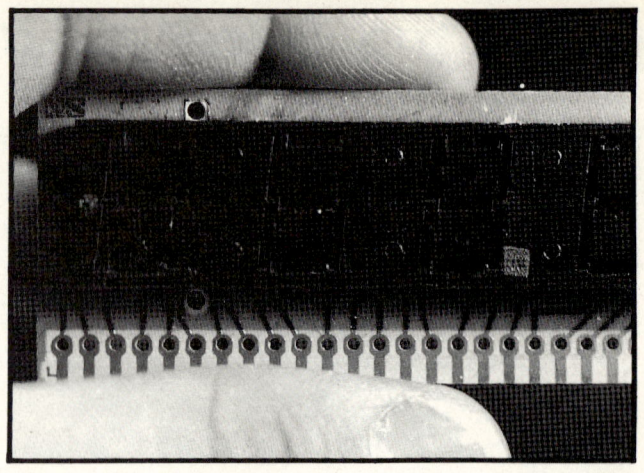

Photo 3b: *The LED display with the lens cover removed.*

Photo 3c: *Because this display has one common cathode line for all 4 digits, I had to cut it at strategic points to isolate the digits.*

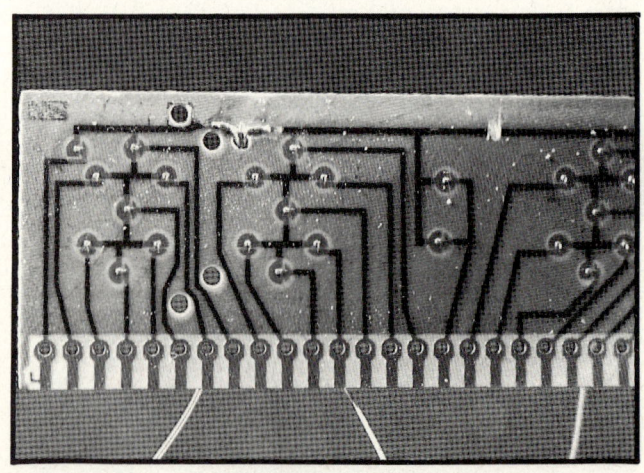

Photo 3d: *Finally, I drilled holes and soldered wires to the individual digit connections.*

At one time (shortly after my article on attaching an X-10 controller to a computer; see reference 1), most of my house was X-10-activated and remotely controlled. Frequently, however, I would find that the switches would reset or change state arbitrarily. I attributed this to power-line transients or electrical noise. Eventually I remedied this situation by retransmitting the intended status of each channel once every minute (obviously a tedious task suitable only for a computer).

Finally, I had a complete falling out with the BSR X-10 system shortly after a thunderstorm a few summers ago. While nothing else (computers, television sets, printers, etc.) in the house was affected, eleven X-10 remote receivers were blown out all at once. I expected to lose a couple receivers now and then, but I didn't expect to replace all of them. And I learned that when a light bulb burned out on a lamp connected to a lamp module, the module was often destroyed. In later production, BSR supposedly installed heavier SCRs (silicon-controlled rectifiers), but this has not completely eliminated the problem. Consequently, I've switched back to the old reliable copper wire and heavy-duty relays. But if I find any better methods I'll be sure to let you know.

A Trick Up My Sleeve

Some of you who are audiophiles might have been wondering how I have managed to record a 90-minute radio program unattended, due to a well-known property of ordinary Philips-type tape cassettes. Strictly speaking, a 90-minute program should just fit on a C-90 cassette. But, of course, a C-90 cassette can record only 45 minutes on one side; people are accustomed to turning the tape over halfway through. (While there do exist C-180 cassettes, they are expensive, and the thin tape doesn't handle the rigors of the automotive environment very well.)

But luckily, when I was still recording manually, I solved this problem. It was hard enough for me to turn the recorder on at all, let alone be there at 6:15 to turn the tape over, so after searching through most of the stereo shops in New England I eventually found a tape recorder that automatically reverses and records on both sides of the tape without turning the cassette over (a Pioneer CTF-750). Now, everything is automatic. But perhaps I should have taken this as an opportunity to build a robot.

In Conclusion

The RTC-4 owes its intelligence to the TMS1121C microcomputer chip. While I used it only in its off-the-shelf configuration, Texas Instruments would like you to know that there are other packaging and functional configurations that can be specifically mask-programmed for high-volume applications. For myself, I'll be satisfied with my somewhat-automatic tape recorder. Of course, to regularly record from radio broadcasts, I had a chat with the program director of the radio station, who gave me permission to record this particular news program. As an author, I'm very careful about copyright infringement.

Acknowledgments

Special thanks to Jeff Bachiochi for his help on this project.

Diagrams pertaining to the TMS1000-series devices are reprinted here through the courtesy of Texas Instruments Inc.

References

1. Ciarcia, Steve. "Computerize a Home." January 1980 BYTE, page 28. Reprinted in *Ciarcia's Circuit Cellar, Volume II*, page 137.
2. Staehlin, David C. "An 8080-Based Remote Appliance Controller." January 1982 BYTE, page 239.

The following items are available from

The Micromint, Inc.
25 Terrace Drive
Vernon, CT 06266
For orders: (800) 635-3355
For information: (203) 871-6170

1. Texas Instruments TMS1121C Universal Timer Controller (pre-programmed 4-bit single-chip microcomputer).............$32
2. Complete kit for the RTC-4 realtime controller. All necessary components provided, including solder-masked and silk-screened printed-circuit board, a TMS1121C, an MAN74A LED display, 20 push-button switches (Panasonic or equivalent), power supply, four 2-ampere 115-VAC relays, and assembly instructions........$119
3. Assembled and tested RTC-4 realtime controller.............$149

On all orders within the continental United States, please include $3 for shipping and handling. Elsewhere, please include $20. New York residents please add 7½ percent sales tax.

Build a Power-Line Carrier-Current Modem

Communicate using electrical power wiring

"Jiggle the printer cable, Jeanette." My assistant reached through the rat's nest of wires behind the computer and grabbed the one connected to the printer. As she moved it, I identified its other end from my cramped vantage point beneath the workbench and pulled it through a slot to attach it to my latest project. I was glad that what we were doing would keep us from having to run cables around the Circuit Cellar so often.

I have long had video terminals, printers, and other data-communicating equipment located at various places in the Circuit Cellar and around the upper stories of my home (see reference 3). Eventually the pain of rerouting cables whenever I moved a peripheral device got to me, so about a year ago I designed a communication system that would save having to string new wires every time. My system revolved around a *carrier-current modem*, which operates in much the same manner as the familiar telephone modem but sends its signals over electrical power wiring instead of over a telephone line.

After I pressed the carrier-current modems into service (with a little help), they served faithfully and I turned my attention to other projects, some of which have appeared in this column. But as of late more and more of my readers have written to me asking for help on how to send data through the AC power line. Apparently the widespread use of and media attention to the BSR X-10 Home Control System and similar products have given many people the idea of using the generally unexploited carrier-current modem for communication. Indeed, about five years ago I published a project on building a remote-control system that communicated through the AC power wiring of a building (see reference 4). It worked very much like the BSR X-10 as a carrier-current remote controller.

I hesitated to present the carrier-current modem as a Circuit Cellar project until now because I feel there is more to general-purpose carrier-current communication than meets the eye.

Simple on/off remote control is different. In most control applications, the communication is generally half-duplex or simplex; the transmission is limited to an intermittent tone or pulse burst that merely triggers a specific receiver into a binary control state. If the receiver is not activated properly by a single transmission because of interference, it's easy to send the control burst more than once. (Many computer control systems that use the BSR X-10 receivers send the same control code 10 times to make sure it is received.) But in general-purpose serial data communication, proper reception of every bit may be necessary, and errors in reception of the data may negate the usefulness of carrier-current operation.

To successfully use a carrier-current modem and the AC power wiring for data communication, we must either tolerate a dropped bit now and then or implement an intelligent protocol of error checking, redundant transmission, and handshaking. A really dependable power-line communication system has the physical link (AC-line transmission and reception) as only one of its components.

I was going to wait until I had perfected the control and error-checking protocols for use with the carrier-current modem, but the increasing interest indicated by my mail suggested that many experimenters might benefit from building a simple carrier-current modem; at least the physical part of the connection could be set up, even if the protocols and software are not ready.

This month's project, a modem for data communcation using the AC

Photo 1: *Prototype of the Circuit Cellar CCM-1 carrier-current modem, which transmits serial data over the AC power line at 1200 bits per second. When in originate mode, the modem transmits mark signals at 90 kHz and space signals at 95 kHz; the answer mode transmits marks at 80 kHz and spaces at 85 kHz. The receive unit, shown here, differs from the transmit unit only in the frequency-selecting passive components.*

power line, is mostly an analog circuit. Successful operation of the modem, therefore, depends much more on tweaking and tuning the components than do digital computer-related projects. I am presenting this two-chip modem chiefly to discuss the principles involved, with some emphasis on selecting components for this application. Because the principles are susceptible to broad application, this knowledge should also be useful in understanding other modem designs as well.

All Modems Are Not Alike

The modem, named after a contraction of the words "modulator" and "demodulator," is a fairly common piece of computer equipment. You've probably seen modems built for sending data over telephone lines, and you may have read my March Circuit Cellar article about a low-cost modem (see reference 2). A modem allows two pieces of digital equipment to communicate with each other over long distances without having a direct hard-wired connection between them. With a telephone modem, the telephone lines form the communication path.

Modems of the usual type translate the voltage levels of the digital input signal (usually RS-232C levels) to tones at two frequencies, one of which signifies a logic 0, the other, a logic 1. The process of shifting the frequency of the output tone as the logic levels change is called *frequency-shift keying*, and the modems are called frequency-shift keyed, or FSK, modems.

To allow communication in two directions at once (full-duplex mode), rather than in only one (half-duplex), two pairs of frequencies are used, avoiding conflict when both ends of the connection talk at the same time. (By convention, one pair of tones is called the "originate" set, and the other is called the "answer" set. The two terms merely signify which set of frequencies each unit is using; no implication is intended regarding the content or origin of the data itself.) For compatibility, modems are built to adhere to certain standards of operation; the most common system in North America for low-speed modems was first used in the Bell System's Model 103 modem, so Bell-103-type modems abound.

Carrier-Current Systems

The AC power line is similar in some respects to the telphone line. One similarity is clear: we can send data through the power line by using an FSK modem.

Obviously, in addition to the data we want to transmit, the power lines

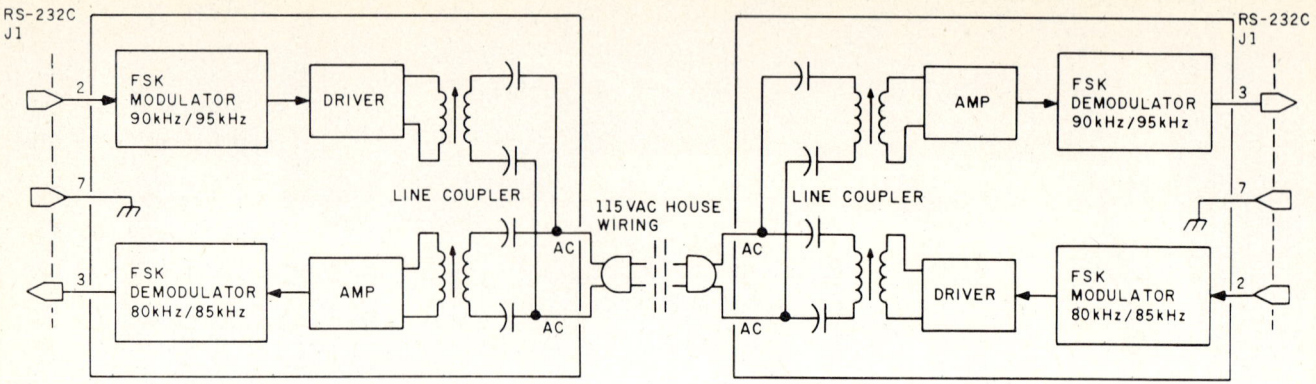

Figure 1: *Block diagram of a data-communication system employing carrier-current modems. The AC power wiring of the building is used to carry the frequency-shift-keyed transmission.*

must continue to carry power or we won't be able to operate the computer equipment. The carrier-current communication system superimposes a high-frequency signal on the 60-Hz power-carrying signal. On an oscilloscope, this is viewed as an additional small voltage carried on or riding atop the 115-V (volt) alternating current. At the receiving end, the modem filters out the 60-Hz signal and any other noise components on the power line, demodulating only the transmitted frequency. Unfortunately, the power line sometimes has an impedance less than 2 ohms, along with thousand-volt noise spikes that make it a hazardous environment and a less-than-optimal communication medium.

There is a price to be paid for the simplicity of this communication system. Unlike the complex digital carrier-current systems, which transmit around the zero-crossing interval of the power signal, the analog FSK carrier-current modems are more sensitive to line peculiarities and noise. However, the digital species is much more complex, and after all, my intention was to present a build-it-yourself project. Learning a little black art for the sake of simplicity can't hurt.

Carrier-Current Modem Circuit

Figure 1 is the block diagram of a carrier-current modem, which consists of three basic components: modulator/driver, amplifier/demodulator, and AC-line coupler. The simplest usable system consists of two modems, one attached to each of two pieces of data-communicating equipment. One of the two modems is arbitrarily designated as the originating modem and the other as the answering modem. As in the case of telephone communication, two sets of FSK frequencies are defined, although the power-line modems operate at much higher frequencies than the telephone-line type. The connections from the communicating equipment to the modulator and demodulator on each modem are through an RS-232C DB-25 connector. The driver and amplifier sections are in turn connected to the AC line through the coupler, the crucial component.

In a direct-connect telephone modem, the coupler is usually a 600-ohm isolation transformer, and the characteristics of the line are well defined. But in a carrier-current modem, the coupling transformer is very often a tuned circuit selected to resonate within the passband of the FSK tones to improve the signal-to-noise ratio in this particularly noisy environment. While tuned couplers are not aways used, most carrier-current driver circuits do employ them to increase the transmission range and receiver selectivity. For most experimenters, the driver and coupler are the hardest sections to construct because so much depends on selecting, balancing, and adjusting the components.

Taming the AC Line

Figure 2 shows two typical carrier-current driver-coupler circuits. Both consist of a transformer capacitively isolated from the AC line; 0.22-μF (microfarad) 600-V capacitors are recommended. Any 0.1-mH (millihenry) slug-tuned transformer will probably work, but I have had best success with standard low-Q (low-resonance) miniature IF (intermediate-frequency) transformers used in transistor radios. In practice, the circuit of figure 2b is less sensitive to component selection and more easily tuned.

If you have any doubts about whether a particular transformer will work, a few brute-force tests can help you tame the AC line and give you the confidence to build the rest of the modem circuit. *Just remember that working directly with the AC power line is dangerous if you aren't careful.*

Begin by building two couplers, one driver, and one receiver, using the component values and circuit layout in the schematic diagram of figure 3. (I'll get around to discussing it shortly.) Temporarily apply power to the driver section and attach it (carefully!) through its coupler to the AC line. The receiver should be powered and connected through its coupler across the AC line at some other nearby location.

Use an oscilloscope (connected to the AC line through an isolation transformer for safety) to monitor the signal present across the secondary coil of the receiver transformer (or from the collector of transistor Q2 to ground), while you inject a signal for transmission using a sine-wave func-

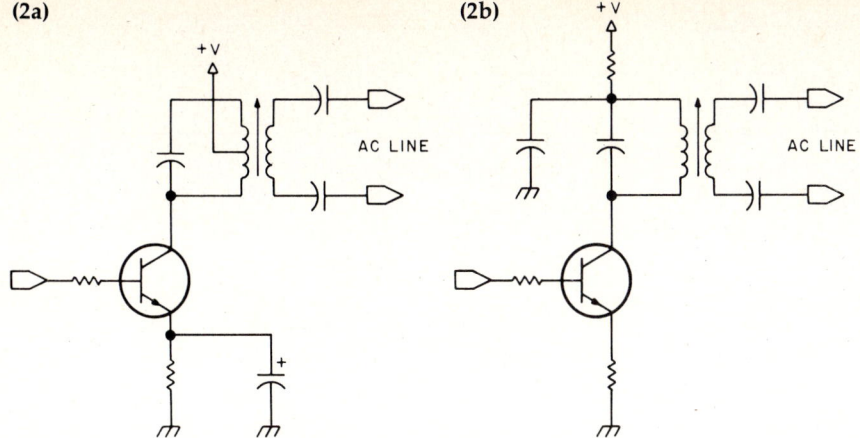

Figure 2: *Two possible schemes for coupling the modulator/driver portion of the carrier-current modem to the AC line. The circuit of figure 2b is the more stable.*

tion generator attached to the base of the driver transistor, Q1. Sweep the frequency between 50 kHz and 150 kHz until you detect the same frequency (at greater than 10 mV [millivolts]) at the receiver. Take care that you are receiving the fundamental frequency and not a harmonic. Don't be too surprised at the strange electrical noise you'll no doubt observe.

You can shift the detection band somewhat by adjusting the tuning slug in and out of the transformer windings or by changing the capacitor across the transformer secondary. The objective is to find the frequency band where the signal level at the receiver is highest. The band should be about 20 kHz wide; the frequency can go as high as 300 kHz (the upper frequency limit of the demodulator) if necessary.

In my case the best results were obtained between 80 kHz and 100 kHz, so I arbitrarily set two originate and answer frequency pairs within this band. One modem transmits on 90 and 95 kHz and receives on 80 and 85 kHz; conversely, the other modem transmits on 80 and 85 kHz and receives on 90 and 95 kHz. In a simple system, any originate and answer frequency pairs that work are acceptable because each frequency pair has its own tuned coupler. I recommend that the frequency separation between the mark and space tones be 5 kHz or less to facilitate easy demodulation. I only caution you not to set any frequency that is a multiple (or submultiple) of another one used in the system.

Remember that, in an analog FSK carrier-current communication system, success largely depends on your peaking the resonance of the coils and finding the proper transmission bands. I can't provide a parts list of components values that will be guaranteed to work because the behavior of parts in the list could be other than that predicted, due to performance and tolerance variations. The most important "component" in the coupler sections is your understanding the objective and knowing how to pursue it through testing and adjustment.

Fortunately, component selection in the FSK modulator and demodulator sections is much more straightforward and follows some basic formulas defined by the frequency and application. However, because it is possible that you might choose frequency pairs different from those in my design, I'll discuss the derivation of the component values rather than just the results.

Exar XR-2206 Modulator

First, let's consider the modulator. The XR-2206 is a function-generator integrated circuit, made by Exar Integrated Systems, which can produce sine, square, and triangular output waveforms at frequencies ranging from 0.01 Hz to 1 MHz. It is ideally suited for FSK applications because it can be set for two different time bases and digitally switched between them. A functional block diagram of the XR-2206 and typical FSK circuit is shown in figure 4.

The mark and space frequencies can be independently set by the choice of timing resistors R2 and R3 and the capacitor between pins 4 and 5. The FSK input signal is applied to pin 9. A high logic-input signal to pin 9 produces the frequency:

$$f_{high} = \frac{1}{R2 \times C}$$

and a low-level input signal produces the frequency:

$$f_{low} = \frac{1}{R3 \times C}$$

where R2 and R3 are in ohms and C is in farads.

R2 and R3 should be between 10 kilohms and 100 kilohms, and the capacitor should be polycarbonate, polystyrene, or Mylar for temperature stability. I chose to use a 0.001-μF capacitor, which produces the following resistor values for the frequency pairs I chose:

R2: 85 kHz, 11.76 kilohms
    95 kHz, 10.53 kilohms

R3: 80 kHz, 12.50 kilohms
    90 kHz, 11.11 kilohms

In the case of R2 and R3, you can use the nearest 1-percent-tolerance resistor or use a potentiometer in combination with the closest 5 percent fixed value.

You must also consider the settings of resistors R1 and R5, which adjust for minimum total harmonic distortion. In our case, where a few-tenths-percent distortion is irrelevant, pins 15 and 16 may be left open and R1 can be replaced by a fixed 200-ohm resistor. With R1 installed (same effect as closing switch S1), the output at pin 2 is a sine wave with an output impedance of 600 ohms and amplitude set by R4. The remaining components serve to stabilize operation and are the same for all frequencies.

XR-2211 Demodulator

The Exar XR-2211 is a phase-locked-loop (PLL) integrated circuit especially designed for data communication

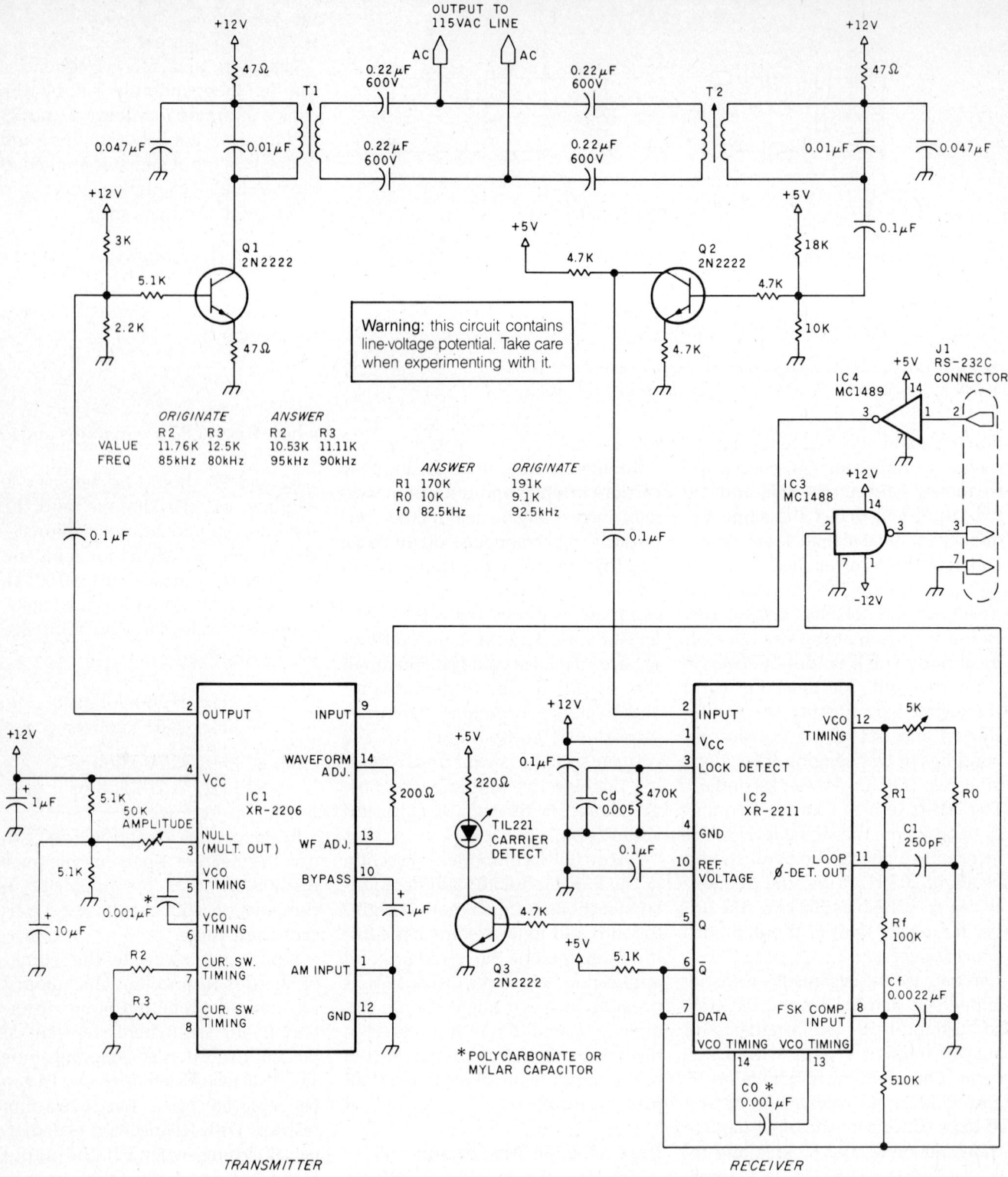

Figure 3: *Schematic diagram of a complete carrier-current modem. The originate-mode modem transmits mark signals at 90 kHz and space signals at 95 kHz; the answer modem transmits marks at 80 kHz and spaces at 85 kHz. This diagram shows the circuit for one end of the link; two such units are needed in the system with the proper component values differing between them. In one unit, the wiring of pins 2 and 3 of J1 should be reversed.*

and particularly suited for FSK applications. It operates over a frequency range of 0.01 Hz to 300 kHz and can accommodate analog input signals between 2 mV and 3 V. An XR-2211 functional block diagram and typical FSK demodulator circuit are shown in figure 5. Frequency-shift-keyed input signals fed into pin 2 of the XR-2211 must be capacitvely coupled through a 0.1-μF capacitor. The internal impedance is 20 kilohms, and the minimum recommended input signal is 10 mV.

The first order of business is to set

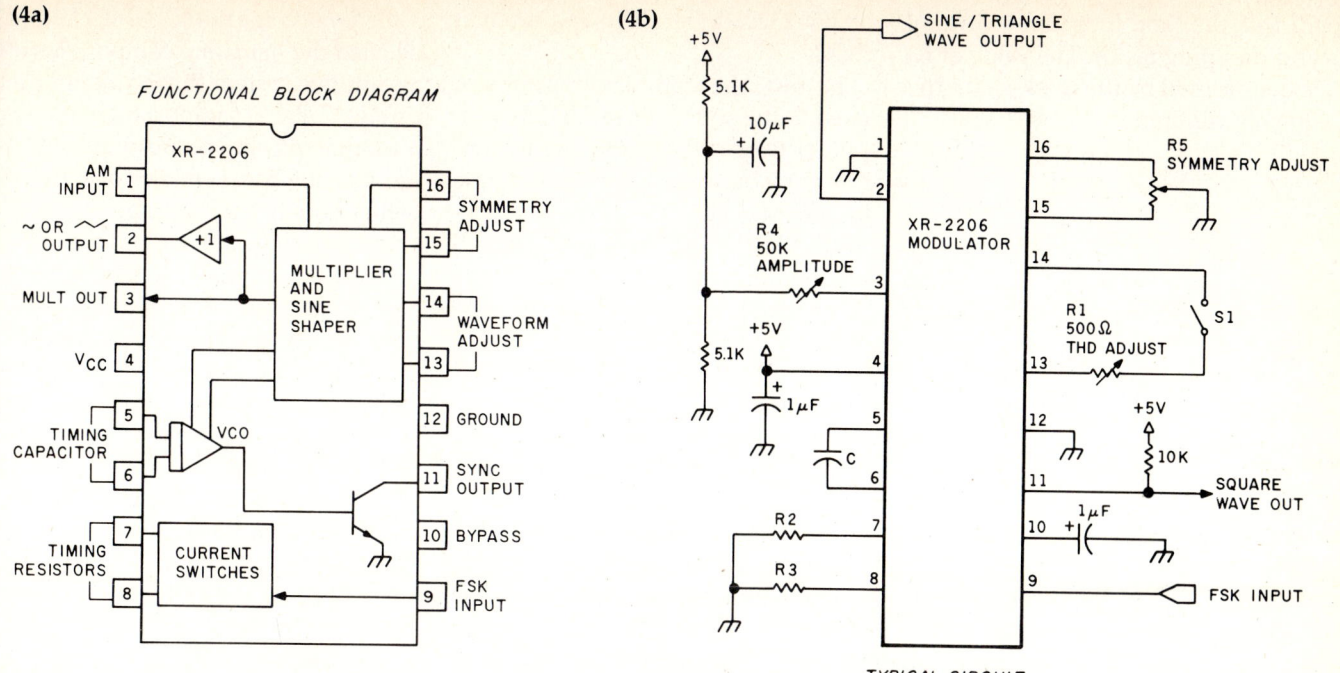

Figure 4: A functional block diagram and pin-out specification of the XR-2206 (4a) and typical FSK circuit (4b).

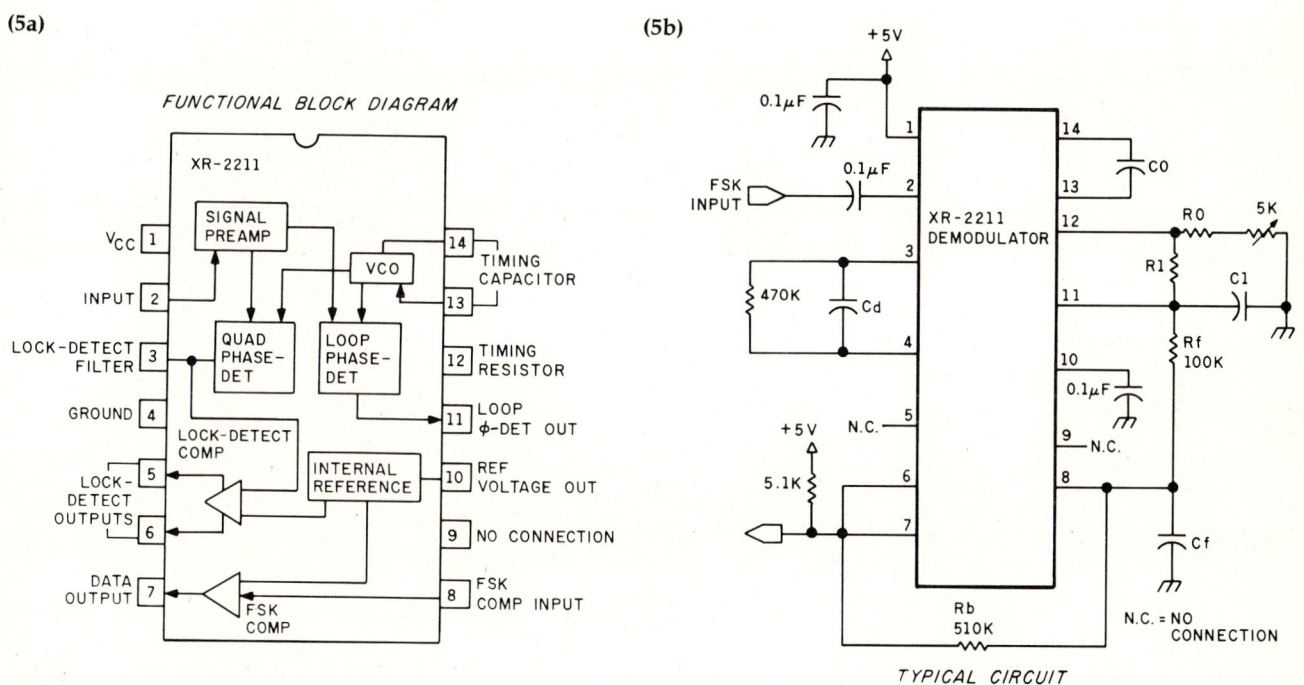

Figure 5: An XR-2211 functional block diagram and pin-out specification (5a) and typical FSK demodulator circuit (5b).

the center frequency of the demodulator passband at the center of the frequency band that we wish to detect. In my case, the passbands are defined by the tone pair at 80/85 kHz and the other pair at 90/95 kHz. The center frequencies for the two demodulators would then be 82.5 kHz and 92.5 kHz, respectively. The component values are computed as follows:

$$f_0 = \frac{1}{R0 \times C0}$$

where R0 is in ohms and C0 is in farads; f_0 is the center frequency.

Generally, R0 is in a range of 10 kilohms to 100 kilohms, but the choice is arbitrary. Often it is more convenient to choose a value for C0 and trim the value of R0 with an adjacent potentiometer. Using 0.001-μF value (Mylar, polycarbonate, or polystyrene) for C0, the computed R0 values are 12.12 kilohms (f_0 = 82.5 kHz) and 10.81 kilohms (f_0 = 92.5 kHz). With a 5-kilohm trim pot in series, more convenient resistors of 10 and 9.1 kilohms can be used instead.

R1 sets the system bandwidth and

C1 sets the loop-filter time constant and damping factor. The value of R1 is determined by the mark/space frequency difference:

$$R1 = \frac{R0 \times f_0}{(f_1 - f_2)}$$

The deviation is 5 kHz by design, and the values for R1 are 170 kilohms (f_0 = 82.5 kHz) and 191 kilohms (f_0 = 92.5 kHz).

While the equation for computing the loop-damping factor associated with C1 is complex, there is a convenient rule of thumb. The damping factor should be approximately ½, and a value of C1 = C0/4 will produce this. With C0 equal to 0.001 μF, C1 equals 250 pf (picofarads).

Resistor Rb provides positive feedback across the FSK comparator and facilitates rapid transition between output logic states. A value of 510 kilohms is used in most applications.

Cf and Rf form a single-pole post-detection filter for the FSK data output. Rf is most often set at 100 kilohms. Cf smooths the data output; its value is roughly calculated: Cf = (3/data rate in bits per second) where Cf is in microfarads. Because this modem is designed for operation at 1200 bps (bits per second), a value of 0.0022 μF or 0.0033 μF is acceptable.

The final area requiring calculation is the lock-detect section of the XR-2211, which is used here in a carrier-detect function. The open-collector lock-detect output, pin 6, is connected to the data output, pin 7. This will disable any output created by noise unless a carrier signal is present within the detection passband of the PLL. Presuming a parallel resistance of 470 kilohms, the minimum value of the lock-detect filter capacitor, Cd, is $16/(f_1 - f_2)/2$. In this case 0.005 μF is adequate.

Testing the Completed Unit

I built the complete Circuit Cellar carrier-current modem the way shown in figure 3, with component values for 80/85 kHz and 90/95 kHz tone pairs, but you may substitute other values as previously discussed. In addition to the three functional sections we have looked at, I have added a carrier-detect indicator and an RS-232C driver (IC3) and receiver (IC4).

To test the completed unit you need some source of serial data output. (I used a full-duplex video terminal.) The easiest test is a simple loop-back circuit. The terminal is connected to the originate modem and plugged into the power line. The answer modem is plugged in some distance away, with pins 2 and 3 jumpered together on J1, its RS-232C connector. As you type on the terminal, the data is transmitted to the answer modem where it is looped back through the jumper and retransmitted to the originate modem where it appears on the terminal's screen.

You should be able to place the modems anywhere within your home or office, or even an adjacent home or apartment. The ultimate range is limited by the power company's step-down transformer and the cross-coupling between the two 115-V legs of a multiphase 230-V distribution system. But you can arrange communication between the latter by attaching a fused capacitor between the two 115-V legs.

In Conclusion

Using this modem I was able to successfully communicate at 1200 bps for extended periods of time without loss of data. I've found FSK carrier-current communication to be fairly reliable; it's best at the lower data rates. Occasionally a few characters have been lost when my air-conditioner compressor or water pump turned on. These are occasions where an intelligent control system might be of significant help. I had intended that the intelligence necessary for error checking and redundant transmissions be part of this project, but as I explained, such a control system is much more involved than the modem itself. Given the excess computing power available in most personal computers, it would certainly be feasible in most cases for error-checking to be performed by applications software, perhaps using something like the well-known file-transfer protocol developed by Ward Christensen for use with his CP/M-based Modem-7 program.

Generally speaking, while I have detailed the hardware components of a complete system that works in the Circuit Cellar, it's important to recognize that AC line conditions differ significantly between locations. Complete frequency bands may be unusable due to interference produced by machinery, digital clocks, microcomputers, and fluorescent lights. For this reason, you should understand how the modem components and coupler are designed. Your ability to customize a basic modem design to the particular electrical environment of your home or office can make or break the project.

References

1. *Applications Data Book.* Sunnyvale, CA: Exar Integrated Systems Inc., 1981.
2. Ciarcia, Steve. "Build the ECM-103, an Originate/Answer Modem." March 1983 BYTE, page 26.
3. Ciarcia, Steve, "Come Upstairs and Be Respectable." May 1977 BYTE, page 50.
4. Ciarcia, Steve. "Tune In and Turn On: A Computerized Wireless AC Control System." Part 1, April 1978 BYTE, page 114. Part 2, May 1978 BYTE, page 97.
5. Edward, Harry J., Jr. *Residential Electrical Wiring: A Practical Guide to Electrical Wiring Practices in Residences.* Reston, VA: Reston Publishing Company, 1982.

Build the Micro D-Cam Solid-State Video Camera

Part 1: The IS32 Optic RAM and the Micro D-Cam Hardware

A 64K-bit dynamic RAM chip is the visual sensor in this digital image camera

If you've followed the activities in the Circuit Cellar for any period of time, you have probably realized that my writing a monthly column is just an excuse to investigate and experiment with whatever I find currently fascinating. One of my longtime fascinations has been the input of visual data to a computer. While I've presented interfaces that allow computers to determine direction and measure distances, receive a variety of sensory inputs via touch or remote signal, and even speak their minds through voice synthesis, until now I have not presented a project that enables a computer to see.

There was a time when most computers communicated via klunky teletypes at 10 characters per second. Improving output technology—high-speed video graphics displays, dot-matrix printers, and voice synthesis—has vastly improved the computer's ability to communicate results and conclusions to its user. Except in specialized applications, however, input technology has been discouragingly static. We still plod along using keyboards or mice as the primary input device even when the input data may be graphical.

So much of our existence involves visual recognition that it only stands to reason that the potential applications of computers would be enhanced if a versatile sensory-input channel were available to the machines. Then, instead of spending hours entering digital coordinates from a picture or map into a computer using a keyboard, you could easily use a "computerized camera" to make a visual snapshot of the material, instantly producing a digitized picture. Once you had such a picture in the computer, it could be interpreted, enhanced, or stored as the application might dictate.

Photo 1: *In this prototype, the Optic RAM is mounted inside a light-tight box with a C-mount lens focusing light onto one of its cell arrays. The ribbon cable leads straight from the Optic RAM to the interface card in the Apple.*

Computers and Vidicon Cameras

Computerized-image cameras are not new, but up to now they have always been too expensive for widespread practical use and casual experimentation. Most of the computer image-input devices currently available use a conventional black-and-white television camera as the image sensor. The camera's video output must be converted to digital logic levels for the computer: a difficult task, because the output, produced using a Vidicon-type pickup tube, is a high-frequency analog signal divided into 30 complete frames of picture information transmitted and scanned each second (or 25 frames for most TV systems outside North America).

Most high-quality TV-camera interfaces convert the analog signal for computer processing through "frame grabbing," in which one of the frames is sampled, digitized, and stored during a 1/30-second frame-scan interval. In these sophisticated visual sensing systems, a high-speed A/D (analog-to-digital) converter digitizes the analog signal in real time at sampling rates exceeding 5 megahertz (MHz) and stores the PCM (pulse-code modulated) data in a high-speed buffer made of semiconductor memory. Because they operate so fast, such units are insensitive to camera motion and fast scene changes.

In less sophisticated TV-camera interfaces, the designer has assumed that the camera and the object in its view will remain still long enough for the picture to be processed by slower, cheaper circuitry. When the TV picture is stationary, all frames in the signal are identical, so a sequential line-sampling technique is often employed. In units of this type, a low-speed A/D converter (sampling from 100,000 to 1,000,000 times per second) operates in bursts of activity shorter than a frame interval, with each successive period of activity, or *sampling window*, triggered at a slightly later time during the frame interval by line and pixel (picture element) position counters. In between the sampling windows, the support circuitry has time to store the digitized information and get ready for the next burst of activity. As a result, the interface assembles the single image from pieces snatched from many frames.

High-speed frame grabbers generally cost more than $10,000, while the slower units cost somewhat less, depending upon speed and resolution. You can expect a 256 by 256 pixel-resolution low-speed interface, the kind used in the computer systems often seen at conventions printing images on T-shirts, to cost between $500 and $1000; half of that price is for the camera and lens.

Solid-State Arrays to the Rescue?

The problem with the Vidicon-type camera is that it is an analog device, which must be adapted to work in digital applications. It would be far better to have a computer video camera that is inherently digital and dispense with the analog-to-digital conversion. Why not use semiconductor devices, the outputs of which are digital signals that change as a function of light level?

The barrier has been price. A variety of semiconductor optical sensors, such as photodiode and charged-coupled-device (CCD) arrays, fill the bill nicely. When I first started to think about building a solid-state image camera, I thought I could just order a 256 by 256 pixel CCD array and add a few binary-counter chips for a quick project. This idea evaporated quickly when I discovered that CCD arrays cost from $800 to $2000, depending upon the number of bad pixels you get.

My success with photodiode arrays wasn't much better. It seems that for about $100 you can buy a 128 by 1 or 256 by 1 array, but arrays more than one element wide are hard to find. To create a 256 by 128 or 128 by 128 picture, I would have needed to devise an optical, mechanical, or electronic way to move the array across the image plane, or move the

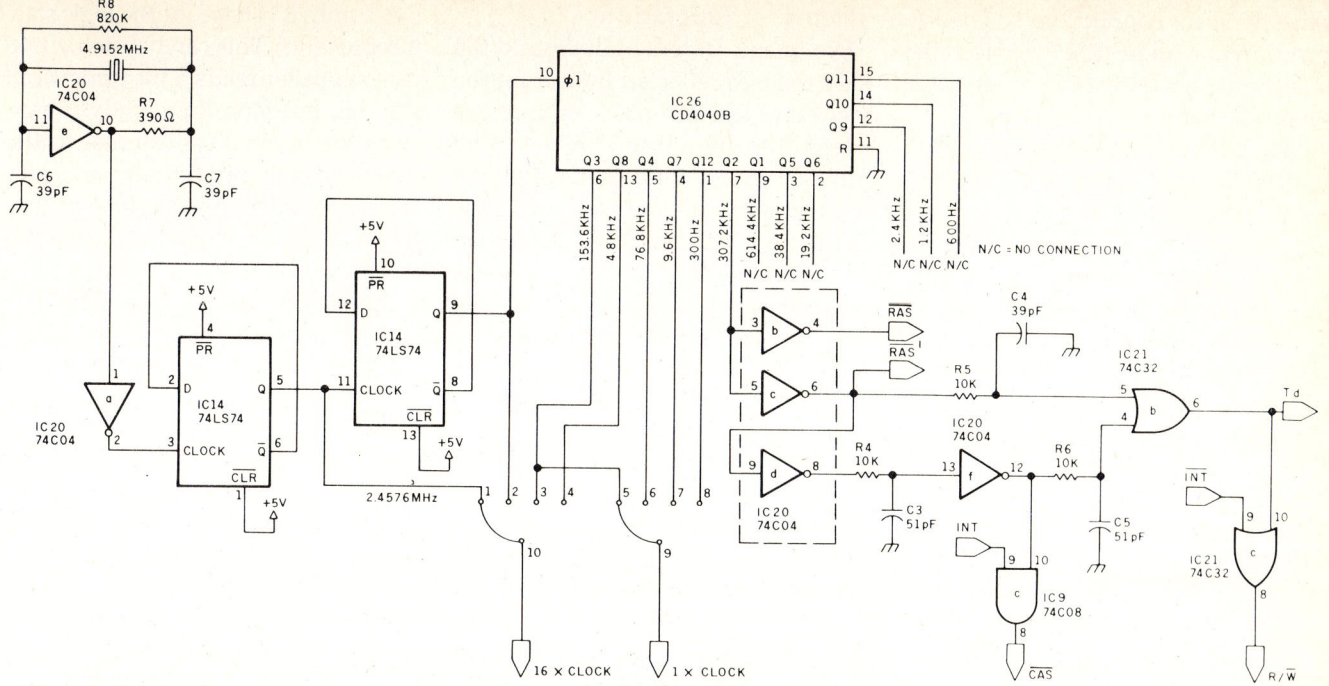

Figure 1: *A schematic diagram of the timing-generator circuit, which contains a CMOS oscillator circuit to generate the fundamental clock rate. This signal is divided down to produce the frequencies for various possible output data rates and controlling the IS32. The data-rate-clock signals control the sequence of operation of the interrupt generator and the transmitter circuit.*

image across the array, stopping periodically for the values of the picture elements to be registered and stored. After tentatively sketching a screw drive and its associated control electronics, I gave up in disgust at the thought of how hard it would be to build.

If Memory Serves...

I had almost decided to join the bandwagon and use a Vidicon camera when I remembered experimenting with the optical sensitivity of dynamic RAM (random-access read/write memory) chips back in 1976. I had even seen a design published for a 64 by 64 pixel resolution camera that used a type-1103 1K-bit dynamic RAM chip as an optical sensor. If the bit-storage cells in those early dynamic RAMs were light-sensitive, could the newer ones be also? If I popped the top off a 64K-bit dynamic RAM chip, wouldn't I find a 256 by 256 array?

Well, yes and no. I took the lids off a few brands of 64K-bit dynamic RAM chips, and it does appear that they are light-sensitive. The problem, however, is that they were designed only as memory devices and not op-

tical arrays. To my knowledge, none of the 64K-bit dynamic RAMs on the market are configured as an orthogonal array laid out 256 elements long by 256 wide. In fact, most have 4 or 8 sections of 16K or 8K bits, and many include redundant sections that can be wired in to replace bad sections on the chip. The bit addresses don't proceed linearly through the chip either; one bit may be in the upper-left corner and the next bit in the lower-right corner.

Just as I was about to abandon all hope, I found an unconventional dynamic-RAM manufacturer that has recognized the light-sensing potential of its 64K-byte device. Micron Technology Inc., of Boise, Idaho (certainly an unconventional place to make integrated circuits), produces a dynamic RAM chip that has its memory cells laid out in only two sections, both of which are 256 by 128 cells, as shown in figure 7 on page 30. With this configuration, the chip can easily be used as an optical sensor. One specially tested 64K-bit dynamic memory device, called the IS32 Optic RAM, comes in a package with a see-through quartz lid (see photo 4 on page 29).

Per pixel, the Micron Technology IS32 Optic RAM costs 1000 times less than the earlier generation of image-sensing chips such as the CCD. The Optic RAM's spectral sensitivity is generally the same as that of other silicon-based light-sensing media, but its bit-for-bit uniformity is not as good as CCDs. Nevertheless, the Optic RAM can bring capabilities to your computer that were previously available only to large industrial users.

Build the Micro D-Cam

This month, using the IS32, I'll show you how to build a relatively low-cost digital image camera I call the Micro D-Cam (see photo 1). Its resolution of 256 by 128 pixels is adequate for many applications in graphics, pattern and character recognition, robotics, process control, and security. (Of course, the output of the Micro D-Cam is a digital signal; it cannot be used to directly drive a composite-video monitor.) I've put together versions of the Micro D-Cam for use with the Apple II computer (II-Plus and IIe, see photo 3, page 29) and the IBM Personal Computer; however, the Micro D-Cam is serially interfaced and requires only five wires for connection, so I'm also working on an RS-232C version that can be attached to any computer that has a serial port.

The Micro D-Cam project is rather complex, so I'll present it in two parts. This month I'll explain how the IS32 Optic RAM and the Micro D-Cam hardware work. Because appropriate software is vital to the success of this project, next month I'll include a lengthy listing of a typical control program for use with the Apple II-Plus version of the Micro D-Cam. We'll also look at some of the Micro D-Cam's capabilities.

IS32 Optic RAM

The IS32 from Micron Technology is an all-digital image-sensing device. Its pertinent characteristics are shown in table 1.

The IS32 contains 65,536 (64K) light-sensitive memory cells laid out in two planar, rectangular arrays of 32,768 elements, each a matrix of 128 rows and 256 columns. The two arrays are separated by an optically nonsensitive "dead" zone about 25 elements wide. To avoid having a gap in the image or using complicated optical systems to eliminate it, only one of the arrays is usually used as an image sensor. Each of the memory elements in the matrix can be accessed randomly when the control circuitry strobes in the appropriate row and column address of the element being accessed.

Theory of Operation

An image camera built around the IS32 focuses reflected light from the viewed object and passes it through a lens onto one of the 32,768-element arrays. When an individual element is struck by photons of light, the capacitor in the cell, which is initially precharged to a fixed voltage, begins to discharge toward zero volts. The capacitor discharges at a rate proportional to the light intensity throughout the duration of the exposure.

After the exposure interval has elapsed, the circuitry reads the element by addressing it as a memory cell. During the cell access, sense amplifiers within the IS32 read the capacitor's voltage value and compare it to a fixed threshold voltage. If the potential is above the threshold, the picture element is deemed to be black; if the potential is too low, the picture element is declared white. The D_{out} pin of the Optic RAM is set to a logic 1 or 0 during the corresponding bit interval as a result of this decision. The raw "gray scale" of the IS32, therefore, has only two shades, white and black. (We'll see how to compensate for this shortly.)

All dynamic-memory devices require refreshing for operation; the charge representing the data stored in each cell capacitor will leak away if left alone (exposure to light merely hastens the leakage). The charge must be sensed and brought back to the nominal voltage for the logic state it represents. This can happen when the computer reads a bit value from the cell, but more frequently it happens when circuitry external to the memory chip periodically activates the cell's address just for the purpose. (Many memory chips, including the IS32 Optic RAM, can refresh their cells a whole row at a time.) The IS32 can be used this way as a regular memory device can, but in optical service there is a twist in the refreshing. The chip is light-sensitive only when it is not being refreshed; the key to using it in a camera is to carefully control its sensitivity by performing the refresh operation in a special way.

In the beginning of an image-sensing cycle, the Micro D-Cam's circuitry addresses all the cells in the active array, filling them with the positive voltages that represent logic 1s. The exposure begins with the receipt of a SOAK command, which is the equivalent of opening the shutter (to allow the array to "soak" in light). Then, after the appropriate exposure interval has elapsed, the control circuitry issues the refresh command, which freezes the states of the memory cells (or pixel cells, if you will). Then the control circuitry activates the interrupt state, during which the value of one cell is fetched and transmitted. Interrupt cycles are continued until all the bits in the array have been transmitted. (The interrupt

Photo 2a: *The Micro D-Cam can focus on UPC bars. Both photos shown here used the Apple II's high-resolution graphics routines to reproduce the camera's output.*

Photo 2b: *The Apple II's display represents gray levels as different densities of white dots. Several different-length exposures are combined to form the gray-scale image.*

mode is not maintained constantly because the IS32 cannot be refreshed during the interrupt.) The Micro D-Cam then causes all the cells to again be set to 1, and the image-sensing cycle starts over.

A white pixel (logic 0) in the output indicates that the capacitor was exposed to a light intensity sufficient to discharge it past the threshold point. A black pixel (logic 1) indicates the light intensity was not enough to discharge the capacitor past the threshold point.

How fast the camera can scan the image varies according to the light intensity. The faster the elements are scanned, or read, the greater the light intensity required. The Micro D-Cam can scan approximately 15 frames per second at maximum speed.

The Optic RAM Resembles Film

The operation of the digital image sensor can be compared to that of a black-and-white film emulsion in a conventional photographic camera. Like the film, the IS32 contains many light-sensitive elements lying in a single plane. The image is focused (optically) on the plane, the user can adjust the aperture (measured in f-stops) and the length of exposure. The aperture is an adjustment of the size of the opening through which the light is allowed to pass on its way to the light-sensitive medium (changed in both cases by mechanically opening or closing an iris). The length of exposure (corresponding to photographic shutter speed) is adjusted in the Optic RAM by the scanning function of the drive electronics. And in the Optic RAM, the film is advanced, so to speak, by refreshing the voltage on all the memory elements.

Also like film, the Optic RAM's elements respond to light in a binary fashion, indicating only black or white, the presence or absence of a certain amount of light during the exposure. However, in a photographic film, the light-sensitive elements

---

1. 128- by 256-element array measuring 5.504 by 1.088 millimeters
2. Element size: 8 microns by 9 microns
3. Vertical center-to-center spacing: 21.5 microns
4. Horizontal spacing: 8.5 microns
5. Spacing between left and right arrays: 150 microns

Table 1: *Specifications of the Micron Technology IS32 Optic RAM, a 64K-bit memory chip that has the extra talent of serving as a digital image detector.*

(grains of silver-halide compounds) come in different sensitivities and respond to different intensities of light, whereas the IS32's cells all respond at about the same intensity for any given condition. To circumnavigate this limitation with the Optic RAM, varying shades of gray can be recorded by making multiple scans of the same optical image, averaging the results obtained from either changing the sensitivity of the cells, using a different threshold voltage for each scan, or varying the scan rate.

By changing the threshold voltage and keeping both the scanned image and light intensity constant, areas on the Optic RAM where intermediately bright portions of the image fall will give differing output levels. The nominal threshold potential, 2.1 volts (V), can be adjusted though pin 1 (Analog Threshold) on the IS32 from 1.5 V to 3 V, but Micron Technology suggests that gray-scale capability be achieved by varying the scan rate rather than by adjusting the threshold voltage. Changes will be exhibited in the response of the pixels where the image is gray (of intermediate brightness) so that the amount of light striking the cell capacitors is near the threshold voltage. Of course, a darker area of the image will generate more logic 1s as output than logic 0s, and a lighter area will generate more logic 0s. By averaging these outputs over a number of scans, the appropriate shade of gray can be produced in a composite image representation.

The Micro D-Cam may not contain a mechanical shutter, but its electronic equivalent is easily controlled by sending the appropriate commands to the control circuitry. The Optic RAM's sensitivity to light varies according to the electrical voltages present on it, allowing for precise continuous control of the Micro D-Cam's exposure values.

Ease of Use

Hooking up the Micro D-Cam to a computer is easy. The unit's control circuitry provides all the requisite timing signals and circuitry to execute commands received from the computer. The Micro D-Cam automatically sequences the Optic RAM so that each image-sensing cell is accessed and the appropriate video information transmitted to the computer for display or processing.

The Micro D-Cam uses a C-mount lens (the type commonly used in 16-millimeter movie cameras and small television cameras) with variable focus. The lens I chose was designed for viewing objects from a distance of at least 18 inches (45 cm); from this distance, the Micro D-Cam can distinguish characters of the size you are now reading. For viewing objects under greater magnification, you can insert a close-up adapter between the lens and its mount to extend the focal length of the lens. (See photo 2.)

The link between the computer and the Micro D-Cam is a TTL- (transistor-transistor logic) level serial interface. The external data-rate clock signal allows the computer to be synchronized to the Micro D-Cam, so the camera can operate at a speed of its own choosing.

Five lines run between the camera and the computer, carrying the transmit, receive, ground, and external clock signals and +5-V power. A general-purpose type-6850 ACIA (asynchronous communication interface adapter) buffered chip performs serial-to-parallel and parallel-to-serial data conversion, mating the Micro D-Cam's nonspecific circuitry to the host computer, as illustrated by the Apple II Plus in this article.

Hardware Details: Timing, Refreshing, and Interrupts

The timing-generator circuit (see figure 1), which generates the timing signals for the operation of the Micro D-Cam, contains a CMOS (complementary metal-oxide semiconductor) oscillator circuit that generates the fundamental clock rate. This signal is divided down to produce the frequencies for various possible output data rates and controlling the IS32. The data-rate-clock signals control the sequence of operation of the interrupt generator and the transmitter circuit.

The oscillator circuit emits a fundamental 4.9152-MHz signal, which is buffered by a type-74C04 inverter section (IC20a). This clock signal is divided again by a type-D flip-flop and brought out to a set of data-rate-selection jumper connections. IC26 divides the frequency by increasing powers of 2; these various subharmonic outputs lead to other data-rate-selection jumpers. Jumper connections 5 through 8 select the data rate used in the transmitter and interrupt-generator circuit (figure 5 on page 27), while connections 1 through 4 are 16× clock signals used in the receiver circuit. The output of IC26's pin 7 drives the Optic RAM's timing circuitry, which generates the familiar $\overline{RAS}$ (row-address strobe), $\overline{CAS}$ (column-address strobe) and R/$\overline{W}$ (read/write) signals as used by most dynamic RAM chips.

When the camera is transmitting data from the Optic RAM, it is in the interrupt mode, and the $\overline{CAS}$ and R/$\overline{W}$ signals are provided to the Optic RAM. When the camera is not transmitting, the interrupt mode is off, and $\overline{CAS}$ and R/$\overline{W}$ are disabled; the active-high interrupt signal INT is low and its complement $\overline{INT}$ is high, so the output of the AND gate driving $\overline{CAS}$ remains high and the OR gate driving R/$\overline{W}$ remains low.

During an interrupt cycle, INT goes high and $\overline{INT}$ goes low, enabling $\overline{CAS}$ and R/$\overline{W}$. The high state of $\overline{RAS}'$ (RAS-bar-prime) passes through a delay line consisting of two inverter sections (IC20d and f) and an R/C (resistance/capacitance) network, and then, combined with INT through an AND gate (IC9c), causes $\overline{CAS}$ to go high. When this happens, the column address is latched into the Optic RAM. At this time the R/$\overline{W}$ signal is still high, so the value stored in the accessed pixel is read out. After another delay period, R/$\overline{W}$ goes low, writing a 1 bit into the accessed cell to restore its charge and make it again able to react to light. When $\overline{RAS}'$ returns low, the interrupt cycle is terminated and $\overline{CAS}$ and R/$\overline{W}$ are disabled.

Command-Receiver Circuit

The serial command line carries commands from the computer to the camera. This data enters the command-receiver section (figure 2) a single bit at a time and is assembled according to the following protocol. The first bit to arrive is the start bit, followed by 8 data bits and then the stop bit. The start bit enables operation of the input shift register and starts the shift-register clock, which is initially low. When the clock goes high, the start bit, always a high level, is latched into the first of eight data positions in the shift register. When the clock goes low, the first data bit arrives at the shift register's input.

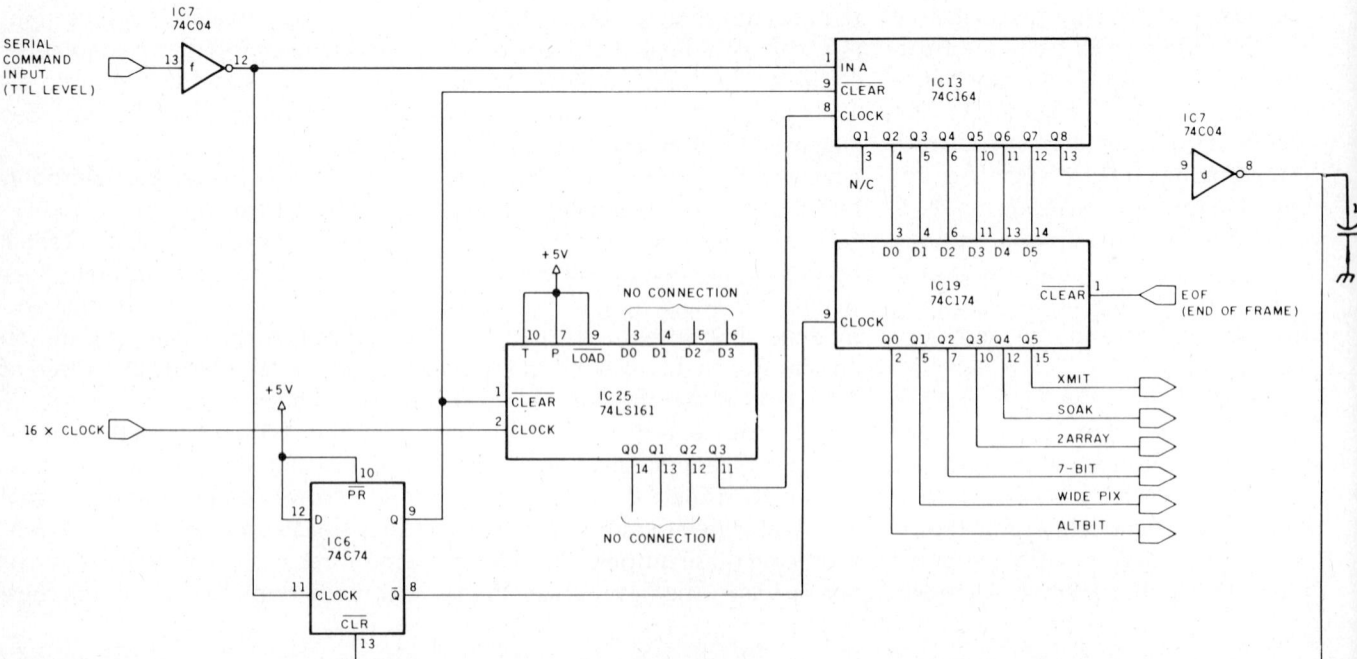

Figure 2: *The serial command line carries commands from the computer to the camera. The data enters the command-receiver section serially and is assembled by the shift register into a decodable word stored in the latch.*

When the rising edge of the clock pulse is detected, the shift register moves the high start bit from position 1 to position 2 and shifts the first data bit from the shift register's input into position 1. As successive bits arrive, each one is shifted into the shift register when the rising edge of the clock pulse is detected.

When the start bit finally reaches position 8, the camera has received the entire command byte, so the first 6 data bits are transferred from the shift register into a latch (a 1-byte memory) called the *command register*. The clock is then disabled and the shift register cleared, leaving the 6 camera-command bits in the command register. The receiver is now ready to accept another command.

Address Registers

The address registers of the circuit (see figure 3) latch the row-address, column-address, and refresh pointers for the Optic RAM addressing. Address registers IC22 and IC16 hold the row and column addresses, respectively, while the third register, IC10, is the *refresh register*.

The first two registers are activated only when the camera is to fetch and transmit a single bit of information from the Optic RAM. (This fetch operation is the interrupt cycle, which, as we saw before, is initiated by the INT signal going high.) The cycle starts on the occurrence of the falling edge of the $\overline{RAS}$ signal and ends on the next falling edge of $\overline{RAS}$. When the camera is not fetching in an interrupt cycle, the refresh register is active. This third address register

---

An unconventional dynamic-RAM manufacturer has recognized the light-sensing potential of its 64K-bit device.

---

continually increments the row-address value from 0 through to 255. Except during interrupts and exposures, this value passes through to the address lines of the IS32, performing a refresh operation. All three address registers have three-state outputs (that is, their outputs can assume a high-impedance condition, not driving the bus either high or low), and only one register is active at any one time.

The selected register drives its data onto a common bus called the *present-address* bus. The present address passes through the descramble-and-soak circuitry (which will be discussed shortly) to the Optic RAM, where it is used to select a row or column. The present-address bus also connects to the address circuit, where a value of 0, 1, or 2 (depending upon software-selected options) is added to the present-address value. The resulting sum is driven out of the adder onto the *next-address* bus, which connects to the inputs of each of the address registers. The value on the next-address bus is latched into the selected address register, and then that register is disabled.

The array-selection circuit simply selects whether one or both of the IS32's cell arrays are to be used. If $\overline{2ARRAY}$ is high, the output of the OR gate (IC21, pin 11) is always high, and the row-register value (IC22) will never be less than 128, so only the second array (rows 128 to 255) will be addressed and transmitted. If $\overline{2ARRAY}$ is low, however, the OR gate will appear transparent and the value on the next-address-bus line D7

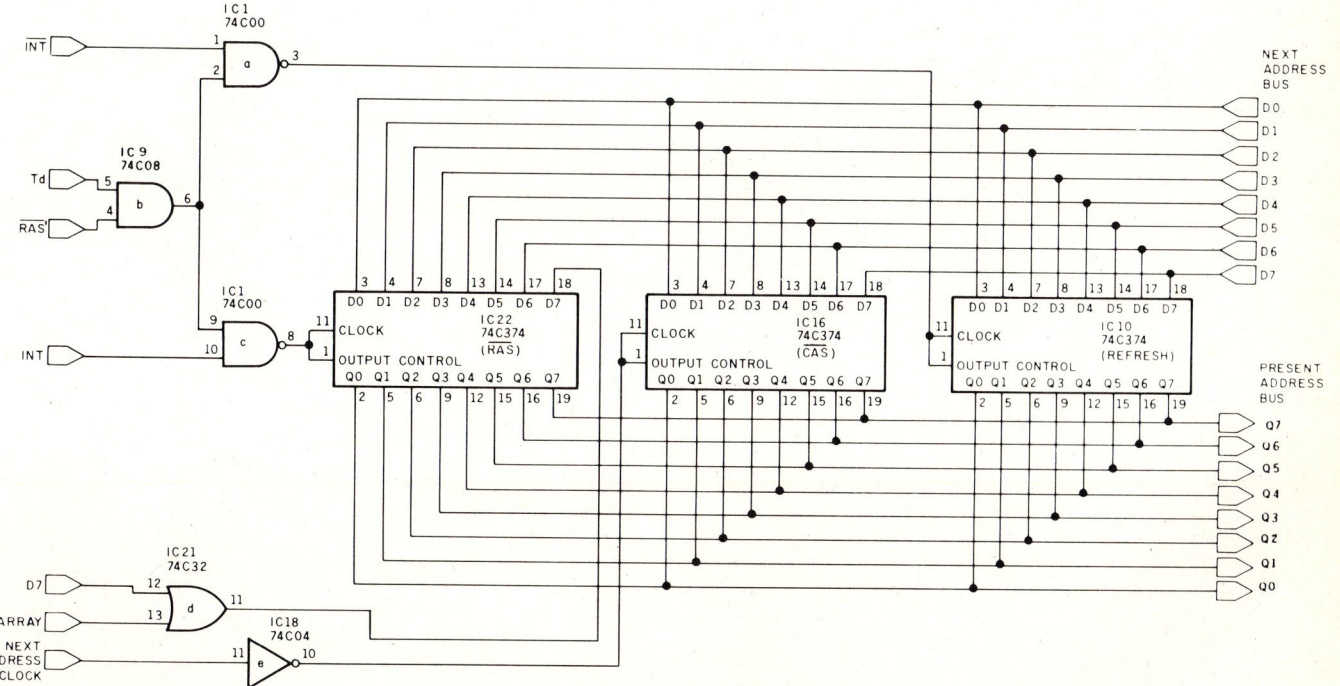

Figure 3: *This section of the circuit latches the row-address, column-address, and refresh pointers for the Optic RAM addressing. Address registers IC22 and IC16 hold the row and column addresses, respectively, while the third register, IC10, is the refresh register.*

will be driven onto IC22. This means all addresses from 0 to 255 will be selected and the values in both arrays will be transmitted.

Address Descramble and Array Soak

The internal circuitry in the Optic RAM scrambles the row and column-address values when accessing a cell. (After all, the IS32 chip was designed for use only as a memory device, not as an optical sensor.) But because element location is a critical issue in optical work, the address-descramble circuit (see figure 4, below) unscrambles the values into a new address, which the Optic RAM decodes to access the desired pixel.

Charged with the task of transforming the data from the address registers into a new address, which the Optic RAM decodes to access the desired pixel, the descrambling circuit consists of two inverters, three exclusive-OR gates, and a multiplexer (IC11). The inverters and exclusive-ORs do the actual descrambling on the row and column addresses; the multiplexer selects between the descrambled row and column addresses at the appropriate times and transmits the address to the Optic RAM.

The multiplexer uses $\overline{RAS}$ to determine which address is selected. If $\overline{RAS}$ is low at the multiplexer's SELECT input (IC11, pin 1), the descrambled row addresses (on the B inputs) are selected. When $\overline{RAS}$ is low, the A inputs, or descrambled column-address inputs, are selected.

The purpose of the $\overline{SOAK}$ circuit is to prevent the refresh addresses from reaching the Optic RAM during the exposure cycles. (Remember, the Optic RAM is light-sensitive only when it is not being refreshed.) During periods when INT is inactive-low (with the refresh register therefore active) and $\overline{SOAK}$ is active-low, the

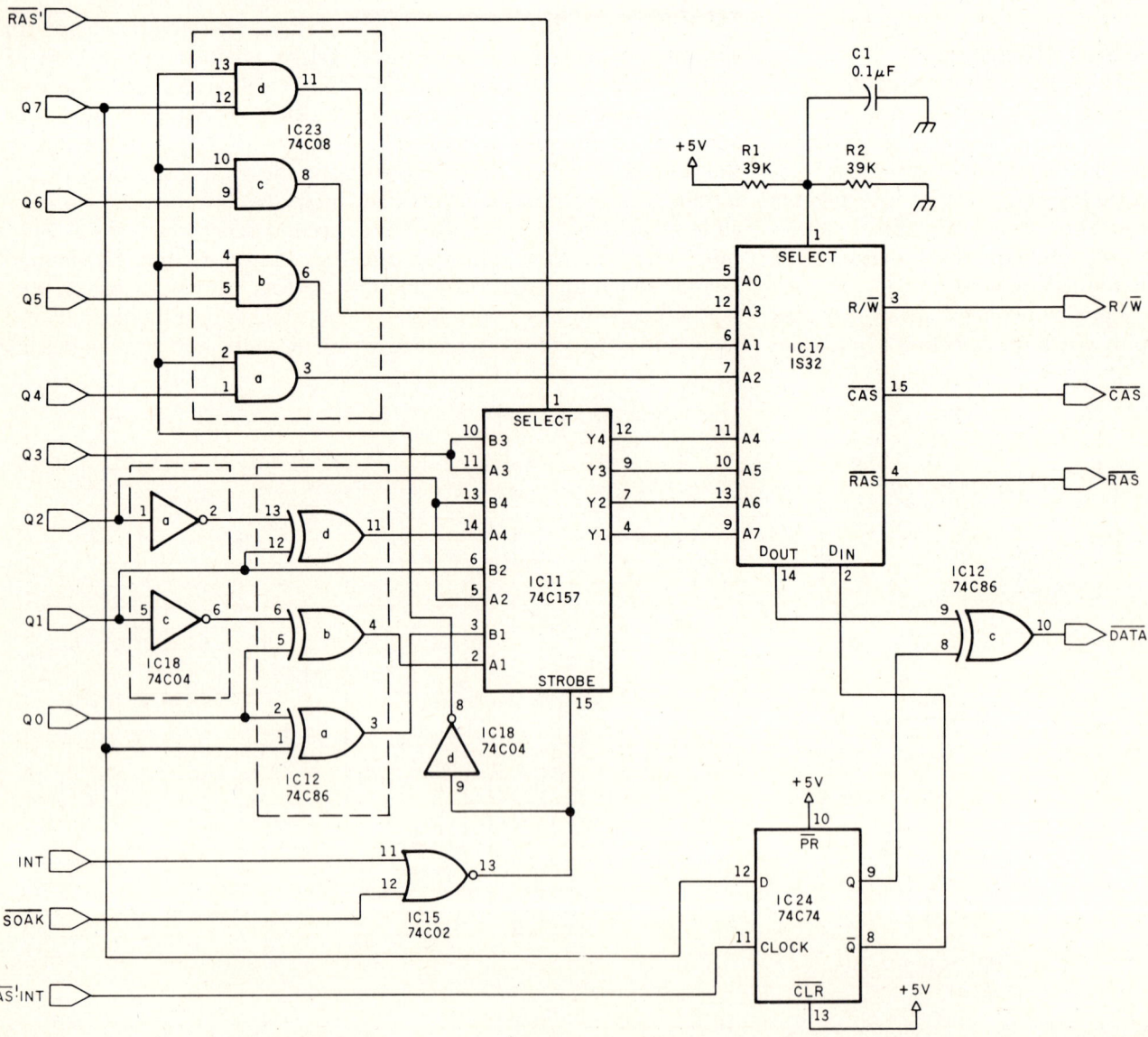

Figure 4: *Consisting of two inverter sections, three exclusive-OR gates, and a multiplexer, the address-descrambling circuitry undoes the internal address scrambling done by the Optic RAM. The soak circuit makes the Optic RAM light-sensitive by depriving it of refresh cycles.*

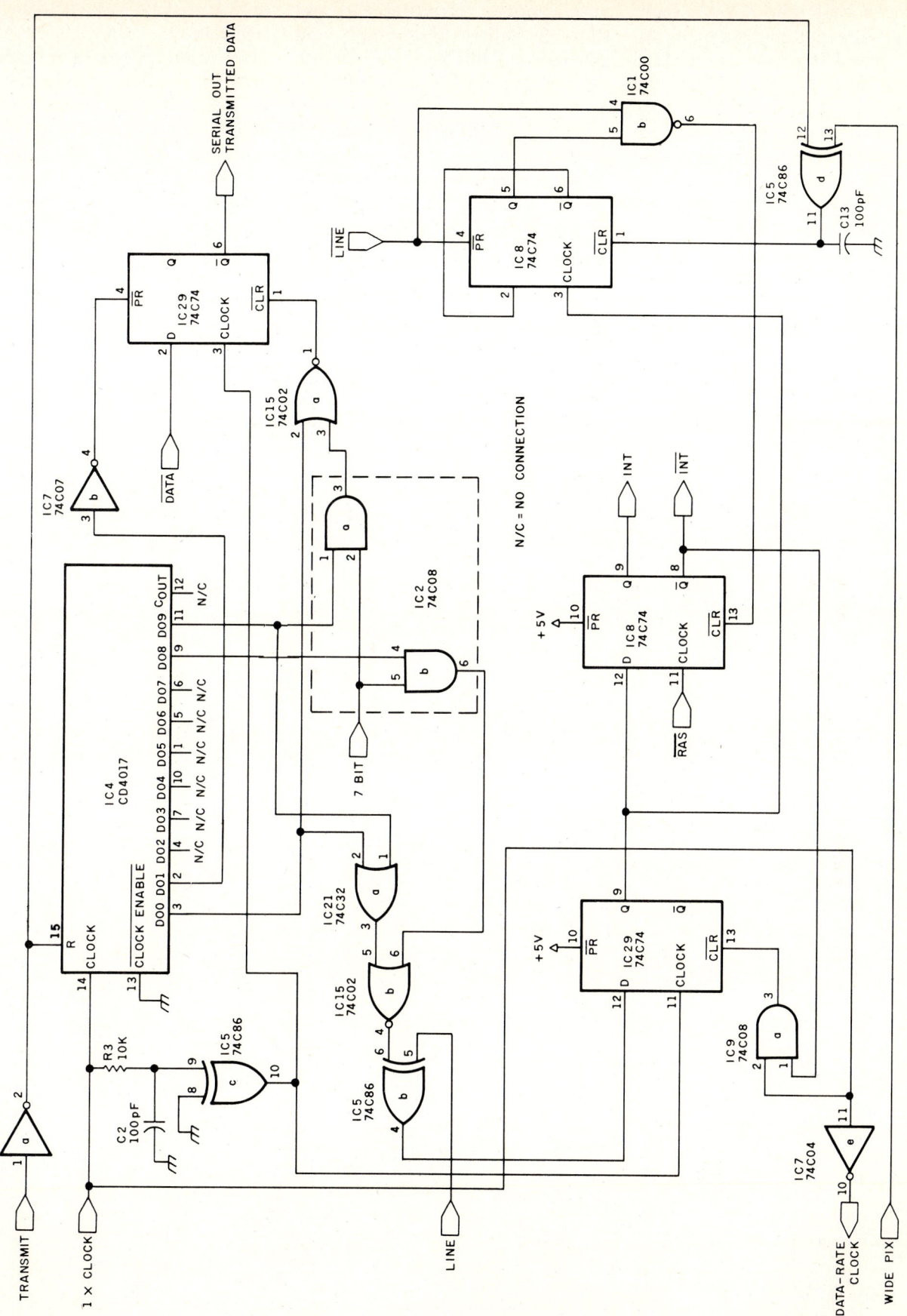

Figure 5: *The circuitry for transmitting the pixel data to the host computer and for generating the interrupt states that allow data to be read from the pixels in the IS32.*

output of NOR gate IC15d is high. This sets the multiplexer's enable input high and drives the multiplexer's outputs low. The high NOR-gate (IC15d) output also forces a low state at the inverter output IC18d, which forces the outputs of the four AND gates IC23a, b, c, and d low. These AND gates stand between the present-address bus and the IS32's four low-order address inputs. Thus, the Optic RAM's address inputs remain low, and the refresh function is performed on only address 0. When $\overline{SOAK}$ goes inactive-high, the multiplexer and AND-gate outputs are enabled and the refresh addresses reach the Optic RAM so that the entire chip is refreshed, making it insensitive to light.

Transmitter and Interrupt-Generator Circuit

This circuit, shown in figure 5, transmits the pixel data serially to the host computer, inserting start and stop bits where appropriate, and generates the INT and $\overline{INT}$ signals for fetching the pixel information from the IS32.

At the heart of this circuit is the ripple counter, IC4, enabled when the

> **The output of the Micro D-Cam is a digital signal; it cannot be used to directly drive a composite-video monitor.**

Micro D-Cam has been commanded to transmit data. It inhibits the interrupt generator when start and stop bits are being transmitted (preventing accessing of the Optic RAM) and enables the interrupt circuit when it is transmitting data. The transmitter's frequency is determined by the data-rate clock. During each clock cycle only one start, stop, or data bit is transmitted.

The interrupt generator is enabled by both the ripple counter (IC4) and the data-rate clock, but the interrupt cycle itself is clocked by $\overline{RAS}$. Because the purpose of the interrupt cycle is to fetch a single pixel for transmission, only one pixel can be transmitted on each clock cycle. The rising edge of the data-rate clock enables the interrupt circuit. The next falling edge of the $\overline{RAS}$ waveform initiates the interrupt cycle, causing a pixel to be read from the Optic RAM. The $\overline{INT}$ signal feeds back into the interrupt circuit, resetting the interrupt enable.

When $\overline{RAS}$ goes low again, the interrupt cycle is terminated. The next falling edge of the data-rate clock enables the interrupt circuit again (unless a start or stop bit is to be transmitted). Thus, only one pixel is transmitted during each data-rate-

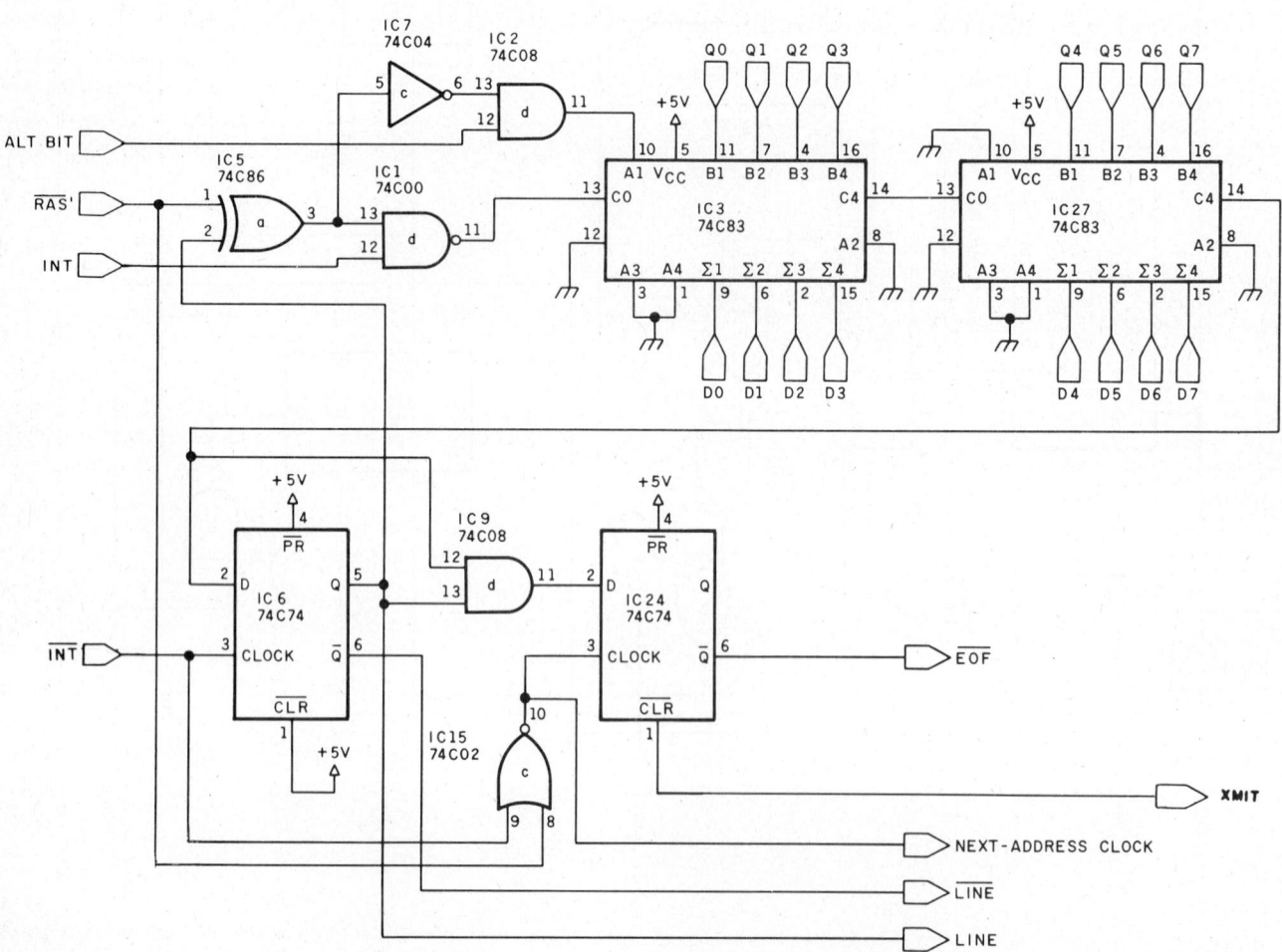

Figure 6: *The adder circuit allows the Micro D-Cam to keep track of the proper values for the row, column, and refresh registers.*

Photo 3: *The control circuitry for the Micro D-Cam image camera is shown here in prototype form mounted in an input/output slot in an Apple II Plus computer.*

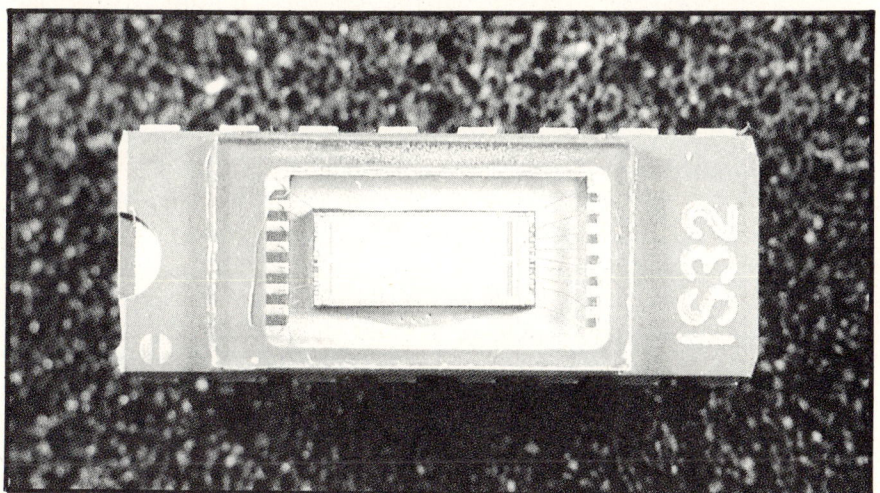

Photo 4: *The IS32 Optic RAM, a 64K-bit dynamic memory device specially packaged with a quartz lid for optical use in image-sensing applications, is made by Micron Technology Inc., 2805 East Columbia Rd., Boise, ID 83706, (208) 383-4000.*

Number	Type	+5V	GND	Number	Type	+5V	GND
IC1	74C00	14	7	IC17	IS32	16	8
IC2	74C08	14	7	IC18	4069	14	7
IC3	74C83	5	12	IC19	74C174	16	8
IC4	4017	16	8	IC20	4069	14	7
IC5	74C86	14	7	IC21	74C32	14	7
IC6	74C74	14	7	IC22	74C374	20	10
IC7	74C04	14	7	IC23	74C08	14	7
IC8	74C74	14	7	IC24	74C74	14	7
IC9	74C08	14	7	IC25	74LS161	16	8
IC10	74C374	20	10	IC26	4040	16	8
IC11	74C157	16	8	IC27	74C83	5	12
IC12	74C86	14	7	IC28	74C00	14	7
IC13	74C164	14	7	IC29	74C74	14	7
IC14	74LS74	14	7	IC30	MC6850	12	1
IC15	74C02	14	7	IC31	74LS245	20	10
IC16	74C374	20	10				

Table 2: *Power wiring for integrated circuits in the Micro D-Cam.*

clock cycle.

The WIDEPIX circuit is used to help compensate for the mismatch in aspect ratios of the Optic RAM and most computer graphics screens. The Optic RAM has a ratio of 2.5:1, compared with the 4:3 aspect ratio of most cathode-ray-tube (CRT) displays, and the pixels are not square. If the image data is displayed on a screen with an aspect ratio that close to 1:1, the image will appear to have been squeezed horizontally. The WIDEPIX circuit helps compensate for this by causing each pixel to be transmitted twice, doubling the width of the image. The circuit is enabled when the Micro D-Cam is transmitting and the WIDEPIX command line is high. This causes the flip-flop IC8a's output to toggle on every data-rate-clock cycle. This flip-flop inhibits the interrupt cycle on alternate data-rate clock cycles. During data-rate-clock cycles in which the interrupt is inhibited, the pixel from the previous interrupt cycle is transmitted again.

The LINE and $\overline{\text{LINE}}$ signals indicate that the column-address register has reached terminal count, meaning that the last column has been scanned. These signals, when active, inhibit the occurrence of further interrupts during the transmission of the current byte so that the value of the last accessed data bit is repeated to fill out the bits in the byte. This guarantees that the next byte transmitted will start off with information from the next row, i.e., that no single byte will contain picture information from two rows. When the stop bit is to be transmitted, assertion of the LINE signal at one input of exclusive-OR gate IC5b causes an interrupt request, and assertion of $\overline{\text{LINE}}$ at the input of the NAND gate IC1b ensures that the interrupt flip-flop is enabled. This "dummy" interrupt is used to increment the row-address register. The pixel that is accessed during this cycle is blanked by the transmission of the stop bit.

Adder and End-of-Frame Circuit

The adder and end-of-frame section, shown in figure 6, adds the proper increments to the row, column, and refresh registers and generates signals indicating end-of-frame (EOF) in the Optic RAM.

When any one of the address registers drives a value onto the present-address bus, the adder circuit receives this value, adds a 0, 1, or 2 to it (depending on the control inputs $\overline{RAS}$, LINE, ALTBIT and INT), and places the sum onto the next-address bus. When the refresh register is active, the INT line causes a 1 to be added each cycle. During interrupt cycles, the row and column registers are active. The adder sequences these registers through the Optic RAM in a "column-fast" mode, i.e., the adder adds 0 to the row address and 1 to the column address until the end of the column (or end of the line) is reached. The adder then adds a 1 to both the row and column, thus incrementing the row register and resetting the column register to 0.

The ALTBIT input simply adds an extra 1 to the value on the present-address bus during interrupt cycles; thus the row and column registers are incremented by a total of 2 rather than 1.

Control and Use

The software routines that control the Micro D-Cam are menu-driven. While the camera is running, several real-time commands are available to alter the operation of the camera from frame to frame. The real-time options are displayed on the screen.

When the camera is first turned on, you start the image-gathering process by selecting one of the options from the menu offered by the software, which I'll discuss in detail next month. If everything is working properly, an image of what the Micro D-Cam is seeing is shown on the computer's video-display screen. If the display screen remains dark, the exposure interval may be insufficient; this situation may be remedied by increasing the exposure time. If the exposure time is excessive, the screen will be white. This situation may be remedied by decreasing the exposure time or changing the aperture on the lens. Eventually, a clear picture will appear on the computer's screen as you reach the proper adjustments.

Special thanks to Carl Baker and Jim Herrud of Micron Technology for their contributions to this project.

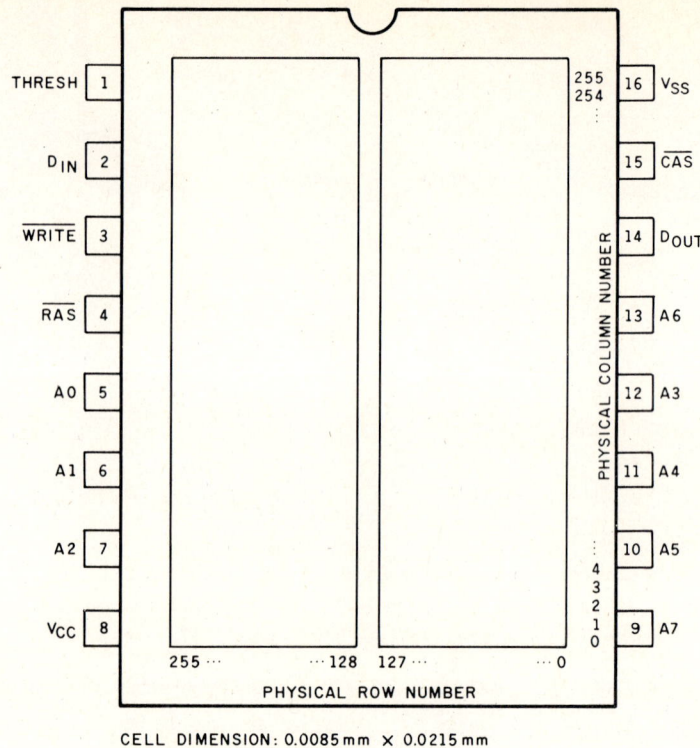

Figure 7: *A diagram of the topology and pinout configuration of the IS32 Optic RAM (not to scale). Each of the two cell regions, visible through the quartz package lid, contains a 128-by 256-cell array.*

The following items are available from:
 The Micromint, Inc.
 25 Terrace Drive
 Vernon, CT 06266
 For orders: (800) 635-3355
 For information: (203) 871-6170

1. *Complete Micro D-Cam unit including interface card, extension cable, IS32 Optic RAM, lens, remote housing, operators manual, and utility software. Specify Apple II (Plus or E), or IBM Personal Computer.*
   Assembled and tested $295
2. *Same as Item 1 except in kit form. Specify Apple II or IBM Personal Computer version.*
   Complete kit $260
3. *IS32 Optic RAM sold separately IS32 each $42*
4. *RS-232C-interfaced Micro D-Cam for general use. Call for price and delivery.*

Please add $4 shipping and insurance in continental United States, $20 overseas. New York residents please include 7 percent sales tax.

Build the Micro D-Cam Solid-State Video Camera

Part 2: Computer Interfaces and Control Software

Serial interfaces for the Apple II and the IBM Personal Computer and versatile software for the Apple II

Last month I introduced you to the Micro D-Cam, a relatively low-cost direct-output digital camera that you can build, either from scratch or from a kit distributed by The Micromint. Using a 64K-bit dynamic memory chip as its optical sensor, it has a resolution of 256 by 128 pixels (picture elements), which is adequate for many applications, including input of graphic images, pattern and character recognition, robotics, process control, and security.

In part 1 I explained the principles of operation of the IS32 Optic RAM (random-access read/write memory) and the rest of the Micro D-Cam's hardware. (Table 1 may help you recall some of the IS32's characteristics.) This month I'd like to finish the project by discussing how the camera can be attached to the expansion buses of the Apple II Plus and the IBM Personal Computer and how the camera is programmed to work.

The amount of software included with this article is somewhat more than you've come to expect from a hardware-type fellow like me, but I feel it is necessary to properly show how software can be used to enhance the final picture. In particular, some of you may be interested in the method used to present a gray scale on an Apple II computer.

A Quick Review

The IS32 Optic RAM from Micron Technology Inc. is a memory chip specially packaged to function as a digital image-sensing device. (Because its output is a pure digital signal, it cannot be used to directly drive a composite-video monitor.) The IS32 contains 32,768 usable light-sensitive elements arranged in a matrix of 128 rows and 256 columns. Each of the elements in the matrix is a light-sensitive capacitor, a memory cell that can be accessed randomly by simply reading in the appropriate row and column address. Light striking a particular element causes the capacitor, which is initially precharged to a fixed voltage, to discharge toward 0 volts (V). The capacitor discharges at a rate proportional to the light intensity throughout the duration of the exposure. When the cell's content is read, a logic 0 remaining in the cell indicates a bright pixel—the capacitor was exposed to a light intensity sufficient to discharge the capacitor past the threshold point. A dark pixel is indicated by a logic 1 remaining in the cell, which happens when the light intensity is not sufficient to discharge the capacitor past the threshold point.

The operation of the image sensor can be compared to the function of film in a camera. The user can regulate the exposure by two adjustments: aperture (f-stop) and shutter speed. The aperture adjustment controls the amount of light that is allowed to expose the light-sensitive medium (either the IS32 or the film emulsion) by mechanically widening or narrowing the hole through which the light passes. The shutter speed (or scanning speed in the case of the IS32) dictates the amount of time the sensitive medium is exposed.

---

1. two 128- by 256-element arrays each measuring 5.504 by 1.088 millimeters
2. element size: 8 microns by 9 microns
3. vertical center-to-center spacing: 21.5 microns
4. horizontal spacing: 8.5 microns
5. spacing between left and right arrays: 150 microns

Table 1: *Specifications of the Micron Technology IS32 Optic RAM, a 64K-bit memory chip that has the extra talent of serving as a digital image detector.*

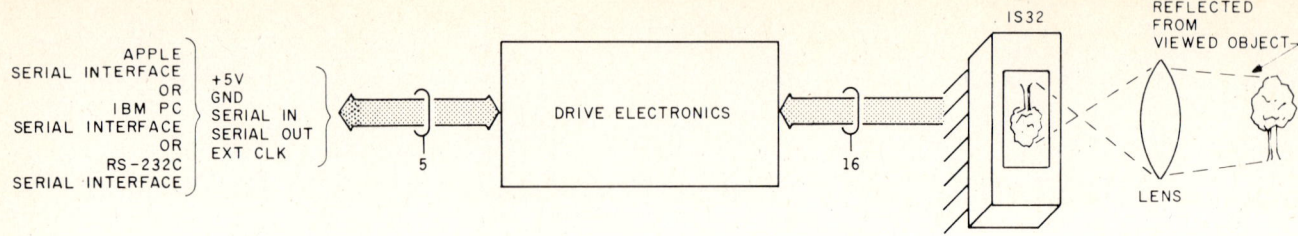

Figure 1: *A block diagram of the Micro D-Cam system.*

The Micro D-Cam's equivalent of an electronic shutter is controlled by commands transmitted to the interface. Sending a SOAK command to the Micro D-Cam has the effect of opening the shutter. After the appropriate period of exposure has elapsed, two commands, REFRESH and SEND, stop the exposure (close the shutter) and transmit the image to the host computer.

Interfacing the Micro D-Cam

Last month, when we looked at the control and driver electronics of the basic Micro D-Cam, we found that it communicates with its host computer serially, one bit at a time. In its minimal configuration, it requires four wires to be connected to the host computer: two supplying +5 V and ground potential and one each for serial data in and out. In a nonspecific configuration, it can operate asynchronously over an RS-232C link (at a data rate of up to 19,200 bps or bits per second), but I have devised serial interfaces for the camera that can be attached directly to the IBM PC and Apple II computers' buses (although still communicating serially). Using a fifth signal, an additional external clock signal provided to the bus interfaces by the drive electronics, the Micro D-Cam can then function at data rates up to 153,600 bps. The complexity of interface circuits of this type depends upon the host computer's bus structure and address range. The general scheme of connection is shown in figure 1.

Figure 2 is a schematic diagram of the circuit that forms the interface from the Micro D-Cam circuitry (shown in part 1) to the expansion bus of the Apple II Plus computer. It owes its simplicity to the predecoding of the I/O (input/output) slot address already provided on the Apple's main circuit board. The address decoders usually required in a peripheral interface are eliminated, and the complete serial interface can be built with only two integrated circuits. The 74LS245 octal bus transceiver buffers the TTL- (transistor/transistor logic) level serial data into and out of the MC6850 ACIA (asynchronous communication interface adapter). The serial bit rate is controlled by the external clock output from the Micro D-Cam drive electronics. For maximum speed, the clock frequency should be set for 153,600 Hz.

Figure 3 on page 33 shows the serial interface circuit for the Micro D-Cam configured for the IBM PC's bus. Due to the greater complexity of the Intel 8088 processor as compared with the Apple's 6502 and the PC's larger memory-address space, the in-

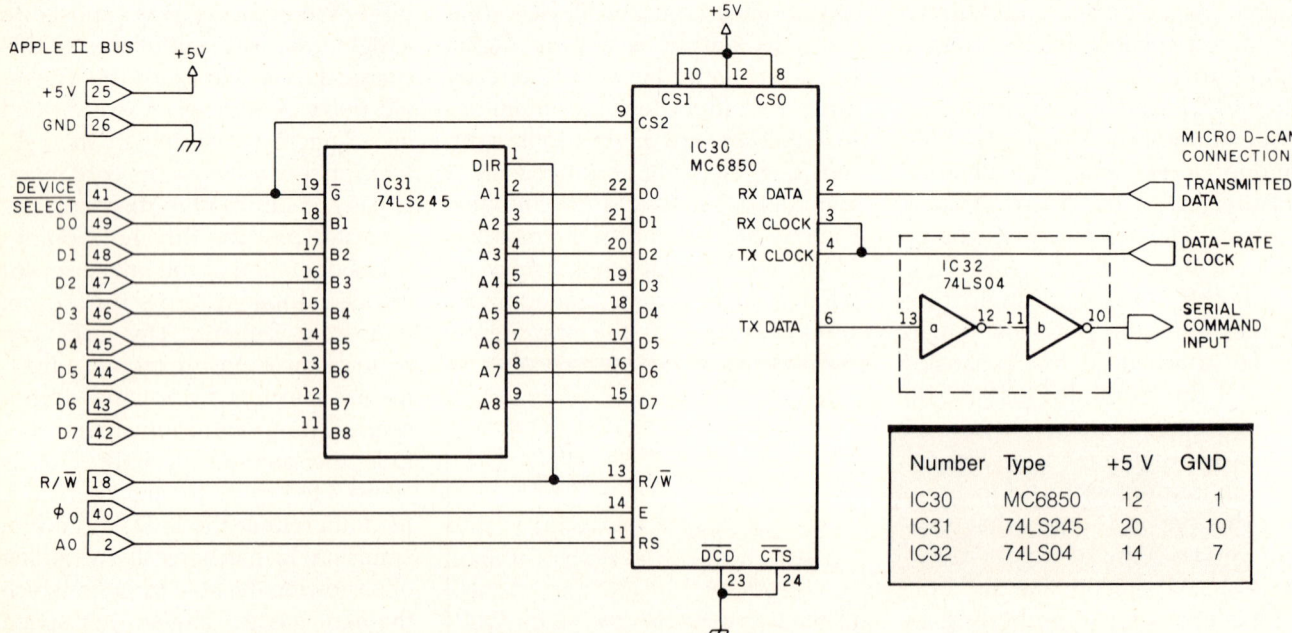

Figure 2: *A schematic diagram of an Apple II Plus or Apple IIe interface for the Micro D-Cam. The serial data stream from the Optic RAM is converted to parallel bytes and placed on the Apple's data bus by the ACIA and bus transceiver. Although operating asynchronously, high data rates (up to 153,600 bps) are possible because of the external data-rate clock input from the camera-control circuitry.*

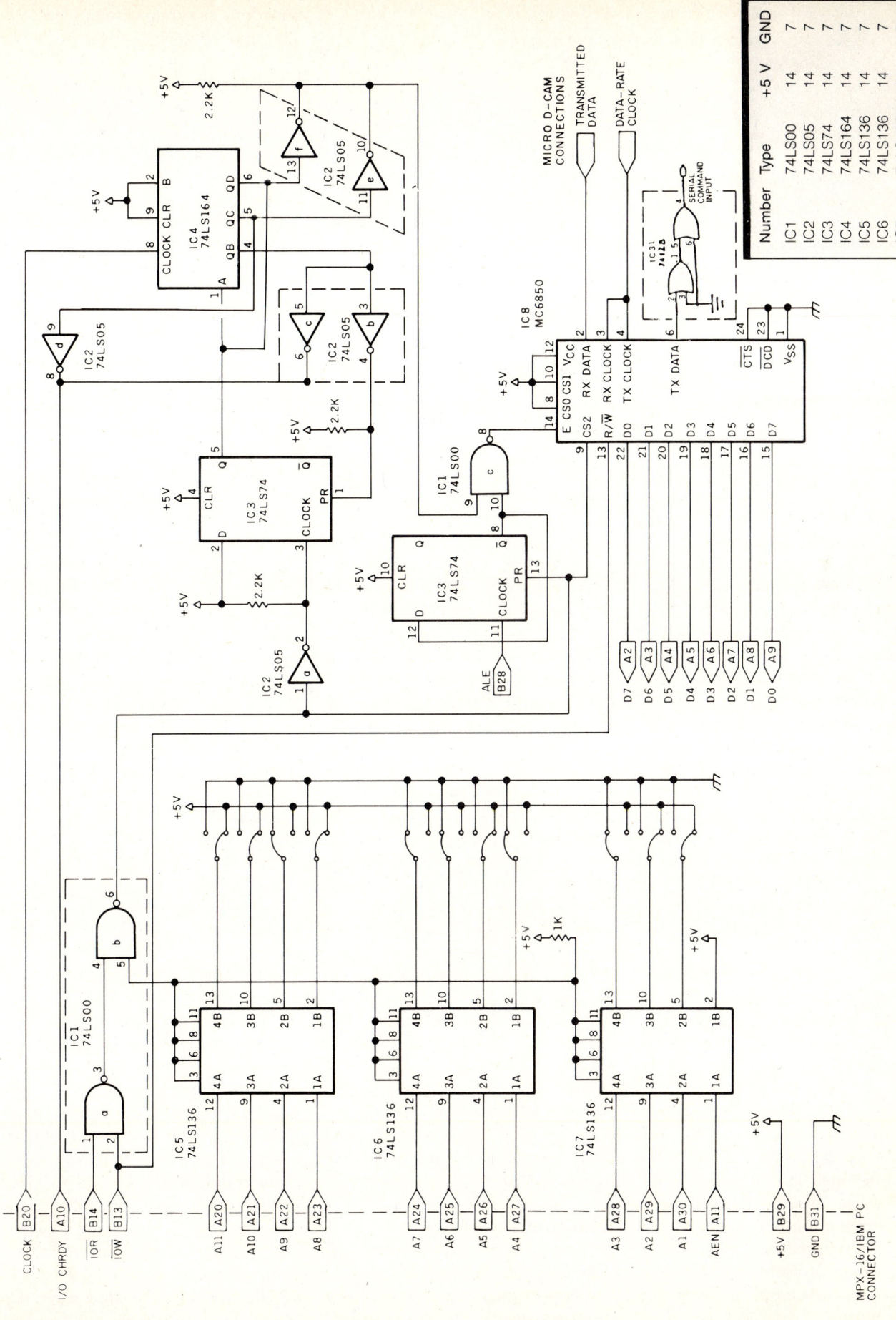

Figure 3: *An interface from the Micro D-Cam to an IBM Personal Computer (or a bus-compatible unit like the MPX-16). A set of jumper connections on the inputs of exclusive-OR gates determines the I/O-bus address of the interface, shown here set to hexadecimal xD26 and xD27.*

Table 2:

(2a)	Status Bit	Meaning When Set to 1
	0	data has been received from the camera
	1	a command may be sent to the camera
	2	unused
	3	unused
	4	received data was improperly framed
	5	data received before previous byte read

(2b)	Command Bit	Meaning When Cleared to 0
	7	none (always 1)
	6	none (always 1)
	5	alternating-bit mode (ALTBIT)
	4	wide-pixel mode (WIDEPIX)
	3	7-bit data bytes (7BIT)
	2	transmit one frame instead of two (1ARRAY)
	1	refresh instead of soak (REFRESH)
	0	send the requested image (SEND)

Table 2: *Meanings of bits in the status register (2a) and command word (2b) for the Apple II/Micro D-Cam interface.*

terface requires three times as many integrated circuits. In the IBM, the Micro D-Cam's two port addresses are decoded by three chips: IC5, IC6, and IC7. These are 74LS136 open-collector exclusive-OR gates connected together in a "wired-OR" configuration. The voltages wired to the 11 inputs of the address decoder determine the interface board's addresses. As shown in figure 3, the addresses I used were xD26 and xD27 (where x can take on any hexadecimal value from 0 to F). The 6850 ACIA (IC8) functions as previously described except that IC2 and IC4 are configured as a wait-state generator to facilitate timely access to the bus.

Data and Command Format

The 6850 ACIA comprises a data register and a status register. You can configure operating parameters (such as parity, stop bits, start bits, clocking, etc.) by writing values into the status register. Before the host computer can access the Micro D-Cam, the ACIA has to be initialized to the proper configuration. The control software does this by writing two bytes, a hexadecimal 03 followed by a hexadecimal 14, into the status register. The first byte performs a master reset on the ACIA, while the second byte specifies that the serial transmission protocol is 1 start bit, followed by 8 data bits, followed by 1 stop bit.

Reading the status register allows the control program to determine when new data has been received and when the ACIA is ready to send data. The meanings of the status bits, when set, are as shown in table 2a.

In normal use, only bit 0 is checked when seeing if data is available from the camera. Bits 4 and 5 are used only in debugging, as these situations should not normally arise. When designing the program that receives the image from the camera, it is a good idea to incorporate a time-out mechanism in case the camera stops sending bytes before the program expects; otherwise, the program can hang up if the software misses even a single byte.

In the Apple II Plus and IIe, the hexadecimal addresses of the type C0nE access the status register of the ACIA on an interface card plugged into the corresponding slot, while C0nF addresses access the ACIA's data register. The n is the hexadecimal value of the slot number plus 8. For example, suppose the interface card were plugged into slot 3; 3 plus 8 equals B, and so address C0BE will access the status register and C0BF the data register.

Command Functions

While the camera is running, the host computer directs the Micro D-Cam's operating modes by sending it command words. Each command word is composed of 8 bits, with functions as summarized in table 2b. Let's look at each of these in detail:

ALTBIT Mode: When bit 5 is clear (equal to 0), the Micro D-Cam transmits only the pixels from the even-numbered rows and columns in the Optic RAM. This mode usually produces a clearer image than the NOALTBIT mode at the expense of losing resolution.

WIDEPIX Mode: When bit 4 is clear, the Micro D-Cam transmits each pixel in the array twice. Each image-sensing element is rectangular in shape, so by "double-transmitting" the pixels, the proper width-to-height (aspect) ratio is maintained when the image is displayed on the computer's video monitor.

7BIT Mode: The Apple II's implementation of high-resolution graphics is somewhat peculiar. The most significant bit of each byte on the hi-res graphics page is reserved as the color bit for a group of pixels, while each of the other 7 bits stores a 1 or 0 as a bright or dark value for a pixel. In 7BIT mode, the Micro D-Cam transmits data in a format compatible with the Apple's high-resolution format, with 7 bits of pixel values per byte. The 7BIT mode is selected by clearing bit 3 of the command byte to 0. The alternative to 7BIT mode is 8BIT mode, which is achieved by setting bit 3 to 1. The 8BIT mode causes the camera to transmit in normal bit-mapped format, with all 8 bits in the byte containing image data, and is preferred for use with all computers other than the Apple.

1ARRAY Mode: The 1ARRAY mode is selected by clearing bit 2 of the command byte. Using this mode, only data from the image focused on the lower light-sensitive array is transmitted from the Micro D-Cam. By setting bit 2 of the command byte, 2ARRAY mode is selected, which causes data from both arrays to be transmitted from the camera. The 2ARRAY mode causes a split-screen effect because of the space between the two arrays in the image-sensor chip.

REFRESH Mode: In some ways, the Micro D-Cam is like any other

Command Character	Control Effect
>	increase exposure time
<	decrease exposure time
F	fix exposure time to current setting
L	load previously stored image from disk
N	print negative of screen image onto Epson printer
P	print screen output onto Epson (Graftrax option required)
Q	quit and return to main menu
R	toggle display of exposure time and light level
S	save current image to disk
T	use current light level and autotrack the exposure

Table 3: *Options for control of the Micro D-Cam that may be selected in real time through the distributed menu-driven software. See table 4 for the options provided in the GREY16 program.*

camera. It must receive the proper amount of light to make the image develop properly. Too much light will overexpose the image, while too little light will underexpose the image. Exposure time is determined by how long the control program in the host computer allows the Optic RAM to be exposed to light without its cells being refreshed. Refreshing the image sensor is the same process used in any dynamic memory: the existing charge in each cell is sensed, the voltage compared with a threshold potential, and a fresh potential of 0 V (for a logic 0) or +5 V (for 1) is rewritten into the cell. (The only difference in the Optic RAM is that all cells must contain +5 V at the beginning of an image-sensing cycle when refreshing stops.) If the image sensor is not continually refreshed, the light focused on each cell causes the voltage in each cell to leak away at a rate proportional to the intensity of the light. When the image sensor is not being refreshed, we say it is "soaking" (in light). Allowing the image sensor to soak for longer periods of time enables the Micro D-Cam to see better in dimmer light.

When the REFRESH mode is selected (by clearing bit 1 of the command byte) the Micro D-Cam keeps the image sensor's cells refreshed while it is sending an image. When bit 1 is set, SOAK mode is invoked. This causes the camera to soak (and therefore remain sensitive to light) while it is transmitting an image.

SEND Mode: When a command is sent to the camera with SEND mode selected (bit 0 cleared), the camera begins transmitting an image.

Control Software

The software for controlling and displaying pictures is vital to the operation of the Micro D-Cam. Menu-driven versions of the Micro D-Cam control software for both the Apple II and IBM PC are available from The Micromint.

However, some of you may already have the Micron Technology Optic RAM or a similar 64K-bit dynamic RAM device with suitable chip layout, and you may want to build the Micro D-Cam from scratch. Consequently, I have included with this article complete listings of two control programs written for the Apple II. One provides experimenters with a means for testing the Micro D-Cam; the second is a more sophisticated software routine that enhances the image and performs gray-scale ordered dithering (I'll explain this term later). While the Micro D-Cam software includes some additional menu-driven utility programs (some options of which are shown in table 3), all the Micro D-Cam photos printed here and last month can be reproduced using only the two programs in the magazine.

A Sample Control Program

The Micro D-Cam demonstration program (listing 1 on pages 36 through 39) illustrates the simplest possible software needed to receive an image from the camera and display it on the Apple's hi-res screen.

It is not really as long and complicated as it looks; the accompanying flowchart (figure 4 on page 50) should reveal the general scheme of operation. The software consists of two parts: a short BASIC main program (listing 1a) and a set of machine-language subroutines (shown in assembled format in listing 1b). The BASIC program loads the machine-language code from disk, interactively sets the correct I/O-slot number and exposure time, and calls the machine-language code to display the image; upon returning to BASIC, the calling program checks to see if you want to terminate the process.

The hard part of the work is done in the machine-language routines, which were necessary to allow the Micro D-Cam to operate at 153,600 bps. When called, the machine code begins by making sure that the hi-res screen is being displayed. It then initializes the ACIA and sends a command to tell the camera to soak without sending an image. (This effectively clears the Optic RAM and tells the camera to begin the exposure.) The program then waits for the duration of the exposure.

The next step is to read the image from the camera and display it on the screen. To save time and memory, the software sends the picture straight to the hi-res screen memory (rather than reading it into a separate buffer area and then moving it) to minimize the processing of the final image. The mode used is alternate-pixel, wide-pixel, with 7-bit data words. Before any part of the picture is received, a number of memory pointers are set up to facilitate proper placement on the screen. A command is sent to the camera to begin transmitting the image, and the program loops to read in each byte of the image and put it on the screen.

The control software knows how many bytes of image data it should receive from the camera, but a problem can arise from relying on byte-counting to determine when to stop reading data: if the computer misses one, it could hang the system up. To be on the safe side, a time-out loop has been provided in the image-reading routine. If the computer times out

Listing 1a: *The BASIC-language portion of the routines to test and demonstrate the Micro D-Cam.*

```
10  REM MICRO DCAM DEMONSTRATION
15  REM PROGRAM
20  REM
25  REM COPYRIGHT (C) 1983 BY
30  REM CIRCUIT CELLAR, INC.
35  REM
40  HGR : TEXT : HOME
50  PRINT CHR$(4)"BLOAD MICRO D
    CAM"
60  INPUT "ENTER CAMERA SLOT: ";S
    L
70  IF SL < 1 OR SL > 7 THEN 60
80  POKE 770,SL * 16: REM SLOT NU
    MBER
90  POKE 768,0: POKE 769,1: REM E
    XPOSURE TIME
100 POKE 771,0: REM UPPER 1/3 OF
     SCREEN
110 CALL 4096: REM UPDATE SCRN
120 IF PEEK (773) = 0 THEN 130:
    REM CHECK FOR KEYPRESS
130 IF PEEK (772) = 209 THEN TEXT
    : HOME : END : REM CHECK FOR
     'Q'
140 GOTO 110
```

Listing 1b: *Micro D-Cam control subroutines, written in 6502 assembly language, called as a machine-language module by the BASIC routine of listing 1.*

```
:ASM      1         ;************************************
          2         ;* MICRO DCAM SUPPORT ROUTINES
          3         ;*
          4         ;*       COPYRIGHT (C) 1983
          5         ;*       BY CIRCUIT CELLAR, INC.
          6         ;*
          7         ;* WHEN CALLED, THIS ROUTINE READS
          8         ;* AN IMAGE FROM THE CAMERA AND
          9         ;* DISPLAYS IT ON THE APPLE'S HI-RES
          10        ;* GRAPHICS SCREEN
          11        ;************************************
          12            KEYCLR   EQU  $C010
          13            KEYHIT   EQU  $C000
          14            BEEP     EQU  $C030
          15            GMODE    EQU  $C050
          16            TMODE    EQU  $C051
          17            MIXED    EQU  $C053
          18            PAGE1    EQU  $C054
          19            HGR      EQU  $C057
          20            STATUS   EQU  $C08E
          21            DATA     EQU  $C08F
          22        ;
          23            SOAKTIME EQU  $300
          24            SLOTADR  EQU  $302
          25            ROWSTART EQU  $303
          26            KEY      EQU  $304
          27            KEYEXIT  EQU  $305
          28        ;
          29            RADR     EQU  $06
          30            CTR      EQU  $08
          31            YREG     EQU  $19
          32        ;
                                 ORG  $1000
                                 
1000: 20 B9 10   33  NSTART     JSR  SETGR      ;SET UP FOR HIRES PAGE1
1003: 20 BF 10   34             JSR  ACIACLR    ;FLUSH THE INPUT BUFFER
1006: A9 D3      35             LDA  #$D3       ;SEND CMD TO SOAK W/O SEND
1008: 20 C6 10   36             JSR  SENDCMD
100B: 20 DA 10   37             JSR  SOAK
100E: A9 C0      38             LDA  #$C0       ;SEND IMAGE W/O SOAK
1010: 20 C6 10   39             JSR  SENDCMD    ;  (ALT,WIDEPIX,7BIT-256X64)
1013: A2 00      40             LDX  #0         ;INITIALIZE THE ROW INDEX
1015: A0 00      41  NEWROW     LDY  #0         ;INIT COLUMN INDEX (Y)
1017: BD 08 11   42             LDA  ROMPTR,X   ;BUILD BASE ADDRS FOR CUR ROW
101A: 18         43             CLC
101B: 6D 03 03   44             ADC  ROWSTART   ;0-SELECT UPPER 1/3 OF SCREEN
101E: 85 06      45             STA  RADR       ;$28-MID 1/3,$50-BOT 1/3
1020: E8         46             INX             ;RADR HAS ADDRS OF CUR ROW
1021: BD 08 11   47             LDA  ROMPTR,X
1024: 85 07      48             STA  RADR+1
1026: E8         49             INX
1027: 84 19      50             STY  YREG       ;PNT X TO NXT ADDRS IN ROMPTR
1029: AC 02 03   51  GET        LDY  SLOTADR    ;GET NEXT BYTE FROM CAMERA
102C: B9 8E C0   52             LDA  STATUS,Y   ;LOAD OFFSET TO CAMERA SLOT
                 53                             ;CHECK IF NEXT BYTE ARRIVED
```

Listing 1b continued:

```
102F: 4A           LSR
1030: B0 21        BCS  NORM3
1032: A9 00        LDA  #0            ;IF BYTE AVAILABLE BRANCH
1034: 85 08        STA  CTR           ;IF BYTE NOT YET AVAILABLE
1036: A9 15        LDA  #$15          ;SET UP TIMEOUT COUNTER
1038: 85 09        STA  CTR+1
103A: C6 08   NORM1 DEC CTR
103C: D0 0F        BNE  NORM2
103E: C6 09        DEC  CTR+1         ;CHK FOR BYTE UNTIL TIMED OUT
1040: D0 08        BNE  NORM2
1042: AD 30 C0     LDA  BEEP          ;IF TIMED OUT, CLICK SPEAKER
1045: AD 00 C0     LDA  KEYHIT        ;CHECK FOR KEYPRESS
1048: 30 1D        BMI  NDONE         ;IF KEY HIT THEN RETURN
104A: 4C 03 10     JMP  NSTART        ;OTHERWISE, TRY COMMAND AGAIN
104D: B9 8E C0 NORM2 LDA STATUS,Y
1050: 4A           LSR
1051: 90 E7        BCC  NORM1
1053: B9 8F C0 NORM3 LDA DATA,Y       ;WHEN BYTE AVAILABLE GET IT
1056: A4 19        LDY  YREG          ;RESTORE COL POINTER TO Y
1058: C0 28        CPY  #$40          ;IF PAST 40TH BYTE IN CUR ROW,
105A: B0 02        BCS  NORM4         ;DONT PUT ONTO HIRES SCREEN
105C: 91 06        STA  (RADR),Y
105E: C8     NORM4 INY                ;INCREMENT COLUMN POINTER
105F: C0 25        CPY  #$37          ;END OF THE COLUMN?
1061: D0 C4        BNE  GET           ;IF NOT, GET THE NEXT BYTE
1063: E0 80        CPX  #$80          ;OTHERWISE
1065: D0 AE        BNE  NEWROW        ;IF NOT DONE, GOTO NEXTROW
1067: A9 D1  NDONE LDA #$D1
1069: 20 C6 10     JSR  SENDCMD
106C: A9 20        LDA  #$20
106E: 8D 04 03     STA  KEY
1071: A9 00        LDA  #0
1073: 8D 05 03     STA  KEYEXIT
1076: AD 00 C0     LDA  KEYHIT        ;CHECK IF KEY WAS HIT
1079: 10 13        BPL  KEY1
107B: 2C 10 C0     BIT  KEYCLR        ;CLEAR THE KEYBOARD STROBE
107E: EE 05 03     INC  KEYEXIT       ;SET 'EXIT CAUSED BY KEY' FLAG
1081: 8D 04 03     STA  KEY
1084: C9 D1        CMP  #"Q"          ;IF THE KEY WAS A 'Q'
1086: D0 06        BNE  KEY1
1088: 20 A1 10     JSR  GRCLR         ;CLEAR THE GRAPHICS SCREEN
108B: AD 51 C0     LDA  TMODE         ;RETURN TO TEXT MODE
108E: 60           RTS
                KEY1
                *
108F: A9 03  ACIACLR LDA #3           ;MASTER RESET ACIA
1091: 84 19        STY  YREG
1093: AC 02 03     LDY  SLOTADR
1096: 99 8E C0     STA  STATUS,Y
1099: A9 14        LDA  #$14          ;1 START,8 DATA,1 STOP,EXT CLK
109B: 99 8E C0     STA  STATUS,Y
109E: A4 19        LDY  YREG
10A0: 60           RTS
                *
10A1: A2 00  GRCLR LDX  #0            ;CLEAR PAGE1 OF HIRES
10A3: A0 00        LDY  #0
10A5: 84 06        STY  RADR
10A7: A9 20        LDA  #$20
10A9: 85 07        STA  RADR+1
10AB: 8A     CLR1  TXA
10AC: 91 06        STA  (RADR),Y
10AE: C8           INY
10AF: D0 FB        BNE  CLR1
10B1: E6 07        INC  RADR+1
10B3: E8           INX
10B4: E0 20        CPX  #$20
10B6: D0 F4        BNE  CLR1
10B8: 60           RTS
                *
10B9: AD 53 C0 SETGR LDA MIXED        ;SET MIXED MODE
10BC: AD 57 C0     LDA  HGR           ;USE HIRES GRAPHICS
10BF: AD 54 C0     LDA  PAGE1         ;USE PAGE1 OF HIRES
10C2: AD 50 C0     LDA  GMODE         ;SWITCH TO GRAPHICS MODE
10C5: 60           RTS
                *
10C6: 84 19  SENDCMD STY YREG
10C8: AC 02 03     LDY  SLOTADR
10CB: 48           PHA
10CC: B9 8E C0     LDA  STATUS,Y
10CF: 29 02        AND  #2
10D1: F0 F9        BEQ  SEND1
10D3: 68           PLA
10D4: 99 8F C0 SEND1 STA DATA,Y       ;SEND BYTE IN A TO CAMERA
10D7: A4 19        LDY  YREG
10D9: 60           RTS
```

Listing continued

Listing 1b continued:

```
                            138  *
10DA: AD 01 03  139  SOAK   LDA  SOAKTIME+1  ;SOAK FOR NUMBER OF MS
10DD: 85 09     140         STA  CTR+1       ;SPECIFIED BY SOAKTIME
10DF: E6 09     141         INC  CTR+1
10E1: AD 00 03  142         LDA  SOAKTIME
10E4: 85 08     143         STA  CTR
10E6: E6 08     144         INC  CTR
10E8: AD 00 03  145         LDA  SOAKTIME
10EB: D0 05     146         BNE  SOAK1
10ED: AD 01 03  147         LDA  SOAKTIME+1
10F0: F0 0B     148         BEQ  SOAK2
10F2: 20 FE 10  149  SOAK1  JSR  MSEC
10F5: C6 08     150         DEC  CTR
10F7: D0 F9     151         BNE  SOAK1
10F9: C6 09     152         DEC  CTR+1
10FB: D0 F5     153         BNE  SOAK1
10FD: 60        154  SOAK2  RTS
                155  *
10FE: 84 19     156  MSEC   STY  YREG        ;1 MILLISECOND LOOP
1100: A0 C7     157         LDY  #199
1102: 88        158  MSEC1  DEY
1103: D0 FD     159         BNE  MSEC1
1105: A4 19     160         LDY  YREG
1107: 60        161         RTS
                162  *
1108: 00 20 00  163  ROWPTR HEX  0020002400280002C00300034003803C
110B: 24 00 28
110E: 00 2C 00
1111: 30 00 34
1114: 00 38 00
1117: 3C        164         HEX  8020802480028802C8030080348038803C
1118: 80 20 80
111B: 24 80 28
111E: 80 2C 80
1121: 30 80 34
1124: 80 38 80
1127: 3C        165         HEX  0021002500290002D00310003500390003D
1128: 00 21 00
112B: 25 00 29
112E: 00 2D 00
1131: 31 00 35
1134: 00 39 00
1137: 3D        166         HEX  8021802580029802D8031803580039803D
1138: 80 21 80
113B: 25 80 29
113E: 80 2D 80
1141: 31 80 35
1144: 80 39 80
1147: 3D        167         HEX  002200026002A002E0032003600003A003E
1148: 00 22 00
114B: 26 00 2A
114E: 00 2E 00
1151: 32 00 36
1154: 00 3A 00
1157: 3E        168         HEX  8022802680002A802E0032803680003A803E
1158: 80 22 80
115B: 26 80 2A
115E: 80 2E 80
1161: 32 80 36
1164: 80 3A 80
1167: 3E        169         HEX  0023002700280002F0033003700003B003F
1168: 00 23 00
116B: 27 00 2B
116E: 00 2F 00
1171: 33 00 37
1174: 00 3B 00
1177: 3F        170         HEX  8023802780028B002F803380003780038B03F
1178: 80 23 80
117B: 27 80 2B
117E: 80 2F 80
1181: 33 80 37
1184: 80 3B 80
1187: 3F        171         HEX  2820282428282C2830283428382B83C
1188: 28 20 28
118B: 24 28 28
118E: 28 2C 28
1191: 30 28 34
1194: 28 38 28
1197: 3C
1198: A8 20 A8
119B: 24 A8 28
119E: A8 2C A8
11A1: 30 A8 34
11A4: A8 38 A8
```

Listing 1b continued:

```
11A7: 3C                  172  HEX  AB20A824AB28AB2CAB30A834AB38AB3C
11AB: 28 21 28
11AB: 25 28 29
11AE: 28 2D 28
11B1: 31 28 35
11B4: 28 39 28
11B7: 3D                  173  HEX  2B212B252B292B2D2B312B352B392B3D
11BB: 25 AB 29
11BE: AB 2D AB
11C1: 31 AB 35
11C4: AB 39 AB
11C7: 3D                  174  HEX  AB21AB25AB29AB2DAB31AB35AB39AB3D
11CB: 28 22 28
11CB: 26 28 2A
11CE: 28 2E 28
11D1: 32 28 36
11D4: 28 3A 28
11D7: 3E                  175  HEX  2B222B262B2A2B2E2B322B362B3A2B3E
11DB: AB 22 AB
11DB: 26 AB 2A
11DE: AB 2E AB
11E1: 32 AB 36
11E4: AB 3A AB
11E7: 3E                  176  HEX  AB22AB26AB2AAB2EAB32AB36AB3AAB3E
11EB: 28 23 28
11EB: 27 28 2B
11EE: 28 2F 28
11F1: 33 28 37
11F4: 28 3B 28
11F7: 3F                  177  HEX  2B232B272B2B2B2F2B332B372B3B2B3F
11FB: AB 23 AB
11FB: 27 AB 2B
11FE: AB 2F AB
1201: 33 AB 37
1204: AB 3B AB
1207: 3F                  178  HEX  AB23AB27AB2BAB2FAB33AB37AB3BAB3F
```

--END ASSEMBLY--

ERRORS: 0

520 BYTES

SYMBOL TABLE - ALPHABETICAL ORDER:

```
ACIACLR =$108F    BEEP    =C030    CLR1    =$10AC    CTR      =$08
DATA    =$C08F    GET     =$1027   GMODE   =$C050    GRCLR    =$10A1
HGR     =$C057    KEY     =$0304   KEY1    =$108E    KEYCLR   =$C010
KEYEXIT =$0305    KEYHIT  =$C000   MIXED   =$C053    MSEC     =$10FE
MSEC1   =$1102    NDONE   =$1067   NEWROW  =$1015    NORM1    =$103A
NORM2   =$104D    NORM3   =$1053   NORM4   =$105E    NSTART   =$1003
PAGE1   =$C054    RADR    =$06     ROWPTR  =$1108    ROWSTART =$0303
SEND1   =$10CC    SENDCMD =$10C6   SETGR   =$10B9    SLOTADR  =$0302
SOAK    =$10DA    SOAK1   =$10F2   SOAK2   =$10FD    SOAKTIME =$0300
STATUS  =$C08E    TMODE   =$C051   YREG    =$19
```

SYMBOL TABLE - NUMERICAL ORDER:

```
RADR    =$06      CTR     =$08     YREG    =$19      SOAKTIME=$0300
SLOTADR =$0302    ROWSTART=$0303   KEY     =$0304    KEYEXIT =$0305
NSTART  =$1003    NEWROW  =$1015   GET     =$1027    NORM1   =$103A
NORM2   =$104D    NORM3   =$1053   NORM4   =$105E    NDONE   =$1067
KEY1    =$108E    ACIACLR =$108F   GRCLR   =$10A1    CLR1    =$10AC
SETGR   =$10B9    SENDCMD =$10C6   SEND1   =$10CC    SOAK    =$10DA
SOAK1   =$10F2    SOAK2   =$10FD   MSEC    =$10FE    MSEC1   =$1102
ROWPTR  =$1108    KEYHIT  =$C000   KEYCLR  =$C010    BEEP    =$C030
GMODE   =$C050    TMODE   =$C051   MIXED   =$C053    PAGE1   =$C054
HGR     =$C057    STATUS  =$C08E   DATA    =$C08F
```

Listing 2a: *The BASIC portion of the GREY16 program that produces dithered gray-scale output on the Apple II's video screen from the Micro D-Cam's output.*

```
10 REM GREY16BAS
12 REM
13 REM COPYRIGHT (C) 1983
15 REM BY CIRCUIT CELLAR, INC.
17 REM
18 REM "LOAD ROUTINES AND INITIA
       LIZE VARIABLES
19 REM
```

Listing continued

Listing 2a continued:

```
20  HGR : TEXT : HOME : PRINT CHR$
    (4)"BLOAD GREY16-48K"
40  NRM = 4096:FULL = 4099:GY = 41
    02
50  CAM = NRM
60  SOAK = 256:LO = 128:HI = 384
70  INC = (HI - LO) / 15
80  SL = 768:SH = 769:ST = 770:SP =
    771:KEY = 772:KP = 773:EI =
    774
90  HOME : INPUT "ENTER CAMERA S
    LOT: ";SN: IF SN < 1 OR SN >
    7 THEN 90
100 POKE ST,SN * 20: REM "SLOT N
    UMBER
110 POKE SP,0: REM "SCREEN POSIT
    ION
117 REM
118 REM "KEYBOARD PROCESSOR SEC
119 REM
120 POKE SH, INT (SOAK / 256): POKE
    SL,SOAK - INT (SOAK / 256) *
    256: REM "SOAK TIME
130 HOME : VTAB 21: PRINT TAB(
    11);"CURRENT EXP: "SOAK" MS
    ": PRINT : PRINT "LO EXP: "
    ;LO;" MS  "; TAB( 24);"HI EX
    P: ";HI;" MS
140 CALL CAM: REM "READ CAMERA A
    ND DISPLAY PICTURE IN CURREN
    T MODE
150 IF NOT ( PEEK (KP)) THEN 14
    0: REM "CHECK FOR KEYPRESS
160 KEY$ = CHR$ ( PEEK (KEY) - 1
    28)
170 IF KEY$ = "E" THEN 400: REM
    "CHANGED EXPOSURE
180 IF KEY$ = "Q" THEN TEXT : HOME
    : END : REM "QUIT PROGRAM
190 IF KEY$ = "N" THEN CAM = NRM
    : HGR : REM "NORMAL SIZE
200 IF KEY$ = "F" THEN CAM = FUL
    L: HGR : REM "FULL SIZE
210 IF KEY$ = "G" THEN GOSUB 30
    0: REM "CREATE GREYSCALE PIC
220 IF KEY$ = "S" THEN GOSUB 60
    0: REM "SAVE PICTURE
230 GOTO 120
297 REM
298 REM "CREATE GREYSCALE PIC
299 REM
300 HOME : VTAB 22: PRINT "COUNT
    DOWN:";
310 POKE SL,LO - INT (LO / 256)
    * 256: POKE SH, INT (LO / 2
    56): POKE EI,INC: REM "POKE
    STARTING EXPOSURE AND EXPOSU
    RE INCREMENT
320 CALL GY: REM "CREATE PIC
330 HOME : VTAB 22: PRINT "HIT '
    S' TO SAVE,": PRINT " ANY OT
    HER KEY TO CONTINUE.": GET K
    EY$
340 RETURN
397 REM
398 REM "CHANGE EXPOSURE
399 REM
400 HOME : VTAB 22: PRINT "CHANG
    E CURRENT, LO, OR": INPUT "
    HI EXPOSURE? (C,L,H): ";CH$
410 IF CH$ = "" THEN 120
420 IF CH$ < > "C" THEN 460
430 INPUT "ENTER NEW CURRENT: ";
    NE$: IF NE$ = "" THEN 540
440 NE = VAL (NE$): IF NE < 1 OR
    NE > 8000 THEN 430
450 SOAK = NE: GOTO 540
460 IF CH$ < > "L" THEN 500
470 INPUT "ENTER NEW LO: ";NE$: IF
    NE$ = "" THEN 540
480 NE = VAL (NE$): IF NE < 1 OR
    NE > 8000 OR NE > HI THEN 47
    0
490 LO = NE: GOTO 540
500 IF CH$ < > "H" THEN 400
510 INPUT "ENTER NEW HI: ";NE$: IF
    NE$ = "" THEN 540
520 NE = VAL (NE$): IF NE < 1 OR
    NE > 8000 OR NE < LO THEN 51
    0
530 HI = NE
540 INC = (HI - LO) / 15
550 GOTO 120
597 REM
598 REM "SAVE PICTURE ON DISK
599 REM
600 HOME : VTAB 22: PRINT "ENTER
    A NAME FOR": INPUT "THE PIC
    TURE: ";NA$
610 IF LEN (NA$) < = 0 THEN RETURN
620 PRINT CHR$ (4)"BSAVE "NA$",
    A$2000,L$2000"
630 HOME : RETURN
```

Listing 2b: *Assembly-language listing of the machine-code portion of the GREY16 program, called from BASIC.*

```
1   ;*********************************
2   ;*
3   ;*  GREY16-48K
4   ;*
5   ;*  COPYRIGHT (C) 1983
6   ;*  BY CIRCUIT CELLAR, INC.
7   ;*
8   ;*********************************
9   ;*
10          ORG     $1000
11  KEYHIT  EQU     $C000
12  KEYCLR  EQU     $C010
13  BEEP    EQU     $C030
14  GMODE   EQU     $C050
15  TMODE   EQU     $C051
```

Listing 2b continued:

```
              16      MIXED    EQU   $C053
              17      PAGE1    EQU   $C054
              18      HGR      EQU   $C057
              19      STATUS   EQU   $C08E
              20      DATA     EQU   $C08F
              21      PRBLNK   EQU   $F948
              22      PRHEX    EQU   $FDE3
              23      *
              24      TBUFFER  EQU   $300
              25      IBUFFER  EQU   $4000
              26      GRTABLE  EQU   $5000
              27      *
              28      SOAKTIME EQU   $300
              29      SLOTADR  EQU   $302
              30      ROWSTART EQU   $303
              31      KEY      EQU   $304
              32      KEYFLAG  EQU   $305
              33      INCRMENT EQU   $306
              34      *
              35      RADR     EQU   $06
              36      DEST     EQU   $08
              37      DEST2    EQU   $1C
              38      CTR      EQU   $19
              39      YREG     EQU   $1B
              40      TMP      EQU   $1E
              41      COUNT    EQU   $1F
              42      TABLE    EQU   $EB
              43      IMAGE    EQU   $ED
              44      KEEPCNT  EQU   $F9
              45      KEEPFLG  EQU   $FA
              46      *
              47      TEMP1    EQU   $06
              48      TEMP2    EQU   $07
              49      TEMP3    EQU   $1C
              50      TEMP4    EQU   $1D
              51      TEMP5    EQU   $1B
              52      *
1000: 4C 09 10 53      JMP   NORMPIC    ;NORMAL PIC (256X64)
1003: 4C 98 10 54      JMP   FULLPIC    ;FULL SIZE (256X128 ENHANCED)
1006: 4C 15 11 55      JMP   GREY       ;GREYSCALE (16 LEVELS OF GREY)
                  56   *
                  57   *
1009: 20 34 12 58 NORMPIC JSR SETGR     ;SET GRAPHICS ON
100C: 20 0A 12 59 NSTART  JSR ACIACLR  ;FLUSH THE INPUT BUFFER
100F: A9 D3    60         LDA ##D3     ;SEND COMMAND TO SOAK W/O SEND
1011: 20 41 12 61         JSR SENDCMD
1014: 20 55 12 62         JSR SOAK     ;WAIT EXPOSURE TIME
1017: A9 C0    63         LDA ##C0     ;SEND IMAGE W/O SOAK
1019: 20 41 12 64         JSR SENDCMD  ;(ALT,WIDEPIX,7BIT-256 X 64)
101C: A2 00    65         LDX #0       ;INITIALIZE THE ROW INDEX
101E: A0 00    66 NEWROW  LDY #0       ;START NXT ROW, INIT COL INDEX
1020: BD F4 13 67         LDA ROMPTR,X ;GET FIRST ROW ADDRS
1023: 18       68         CLC
1024: 6D 03 03 69         ADC ROWSTART ;0-SELECT UPPER 1/3 OF SCREEN
1027: 85 06    70         STA RADR     ;$28-MID 1/3, $50-BOT 1/3
1029: E8       71         INX          ;RADR HAS ADDRS OF CUR ROW
102A: BD F4 13 72         LDA ROMPTR,X
102D: 85 07    73         STA RADR+1
102F: E8       74         INX          ;POINT X-REG TO NEXT ADDRESS
1030: 84 1B    75 GET     STY YREG     ;GET NEXT BYTE FROM CAMERA
1032: AC 02 03 76         LDY SLOTADR  ;LOAD OFFSET TO CAMERA SLOT
1035: B9 8E C0 77         LDA STATUS,Y ;CHECK IF BYTE HAS ARRIVED
1038: 4A       78         LSR
1039: B0 21    79         BCS C15      ;IF BYTE AVAILABLE THEN BRANCH
103B: A9 00    80         LDA #0       ;IF BYTE NOT AVAILABLE THEN
103D: 85 19    81         STA CTR      ;SET UP TIMEOUT COUNTER
103F: A9 15    82         LDA ##15
1041: 85 1A    83         STA CTR+1
1043: C6 19    84 C0      DEC CTR
1045: D0 0F    85         BNE C1
1047: C6 1A    86         DEC CTR+1
1049: D0 0B    87         BNE C1
104B: AD 30 C0 88         LDA BEEP     ;IF TIMED OUT, CLICK SPEAKER
104E: AD 00 C0 89         LDA KEYHIT   ;CHECK FOR KEYPRESS
1051: 30 1D    90         BMI NDONE    ;IF KEY HIT, RETURN TO BASIC
1053: 4C 0C 10 91         JMP NSTART   ;ELSE, RESTART CMD SEQUENCE
1056: B9 8E C0 92 C1      LDA STATUS,Y
1059: 4A       93         LSR
105A: 90 E7    94         BCC C0
105C: B9 8F C0 95 C15     LDA DATA,Y   ;WHEN BYTE AVAILABLE GET IT
105F: A4 1B    96         LDY YREG     ;RESTORE COL POINTER TO Y-REG
1061: C0 28    97         CPY #40      ;IF PAST 40TH BYTE IN CURRENT
```

Listing continued

Listing 2b continued:

```
1063: B0 02        98           BGE  C3                    ;ROW THEN DONT PUT ON SCREEN
1065: 91 06        99           STA  (RADR),Y
1067: C8          100           INY
              C3  101           CPY  #$37                  ;INCREMENT COLUMN POINTER
1068: C0 25       101           CPY  #$37                  ;CHECK FOR END OF COLUMN
106A: D0 C4       102           BNE  GET                   ;IF NOT, GET THE NEXT BYTE
106C: E0 80       103           CPX  #$80                  ;CHECK FOR END OF IMAGE
106E: D0 AE       104           BNE  NEWROW                ;GOTO NEXT ROW
1070: A9 D1       105  NDONE    LDA  #$D1
1072: 20 41 12 106           JSR  SENDCMD                ;REFRESH W/O SENDING
1075: A9 20       107  DOKEY    LDA  #$20
1077: 8D 04 03 108           STA  KEY                   ;CLEAR 'KEY HIT' VALUE
107A: A9 00       109           LDA  #0
107C: 8D 05 03 110           STA  KEYFLAG               ;CLEAR 'KEY HIT' FLAG
107F: AD 00 C0 111           LDA  KEYHIT                ;LOOK AT KEYBOARD
1082: 10 13       112           BPL  D1                    ;BRANCH IF NO KEYPRESS
1084: 2C 10 C0    113           BIT  KEYCLR                ;CLEAR KEYBOARD STROBE
1087: EE 05 03 114           INC  KEYFLAG               ;SET 'KEY HIT' FLAG
108A: 8D 04 03 115           STA  KEY                   ;SAVE KEYPRESS
108D: C9 D1       116           CMP  #"Q"                  ;CHECK IF 'Q' HIT
108F: D0 06       117           BNE  D1                    ;IF NOT, RETURN
1091: 20 1C 12 118           JSR  GRCLR                 ;CLEAR GRAPHICS SCREEN
1094: AD 51 C0 119  D1       LDA  TMODE                 ;SET TEXT MODE
1097: 60          120           RTS                         ;RETURN TO BASIC
              121  *
1098: 20 34 12 122  FULLPIC  JSR  SETGR                 ;SET GRAPHICS ON
109B: 20 A7 10 123           JSR  TAKEPIC               ;TAKE A PICTURE
109E: 20 83 12 124           JSR  MOVE                  ;MOVE TO CORRECT BUFFER
10A1: 20 A3 12 125           JSR  DISPLAY               ;TRANS IMAGE TO HIRES SCREEN
10A4: 4C 75 10 126           JMP  DOKEY                 ;CHECK FOR KEYPRESS
              127  *
10A7: 20 0A 12 128  TAKEPIC  JSR  ACIACLR               ;CLEAR ACIA
10AA: A9 FB       129  START    LDA  #$FB
10AC: 20 41 12 130           JSR  SENDCMD               ;TELL CAMERA TO SOAK W/O SEND
10AF: 20 55 12 131           JSR  SOAK                  ;WAIT EXPOSURE TIME
10B2: A9 E8       132           LDA  #$E8
10B4: 20 41 12 133           JSR  SENDCMD               ;TELL CAMERA TO SEND IMAGE
10B7: A9 00       134           LDA  #0                   ;(NOALT,WIDEPIX,8BIT-512X128)
10B9: 85 06       135           STA  #<IBUFFER             ;SET UP BUFFER ADDRESS
10BB: A9 40       136           LDA  #>IBUFFER
10BD: 85 07       137           STA  RADR+1
10BF: A9 20       138           LDA  #$20                  ;INIT KEEP COUNTER
10C1: 85 F9       139           STA  KEEPCNT
10C3: A9 01       140           LDA  #$01                  ;INIT KEEP FLAG
10C5: 85 FA       141           STA  KEEPFLG
10C7: A0 00       142           LDY  #0
10C9: AE 07 03 143           LDX  SLOTADR               ;GET SLOT ADDRESS
10CC: A9 00       144  GSTAT    LDA  #0
10CE: 85 19       145           STA  CTR                   ;SETUP TIMER FOR TIMEOUT
10D0: BD BE C0 146           LDA  STATUS,X              ;CHECK IF BYTE READY
10D3: 4A          147           LSR
10D4: C6 19       148           DEC  CTR                   ;LOOP UNTIL CHAR AVAIL
10D6: D0 06       149           BNE  NOHANG                ;OR TIMED OUT
10D8: AD 30 C0 150           LDA  BEEP                  ;IF TIMED OUT, CLICK & RESTART
10DB: 4C AA 10 151           JMP  START
10DE: 90 EC       152  NOHANG   BCC  GSTAT
10E0: A5 FA       153           LDA  KEEPFLG               ;KEEP OR DISCARD?
10E2: F0 08       154           BEQ  IGNORE
10E4: BD BF C0 155           LDA  DATA,X                ;GET BYTE
10E7: 91 06       156           STA  (RADR),Y             ;STORE BYTE IN BUFFER
10E9: 4C EF 10 157           JMP  CONT
10EC: BD BF C0 158  IGNORE   LDA  DATA,X                ;GET BYTE, BUT DISCARD
10EF: C6 F9       159           DEC  KEEPCNT               ;DECREMENT COUNTER
10F1: D0 0A       160           BNE  CONT1
10F3: A9 20       161           LDA  #$20                  ;INIT COUNTER
10F5: 85 F9       162           STA  KEEPCNT
10F7: A5 FA       163           LDA  KEEPFLG
10F9: 49 01       164           EOR  #$01                  ;TOGGLE KEEP FLAG
10FB: 85 FA       165           STA  KEEPFLG
10FD: A5 FA       166  CONT1    LDA  KEEPFLG               ;CHECK KEEP FLAG
10FF: F0 CB       167           BEQ  GSTAT                 ;IF IGNORING, DONT INCR PNTRS
1101: C8          168           INY                        ;INCREMENT POINTER LO
1102: D0 C8       169           BNE  GSTAT
1104: E6 07       170           INC  RADR+1               ;INCREMENT POINTER HI
1106: A5 07       171           LDA  RADR+1
1108: C9 50       172           CMP  #$50                  ;CHECK IF DONE
110A: D0 C0       173           BNE  GSTAT                 ;GET NEXT BYTE
110C: A9 F9       174           LDA  #$F9
110E: 20 41 12 175           JSR  SENDCMD               ;TELL CAMERA REFRESH W/O SEND
1111: 20 F2 12 176           JSR  ENHANCE               ;ENHANCE IMAGE
1114: 60          177           RTS
              178  *
```

42

Listing 2b continued:

```
1115: A9 00       179  GREY      LDA #<GRTABLE    ;INIT TABLE POINTERS
1117: 85 EB       180            STA TABLE
1119: A9 50       181            LDA #>GRTABLE
111B: 85 EC       182            STA TABLE+1
111D: A0 00       183            LDY #$00
111F: 98          184  CLR1      TYA              ;CLEAR GREYSCALE COUNTER TABLE
1120: 91 EB       185            STA (TABLE),Y
1122: E6 EB       186            INC TABLE
1124: D0 F9       187            BNE CLR1
1126: E6 EC       188            INC TABLE+1
1128: A9 90       189            LDA #>GRTABLE+$4000
112A: C5 EC       190            CMP TABLE+1      ;CHECK FOR END
112C: D0 F1       191            BNE CLR1
112E: A9 0F       192            LDA #$0F         ;INIT EXPOSURE COUNTER
1130: 85 1F       193            STA COUNT
1132: A5 1F       194  NEXTPIC   LDA COUNT
1134: 20 E3 FD    195            JSR PRHEX        ;DISPLAY COUNT FOR COUNTDOWN
1137: 20 48 F9    196            JSR PRBLNK       ;LEAVE THREE SPACES
113A: 20 A7 10    197            JSR TAKEPIC      ;TAKE A PICTURE
113D: A9 00       198            LDA #<GRTABLE    ;INIT TABLE AND BUFFER PNTRS
113F: 85 ED       199            STA IMAGE
1141: 85 EB       200            STA TABLE
1143: A9 30       201            LDA #>GRTABLE
1145: 85 EE       202            STA IMAGE+1
1147: A9 50       203            LDA #>GRTABLE
1149: 85 EC       204            STA TABLE+1
114B: A0 00       205            LDY #$00
114D: A2 04       206  NXTBYTE   LDX #$04
114F: B1 ED       207            LDA (IMAGE),Y    ;GET NEXT BYTE
1151: 85 1E       208            STA TMP
1153: 06 1E       209  LOADTBL   ASL TMP          ;TEST EACH BIT IN THE BYTE
1155: A9 00       210            LDA #$00
1157: 90 02       211            BCC ZERO
1159: A9 01       212            LDA #$01
115B: 06 1E       213  ZERO      ASL TMP
115D: 90 03       214            BCC ZERO1
115F: 18          215            CLC
1160: 69 10       216            ADC #$10
1162: 18          217  ZERO1     CLC
1163: 71 EB       218            ADC (TABLE),Y    ;AND INCR APPROPRIATE CNTRS
1165: 91 EB       219            STA (TABLE),Y    ;(TWO BIT-COUNTERS PER BYTE)
1167: E6 EB       220            INC TABLE        ;INCREMENT TABLE POINTER
1169: D0 02       221            BNE TBL1
116B: E6 EC       222            INC TABLE+1
116D: CA          223  TBL1      DEX              ;DONE WITH THIS BYTE?
116E: D0 E3       224            BNE LOADTBL
1170: E6 ED       225            INC IMAGE        ;INCREMENT BUFFER POINTER
1172: D0 D9       226            BNE NXTBYTE
1174: E6 EE       227            INC IMAGE+1
1176: A9 40       228            LDA #>TBUFFER+$1000
1178: C5 EE       229            CMP IMAGE+1      ;CHECK FOR END OF BUFFER
117A: D0 D1       230            BNE NXTBYTE      ;IF NOT, GET NEXT BYTE
117C: 18          231            CLC
117D: AD 00 03    232            LDA SOAKTIME
1180: 6D 06 03    233            ADC INCREMENT    ;INCREMENT EXPOSURE TIME FOR
1183: 8D 00 03    234            STA SOAKTIME     ;NEXT EXPOSYRE
1186: AD 01 03    235            LDA SOAKTIME+1
1189: 69 00       236            ADC #$00
118B: 8D 01 03    237            STA SOAKTIME+1
118E: C6 1F       238            DEC COUNT        ;DONE WITH 15 EXPOSURES?
1190: D0 A0       239            BNE NEXTPIC      ;IF NOT, TAKE NEXT PICTURE
                  240  *
1192: A9 00       241            LDA #<GRTABLE    ;INIT TABLE AND BUFFER PNTRS
1194: 85 EB       242            STA TABLE
1196: 85 ED       243            STA IMAGE
1198: A9 50       244            LDA #>GRTABLE
119A: 85 EC       245            STA TABLE+1
119C: A9 40       246            LDA #>TBUFFER
119E: 85 EE       247            STA IMAGE+1
11A0: A2 00       248            LDX #$00
11A2: A9 20       249  NEXTROW   LDA #$20
11A4: 85 1F       250            STA COUNT        ;INIT COLUMN COUNTER
11A6: A0 00       251  THISROW   LDY #$00
11A8: 84 1E       252            STY TMP
11AA: A9 02       253            LDA #$02
11AC: 85 19       254            STA CTR          ;INIT BIT COUNTER
11AE: B1 EB       255  THISBYTE  LDA (TABLE),Y    ;GET NEXT BYTE
11B0: 29 0F       256            AND #$0F         ;MASK OFF TOP NIBBLE
11B2: DD E4 13    257            CMP VAL1,X       ;COMP WITH DITHER MATRX VAL
11B5: 26 1E       258            ROL TMP          ;ROTATE CARRY BIT INTO BYTE
11B7: B1 EB       259            LDA (TABLE),Y    ;GET BYTE AGAIN
11B8: 29 F0       260            AND #$F0         ;MASK OFF LOWER NIBBLE
11BB: DD E8 13    261            CMP VAL2,X       ;COMP WITH NEXT MATRX VALUE
```

Listing continued

Listing 2b continued:

```
                                                                                    ;GET CAMERA ADDRESS
11BE: 26 1E         262             ROL    TMP           ;ROTATE CARRY BIT INTO BYTE
11C0: C8            263             INY                  ;INCR TABLE INDEX
11C1: B1 EB         264             LDA    (TABLE),Y     ;GET NEXT BYTE
11C3: 29 0F         265             AND    #$0F          ;MASK OFF UPPER NIBBLE
11C5: DD EC 13      266             CMP    VAL3,X        ;COMP WITH THIRD MATRX VALUE
11C8: 26 1E         267             ROL    TMP           ;ROTATE CARRY INTO BYTE
11CA: B1 EB         268             LDA    (TABLE),Y     ;GET NEXT BYTE AGAIN
11CC: 29 F0         269             AND    #$F0          ;MASK OFF LOWER NIBBLE
11CE: DD F0 13      270             CMP    VAL4,X        ;COMP WITH FOURTH MATRX VALUE
11D1: 26 1E         271             ROL    TMP           ;ROTATE CARRY INTO BYTE
11D3: C8            272             INY                  ;INCR TABLE INDEX
11D4: C6 19         273             DEC    CTR           ;DECR BIT COUNTER
11D6: D0 D6         274             BNE    THISBYTE      ;CONTINUE WITH THIS BYTE
11D8: A0 00         275             LDY    #$00
11DA: A5 1E         276             LDA    TMP           ;GET BYTE FOR NXT COL IN IMAGE
11DC: 91 ED         277             STA    (IMAGE),Y     ;PUT INTO IMAGE BUFFER
11DE: 18            278             CLC
11DF: A9 04         279             LDA    #$04
11E1: 65 EB         280             ADC    TABLE
11E3: 85 EB         281             STA    TABLE
11E5: 90 02         282             BCC    NEXT
11E7: E6 EC         283             INC    TABLE+1       ;INCR TABLE POINTER
11E9: E6 ED         284   NEXT      INC    IMAGE
11EB: D0 08         285             BNE    NXT1
11ED: E6 EE         286             INC    IMAGE+1       ;INCR BUFFER POINTER
11EF: A5 EE         287             LDA    IMAGE+1
11F1: C9 50         288             CMP    #>IBUFFER+$1000 ;CHECK FOR END OF IMAGE
11F3: B0 0E         289             BGE    DONE
11F5: C6 1F         290   NXT1      DEC    COUNT         ;IF NOT, DECR COL COUNTER
11F7: D0 AD         291             BNE    THISROW       ;IF NOT END, STAY ON THIS ROW
11F9: E8            292             INX                  ;INCR DITHER MATRX INDEX
11FA: E0 04         293             CPX    #$04
11FC: D0 02         294             BNE    NXT2
11FE: A2 00         295             LDX    #$00          ;IF REACHED 4, RESET TO 0
1200: 4C A2 11      296   NXT2      JMP    NEXTROW       ;DO NEXT ROW
1203: 20 1C 12      297   DONE      JSR    GRCLR         ;CLEAR GRAPHICS SCREEN
1206: 20 A3 12      298             JSR    DISPLAY       ;SEND IMAGE TO SCREEN
1209: 60            299             RTS
                    300   *
120A: A9 03         301   ACIACLR   LDA    #3            ;INIT ACIA
120C: 84 1B         302             STY    YREG          ;SAVE Y-REG
120E: AC 02 03      303             LDY    SLOTADR       ;GET CAMERA ADDRESS
1211: 99 8E C0      304             STA    STATUS,Y
1214: A9 14         305             LDA    #$14          ;8 BITS,1 START,1 STOP,EXT CLK
1216: 99 8E C0      306             STA    STATUS,Y
1219: A4 1B         307             LDY    YREG
121B: 60            308             RTS
                    309   *
121C: A2 00         310   GRCLR     LDX    #0            ;CLEAR GRAPHICS SCREEN
121E: A0 00         311             LDY    #0
1220: 84 06         312             STY    RADR
1222: A9 20         313             LDA    #$20
1224: 85 07         314             STA    RADR+1
1226: 8A            315             TXA
1227: 91 06         316   CLR2      STA    (RADR),Y
1229: C8            317             INY
122A: D0 FB         318             BNE    CLR2
122C: E6 07         319             INC    RADR+1
122E: E8            320             INX
122F: E0 20         321             CPX    #$20
1231: D0 F4         322             BNE    CLR2
1233: 60            323             RTS
                    324   *
1234: AD 53 C0      325   SETGR     LDA    MIXED         ;SET MIXED GRAPHICS & TEXT
1237: AD 57 C0      326             LDA    HGR           ;SET HIRES
123A: AD 54 C0      327             LDA    PAGE1         ;SET PAGE 1
123D: AD 50 C0      328             LDA    GMODE         ;SET GRAPHICS MODE
1240: 60            329             RTS
                    330   *
1241: 84 1B         331   SENDCMD   STY    YREG          ;SAVE Y-REG
1243: AC 02 03      332             LDY    SLOTADR       ;GET CAMERA ADDRESS
1246: 48            333             PHA                  ;SAVE A-REG
1247: B9 8E C0      334   SEND1     LDA    STATUS,Y      ;GET STATUS REGISTER
124A: 29 02         335             AND    #2            ;CHECK IF READY FOR BYTE
124C: F0 F9         336             BEQ    SEND1
124E: 68            337             PLA                  ;RESTORE A-REG
124F: 99 8F C0      338             STA    DATA,Y        ;SEND COMMAND
1252: A4 1B         339             LDY    YREG          ;RESTORE Y-REG
1254: 60            340             RTS
                    341   *
1255: AD 01 03      342   SOAK      LDA    SOAKTIME+1    ;LOAD COUNTER WITH SOAKTIME
1258: 85 1A         343             STA    CTR+1         ; BY SOAKTIME
125A: E6 1A         344             INC    CTR+1
```

Listing 2b continued:

```
125C: AD 00 03  345            LDA SOAKTIME
125F: 85 19      346            STA CTR
1261: E6 19      347            INC CTR
1263: AD 00 03  348            LDA SOAKTIME     ;CHECK FOR ZERO SOAKTIME
1266: D0 05      349            BNE SOAK1
1268: AD 01 03  350            LDA SOAKTIME+1
126B: F0 0B      351            BEQ SOAK2        ;RETURN IF ZERO
126D: 20 79 12  352 SOAK1      JSR MSEC         ;WAIT FOR 1 MS
1270: C6 19      353            DEC CTR          ;DECR COUNTER
1272: D0 F9      354            BNE SOAK1        ;LOOP IF NOT ZERO
1274: C6 1A      355            DEC CTR+1        ;DECR COUNTER HI
1276: D0 F5      356            BNE SOAK1        ;LOOP IF NOT DONE
1278: 60         357 SOAK2     RTS
                 358 $
1279: 84 1B      359 MSEC       STY YREG         ;SAVE Y-REG
127B: A0 C7      360            LDY #199         ;INIT TIMER
127D: 88         361 DLY1       DEY              ;DECR TIMER
127E: D0 FD      362            BNE DLY1         ;LOOP IF NOT DONE
1280: A4 1B      363            LDY YREG         ;RESTORE Y-REG
1282: 60         364            RTS
                 365 $
1283: A0 00      366 MOVE       LDY #$00         ;ZERO INDEX
1285: 84 ED      367            STY IMAGE        ;INIT BUFFER POINTERS
1287: 84 08      368            STY DEST
1289: A9 30      369            LDA #>IBUFFER
128B: 85 EE      370            STA IMAGE+1
128D: A9 40      371            LDA #>IBUFFER
128F: 85 09      372            STA DEST+1
1291: B1 ED      373 LOOP       LDA (IMAGE),Y    ;GET BYTE FROM TEMP BUFFER
1293: 91 08      374            STA (DEST),Y     ;PUT INTO INPUT BUFFER
1295: C8         375            INY              ;INCR INDEX
1296: D0 F9      376            BNE LOOP         ;LOOP IF NOT ZERO
1298: E6 EE      377            INC IMAGE+1      ;INCR POINTERS
129A: E6 09      378            INC DEST+1
129C: A9 50      379            LDA #>IBUFFER+$1000
129E: C5 09      380            CMP DEST+1       ;CHECK FOR END OF BUFFERS
12A0: D0 EF      381            BNE LOOP         ;LOOP IF NOT DONE
12A2: 60         382            RTS
                 383 $
12A3: A2 00      384 DISPLAY    LDX #$00
12A5: A9 00      385            LDA #<IBUFFER    ;INIT BUFFER POINTERS
12A7: 85 06      386            STA RADR
12A9: A9 40      387            LDA #>IBUFFER
12AB: 85 07      388            STA RADR+1
12AD: BD F4 13  389 NXTROW     LDA ROWPTR,X     ;GET STARTING ADDRS OF CUR ROW
12B0: 85 08      390            STA DEST         ;PUT IN SCREEN POINTER
12B2: E8         391            INX              ;INCR INDEX
12B3: BD F4 13  392            LDA ROWPTR,X     ;GET UPPER HALF OF ADDRESS
12B6: 85 09      393            STA DEST+1       ;PUT IN SCREEN POINTER
12B8: E8         394            INX              ;INCR INDEX
12B9: A0 20      395            LDY #$20         ;INIT COLUMN COUNTER
12BB: B1 06      396            LDA (RADR),Y     ;GET BYTE FROM BUFFER
12BD: 88         397            DEY              ;DECR COL COUNTER
12BE: 91 08      398            STA (DEST),Y     ;PUT BYTE ON SCREEN
12C0: C0 00      399            CPY #0           ;CHECK IF COUNTER IS ZERO
12C2: D0 F7      400            BNE MOV          ;IF NOT, DO NEXT BYTE
12C4: A0 00      401            LDY #0
12C6: 08         402 RESHFT     PHP              ;ROTATE EACH BYTE IN ROW SO
12C7: 84 1E      403            STY TMP          ;THE HIGH BIT IS CARRIED TO
12C9: B1 08      404 SHFT       LDA (DEST),Y     ;THE NEXT BYTE
12CB: 28         405            PLP
12CC: 2A         406            ROL
12CD: 91 08      407            STA (DEST),Y
12CF: 08         408            PHP
12D0: C8         409            INY
12D1: C0 20      410            CPY #$20
12D3: D0 F4      411            BNE SHFT
12D5: 28         412            PLP
12D6: A4 1E      413            LDY TMP
12D8: B1 08      414            LDA (DEST),Y     ;SHIFT THE LEADING BYTE SO THE
12DA: 4A         415            LSR              ;TOP BIT IS ZERO
12DB: 91 08      416            STA (DEST),Y
12DD: C8         417            INY
12DE: C0 20      418            CPY #$20
12E0: D0 E4      419            BNE RESHFT       ;IF NOT DONE, ROTATE ALL OF
12E2: 18         420            CLC              ;ROW AFTER LEADING BYTE AGAIN
12E3: A5 06      421            LDA RADR
12E5: 69 20      422            ADC #$20
12E7: 85 06      423            STA RADR
12E9: 90 02      424            BCC DISP1
12EB: E6 07      425            INC RADR+1
12ED: E0 00      426 DISP1      CPX #0           ;IF SCRN NOT DONE, DO NXT ROW
12EF: D0 BC      427            BNE NXTROW
```

Listing continued

Listing 2b continued:

```
12F1: 60              428          RTS
                      429  ;************************************************
                      430  ;
                      431  ; ENHANCE REARRANGES THE IMAGE IN IBUFFER TO AN
                      432  ; IMAGE IN TBUFFER THAT CORRESPONDS MORE
                      433  ; PRECISELY TO THE ACTUAL PICTURE BEING SEEN
                      434  ; BY THE CAMERA.
                      435  ; THE ALGORITHM REARRANGES EACH BYTE AS FOLLOWS:
                      436  ; FOR BYTES IN EVEN ROWS(STARTING WITH ROW 0) --
                      437  ;    BITS 2 AND 6 ARE UNCHANGED
                      438  ;    BIT 0 MOVES TO BIT 3 OF THE BYTE 1 NEXT ROW
                      439  ;    BIT 4 MOVES TO BIT 7 OF THE BYTE 1 NEXT ROW
                      440  ;
                      441  ; FOR BYTES IN ODD ROWS --
                      442  ;    BITS 0 AND 4 ARE UNCHANGED
                      443  ;    BIT 2 MOVES TO BIT 5 OF THIS BYTE IN NEXT ROW
                      444  ;    BIT 6 MOVES TO BIT 1 OF THIS BYTE+1 IN NEXT ROW
                      445  ;************************************************
                      446  ;
12F2: A9 DF           447  ENHANCE  LDA  #<IBUFFER-$21  ;INIT BUFFER POINTERS
12F4: 85 06           448           STA  RADR
12F6: 85 08           449           STA  DEST
12F8: A9 3F           450           LDA  #>IBUFFER-$21
12FA: 85 07           451           STA  RADR+1
12FC: A9 2F           452           LDA  #>TBUFFER-$21
12FE: 85 09           453           STA  DEST+1
1300: A9 20           454           LDA  #$20
1302: 85 1E           455           STA  TMP
1304: A0 01           456  NEVEN    LDY  #1         ;PERFORM OPERATION AS DESCR
1306: B1 06           457           LDA  (RADR),Y   ;ABOVE ON EVEN ROWS
1308: 0A              458           ASL
1309: 0A              459           ASL
130A: 0A              460           ASL
130B: 09 DF           461           ORA  #$DF
130D: 85 1C           462           STA  DEST2
130F: 88              463           DEY
1310: B1 06           464           LDA  (RADR),Y
1312: 2A              465           ROL
1313: 2A              466           ROL
1314: 2A              467           ROL
1315: 09 FD           468           ORA  #$FD
1317: 85 1D           469           STA  DEST2+1
1319: A0 21           470           LDY  #$21
131B: B1 06           471           LDA  (RADR),Y
131D: 09 BB           472           ORA  #$BB
131F: 29 FF           473           AND  #$FF
1321: 25 1C           474           AND  DEST2
1323: 25 1D           475           AND  DEST2+1
1325: 91 08           476           STA  (DEST),Y
1327: E6 06           477           INC  RADR
1329: E6 08           478           INC  DEST
132B: D0 04           479           BNE  NODUB
132D: E6 07           480           INC  RADR+1
132F: E6 09           481           INC  DEST+1
1331: C6 1E           482  NODUB    DEC  TMP
1333: D0 CF           483           BNE  NEVEN
1335: A9 20           484           LDA  #$20
1337: 85 1E           485           STA  TMP
1339: A0 01           486  NODD     LDY  #1
133B: B1 06           487           LDA  (RADR),Y
133D: 0A              488           ASL
133E: 0A              489           ASL
133F: 0A              490           ASL
1340: 09 77           491           ORA  #$77
1342: 85 1C           492           STA  DEST2
1344: A0 21           493           LDY  #$21
1346: B1 06           494           LDA  (RADR),Y
1348: 09 EE           495           ORA  #$EE
134A: 29 FF           496           AND  #$FF
134C: 25 1C           497           AND  DEST2
134E: 91 08           498           STA  (DEST),Y
1350: E6 06           499           INC  RADR
1352: E6 08           500           INC  DEST
1354: D0 04           501           BNE  NODUB2
1356: E6 07           502           INC  RADR+1
1358: E6 09           503           INC  DEST+1
135A: C6 1E           504  NODUB2   DEC  TMP
135C: D0 DB           505           BNE  NODD
135E: A9 20           506           LDA  #$20
1360: 85 1E           507           STA  TMP
1362: A5 08           508           LDA  DEST
1364: C9 DF           509           CMP  #<TBUFFER+$FDF ;CHECK FOR END OF BUFFER
1366: D0 09           510           BNE  NEVENJ
1368: A5 09           511           LDA  DEST+1
136A: C9 3F           512           CMP  #>TBUFFER+$FDF
```

46

Listing 2b continued:

```
136C: D0 03        513              BNE    NEVENJ
136E: 4C 74 13     514              JMP    FILLIN
1371: 4C 04 13     515  NEVENJ      JMP    NEVEN      ;LOOP IF NOT DONE
                   516  ;
                   517  ;************************************************
                   518  ; FILLIN COLORS HOLES IN THE IMAGE IN TBUFFER
                   519  ; EITHER BLACK OR WHITE, DEPENDING ON THE COLORS
                   520  ; OF THE SURROUNDING PIXELS.
                   521  ; THE ALGORITHM USED IS AS FOLLOWS:
                   522  ; FOR BYTES IN EVEN ROWS (STARTING WITH ROW 0)--
                   523  ;     BITS 0,3,4,7 ARE 'HOLES'
                   524  ;     BITS 2 AND 6 ARE UNCHANGED
                   525  ;     BIT 5 COMES FROM BIT 2 OF THE BYTE 1 ROW PREV
                   526  ;     BIT 1 COMES FROM BIT 6 OF THE BYTE 1 ROW
                   527  ;           PREVIOUS LESS ONE BYTE
                   528  ;
                   529  ; FOR BYTES IN ODD ROWS--
                   530  ;     BITS 1,2,5,6 ARE 'HOLES
                   531  ;     BITS 0 AND 4 ARE UNCHANGED
                   532  ;     BITS 3 AND 7 COME FROM BITS 0 AND 4 OF THE
                   533  ;           BYTE 1 ROW PREVIOUS
                   534  ;
                   535  ;************************************************
1374: A9 E0        536  FILLIN      LDA    #<TBUFFER-$20  ;INIT BUFFER POINTER
1376: 85 08        537              STA    DEST
1378: A9 2F        538              LDA    #>TBUFFER-$20
137A: 85 09        539              STA    DEST+1
137C: A2 00        540              LDX    #0
137E: A9 20        541              LDA    #$20
1380: 85 1E        542              STA    TMP
1382: A0 00        543  FILL1       LDY    #0           ;FILLIN EACH ROW AS
1384: B1 08        544              LDA    (DEST),Y     ;DESCRIBED ABOVE
1386: 85 07        545              STA    TEMP2
1388: A0 40        546              LDY    #$40
138A: B1 08        547              LDA    (DEST),Y
138C: 85 1C        548              STA    TEMP3
138E: 25 07        549              AND    TEMP2
1390: 85 06        550              STA    TEMP1
1392: A0 20        551              LDY    #$20
1394: B1 08        552              LDA    (DEST),Y
1396: 4A           553              LSR
1397: 1D E0 13     554              ORA    RMASK,X
139A: 05 06        555              ORA    TEMP1
139C: 85 1D        556              STA    TEMP4
139E: B1 08        557              LDA    (DEST),Y
13A0: 0A           558              ASL
13A1: 1D E2 13     559              ORA    LMASK,X
13A4: 05 06        560              ORA    TEMP1
13A6: 85 1B        561              STA    TEMP5
13A8: A5 07        562              LDA    TEMP2
13AA: 05 1C        563              ORA    TEMP3
13AC: 1D DE 13     564              ORA    CMASK,X
13AF: 31 08        565              AND    (DEST),Y
13B1: 25 1D        566              AND    TEMP4
13B3: 25 1B        567              AND    TEMP5
13B5: 91 08        568              STA    (DEST),Y
13B7: E6 08        569              INC    DEST
13B9: D0 08        570              BNE    FILL2
13BB: E6 09        571              INC    DEST+1
13BD: A9 40        572  FILL2       LDA    #>TBUFFER+$1000
13BF: C5 09        573              CMP    DEST+1       ;CHECK FOR END OF BUFFER
13C1: F0 0F        574              BEQ    FILL3
13C3: C6 1E        575              DEC    TMP
13C5: D0 BB        576              BNE    FILL1
13C7: A9 20        577              LDA    #$20
13C9: 85 1E        578              STA    TMP
13CB: 8A           579              TXA
13CC: 49 01        580              EOR    #1
13CE: AA           581              TAX
13CF: 4C 82 13     582              JMP    FILL1
13D2: A2 1F        583  FILL3       LDX    #$1F
13D4: BD 40 30     584              LDA    TBUFFER+$40,X ;CLEAN UP THE FIRST ROW
13D7: 9D 00 30     585              STA    TBUFFER,X
13DA: CA           586              DEX
13DB: 10 F7        587              BPL    CLEAN
13DD: 60           588              RTS
                   589  ;
13DE: 66 99        590  CMASK       DFB    $66,$99
13E0: EE BB        591  RMASK       DFB    $EE,$BB
13E2: 77 DD        592  LMASK       DFB    $77,$DD
13E4: 01 0D 04     593  VAL1        DFB    $01,$0D,$04,$0F
      0F
13E8: 90 50 C0     594  VAL2        DFB    $90,$50,$C0,$80
      80
```

Listing continued

47

Listing 2b continued:
```
13EC: 03 0F 02          DFB  $03,$0F,$02,$0E
13EF: 0E            595 VAL3
13F0: B0 70 A0          DFB  $B0,$70,$A0,$60
13F3: 60            596 VAL4
                    597 *
                    598 *************************
13F4: 00 20 00
13F7: 24 00 28
13FA: 00 2C 00
13FD: 30 00 34
1400: 00 38 00
1403: 3C            599 ROWPTR HEX 002000240028002C0030003400380 03C
1404: 80 20 80
1407: 24 80 28
140A: 80 2C 80
140D: 30 80 34
1410: 80 38 80
1413: 3C            600        HEX 80208024802880 2C80308034803880 3C
1414: 00 21 00
1417: 25 00 29
141A: 00 2D 00
141D: 31 00 35
1420: 00 39 00
1423: 3D            601        HEX 002100250029002D0031003500390 03D
1424: 80 21 80
1427: 25 80 29
142A: 80 2D 80
142D: 31 80 35
1430: 80 39 80
1433: 3D            602        HEX 80218025802980 2D80318035803980 3D
1434: 00 22 00
1437: 26 00 2A
143A: 00 2E 00
143D: 32 00 36
1440: 00 3A 00
1443: 3E            603        HEX 002200260 02A002E0032003600 3A003E
1444: 80 22 80
1447: 26 80 2A
144A: 80 2E 80
144D: 32 80 36
1450: 80 3A 80
1453: 3E            604        HEX 80228026802A802E8032803680 3A803E
1454: 00 23 00
1457: 27 00 2B
145A: 00 2F 00
145D: 33 00 37
1460: 00 3B 00
1463: 3F            605        HEX 002300270 02B002F003300370 03B003F
1464: 80 23 80
1467: 27 80 2B
146A: 80 2F 80
146D: 33 80 37
1470: 80 3B 80
1473: 3F            606        HEX 80238027802B802F8033803780 3B803F
1474: 28 20 28
1477: 24 28 28
147A: 28 2C 28
147D: 30 28 34
1480: 28 38 28
1483: 3C            607        HEX 2820282428282C28302834283828 3C
1484: A8 20 A8
1487: 24 A8 28
148A: A8 2C A8
148D: 30 A8 34
1490: A8 38 A8
1493: 3C            608        HEX A820A824A828A82CA830A834A838A83C
1494: 28 21 28
1497: 25 28 29
149A: 28 2D 28
149D: 31 28 35
14A0: 28 39 28
14A3: 3D            609        HEX 2821282528292 82D28312835283928 3D
14A4: A8 21 A8
14A7: 25 A8 29
14AA: A8 2D A8
14AD: 31 A8 35
14B0: A8 39 A8
14B3: 3D            610        HEX A821A825A829A82DA831A835A839A83D
14B4: 28 22 28
14B7: 26 28 2A
14BA: 28 2E 28
14BD: 32 28 36
14C0: 28 3A 28
14C3: 3E            611        HEX 2822282628 2A282E2832283628 3A283E
14C4: A8 22 A8
14C7: 26 A8 2A
```

Listing 2b continued:

```
14CA: A8 2E A8
14CD: 32 A8 36
14D0: A8 3A A8
14D3: 3E            612   HEX  A822A826A82AA82EA832A836A83AA83E
14D4: 28 23 28
14D7: 27 28 2B
14DA: 28 2F 28
14DD: 33 28 37
14E0: 28 3B 28
14E3: 3F            613   HEX  28232827282B282F2833283728382B283F
14E4: A8 23 A8
14E7: 27 A8 2B
14EA: A8 2F A8
14ED: 33 A8 37
14F0: A8 3B A8
14F3: 3F            614   HEX  A823A827A82BA82FA833A837A83BA83F
```

--END ASSEMBLY--

ERRORS: 0

1268 BYTES

SYMBOL TABLE - ALPHABETICAL ORDER:

```
ACIACLR =$120A   BEEP    =$C030   C0       =$1043   C1       =$1056
C15     =$105C   C3      =$1067   CLEAN    =$13D4   CLR1     =$111F
CLR2    =$1227   CMASK   =$13DE   CONT     =$10EF   CONT1    =$10FD
COUNT   =$1F     CTR     =$19     D1       =$1097   DATA     =$C08F
DEST    =$08     DEST2   =$1C     DISP1    =$12ED   DISPLAY  =$12A3
DLY1    =$127D   DOKEY   =$1075   DONE     =$1203   ENHANCE  =$12F2
FILL1   =$1382   FILL2   =$13C3   FILL3    =$13D2   FILLIN   =$1374
FULLPIC =$1098   GET     =$1030   GMODE    =$C050   GRCLR    =$121C
GREY    =$1115   GRTABLE =$5000   GSTAT    =$10CC   HGR      =$C057
IBUFFER =$4000   IGNORE  =$10EC   IMAGE    =$ED     INCREMENT=$0306
KEEPCNT =$F9     KEEPFLG =$FA     KEY      =$0304   KEYCLR   =$C010
KEYFLAG =$0305   KEYHIT  =$C000   LMASK    =$13E2   LOADTBL  =$1153
LOOP    =$1291   MIXED   =$C053   MOV      =$12BB   MOVE     =$1283
MSEC    =$1279   NDONE   =$1070   NEVEN    =$1304   NEVENJ   =$1371
NEWROW  =$101E   NEXT    =$11E9   NEXTPIC  =$1132   NEXTROW  =$11A2
NODD    =$1339   NODUB   =$1331   NODUB2   =$135A   NOHANG   =$10DE
NORMPIC =$1009   NSTART  =$100C   NXT1     =$11F5   NXT2     =$1200
```

```
NXTBYTE =$114D   NXTROW  =$12AD   PAGE1    =$C054   PRBLNK   =$F948
PRHEX   =$FDE3   RADR    =$06     RESHFT   =$12C6   RMASK    =$13E0
ROWPTR  =$13F4   ROWSTART=$0303   SEND1    =$1247   SENDCMD  =$1241
SETGR   =$1234   SHFT    =$12C9   SLOTADR  =$0302   SOAK     =$1255
SOAK1   =$126D   SOAK2   =$1278   SOAKTIME =$0300   START    =$10AA
STATUS  =$C08E   TABLE   =$EB     TAKEPIC  =$10A7   TBL1     =$116D
TBUFFER =$3000   TEMP1   =$06     TEMP2    =$07     TEMP3    =$1C
TEMP4   =$1D     TEMP5   =$1B     THISBYTE =$11AE   THISROW  =$11A6
TMODE   =$C051   TMP     =$1E     VAL1     =$13E4   VAL2     =$13E8
VAL3    =$13EC   VAL4    =$13F0   YREG     =$1B     ZERO     =$115B
ZERO1   =$1162
```

SYMBOL TABLE - NUMERICAL ORDER:

```
RADR    =$06     TEMP1   =$06     TEMP2    =$07     DEST     =$08
CTR     =$19     YREG    =$1B     TEMP5    =$1B     DEST2    =$1C
TEMP3   =$1C     TEMP4   =$1D     TMP      =$1E     COUNT    =$1F
TABLE   =$EB     IMAGE   =$ED     KEEPCNT  =$F9     KEEPFLG  =$FA
SOAKTIME=$0300   SLOTADR =$0302   ROWSTART =$0303   KEY      =$0304
KEYFLAG =$0305   INCREMENT=$0306  NORMPIC  =$1009   NSTART   =$100C
NEWROW  =$101E   GET     =$1030   C0       =$1043   C1       =$1056
C15     =$105C   C3      =$1067   NDONE    =$1070   DOKEY    =$1075
D1      =$1097   FULLPIC =$1098   TAKEPIC  =$10A7   START    =$10AA
GSTAT   =$10CC   NOHANG  =$10DE   IGNORE   =$10EC   CONT     =$10EF
CONT1   =$10FD   GREY    =$1115   CLR1     =$111F   NEXTPIC  =$1132
NXTBYTE =$114D   LOADTBL =$1153   ZERO     =$115B   ZERO1    =$1162
TBL1    =$116D   NEXTROW =$11A2   THISROW  =$11A6   THISBYTE =$11AE
NEXT    =$11E9   NXT1    =$11F5   DONE     =$1200   ACIACLR  =$120A
GRCLR   =$121C   NXT2    =$1227   SETGR    =$1234   SENDCMD  =$1241
SEND1   =$1247   SOAK    =$1255   SOAK2   =$1278   MSEC     =$1279
SOAK1   =$126D   DLY1    =$127D   LOOP     =$1291   MOVE     =$1283
DISPLAY =$12A3   NXTROW  =$12AD   MOV      =$12BB   RESHFT   =$12C6
SHFT    =$12C9   DISP1   =$12ED   ENHANCE  =$12F2   NEVEN    =$1304
NODUB   =$1331   NODD    =$1339   NODUB2   =$135A   NEVENJ   =$1371
FILLIN  =$1374   FILL1   =$1382   CMASK    =$13DE   FILL2    =$13C3
FILL3   =$13D2   CLEAN   =$13D4   RMASK    =$13E0   LMASK    =$13E2
VAL1    =$13E4   VAL2    =$13E8   VAL3    =$13EC   VAL4     =$13F0
ROWPTR  =$13F4   GRTABLE =$5000   TBUFFER =$3000   IBUFFER  =$4000
KEYHIT  =$C000   KEYCLR  =$C010   GMODE   =$C050   TMODE    =$C051
MIXED   =$C053   PAGE1   =$C054   HGR     =$C057   STATUS   =$C08E
DATA    =$C08F   PRHEX   =$FDE3   PRBLNK  =$F948
```

49

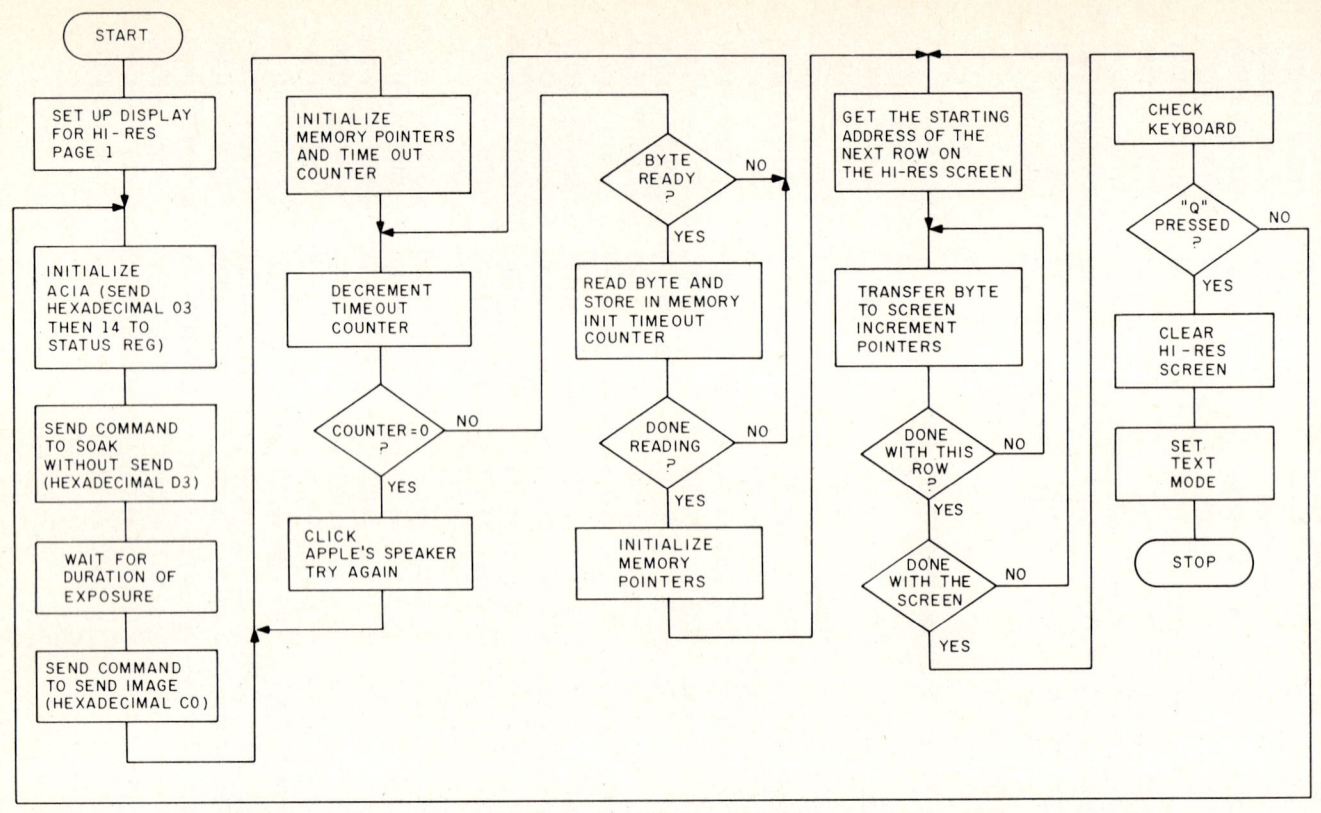

Figure 4: *A flowchart of the Micro D-Cam demonstration program for the Apple II. The program consists of a BASIC main routine, shown in listing 1a, and some 6502 machine-language subroutines, shown in assembly-language form in listing 1b.*

while waiting for the camera, it clicks the speaker, checks for a keypress, and tries the entire command sequence again. In this manner, you are alerted to any possible problems.

Because the Apple's hi-res screen display is mapped nonlinearly into memory space, a lengthy table at the end of the machine-language code provides the starting address for each consecutive row of the hi-res screen. The program gets the address of the beginning of each row and then reads 40 bytes from the camera, placing them consecutively on the screen. The next row, and each row after it, is done in a similar manner.

Once the image is on the screen, a command is sent to the camera to refresh without sending. This gets it ready for the next exposure. Finally, the machine code checks the keyboard and processes any command inputs before returning to BASIC.

Obtaining Gray Scale

A more user-friendly demonstration of the Micro D-Cam that also provides a level of gray-scale capability is the GREY16 program of listing 2 (pages 39 through 49). It has one mode that allows you to do quick aiming and focusing of the camera, another to let you get an idea of what the final picture will look like, and a third to create a 15-intensity-level gray-scale picture on the Apple II. (The processes involved are outlined in the flowchart of figure 5. Unfortunately, space constraints prevent me from showing you a similar program for the IBM PC.)

Using GREY16, you can change the length of exposure for the image being displayed, or you can change the upper and lower exposure limits of the gray-scale image. Once you've obtained a satisfactory picture, you can save it on disk for later use or print it on an Epson MX-80 printer (equipped with Graftrax) using the screen-dump program. A summary of available commands in the GREY16 program is shown in table 4.

When it is first powered up, you start the camera running by selecting one of the options from the GREY16 menu. If the exposure time is insufficient, the screen will be black. If the exposure time is excessive, the screen

Command Character	Control Effect
N	display the image in normal size (256 by 64)
F	display the image in full size (256 by 128)
G	create a picture (256 by 128) with 15 levels of gray (this process takes about 30 seconds and displays a countdown of the number of exposures from F to 0)
E	change the exposure time of the current displayed image, the upper limit of the gray-scale image, or the lower limit of the gray-scale image
S	save to disk the picture currently being displayed (this may be done in any of the three display modes: normal, full, or gray)
Q	quit the program and return to BASIC

Table 4: *A summary of user commands implemented in the GREY16 program of listing 2.*

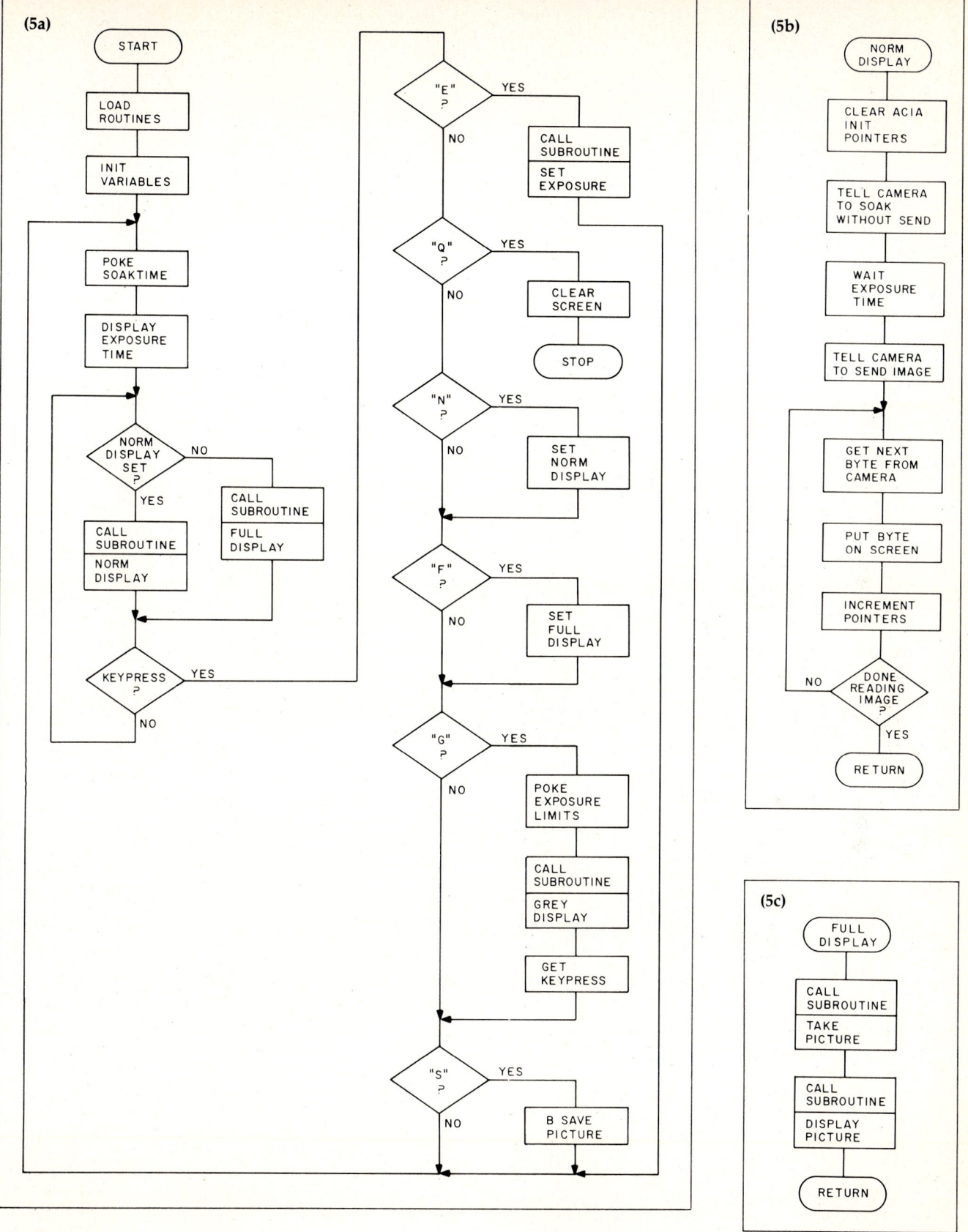

Figure 5: *Flowcharts of the GREY16 program for the Apple II (the figure continues on pages 80 and 82). The BASIC portion appears as listing 2a, the machine-language portion as listing 2b. The main routine (5a) calls various subroutines: NORM DISPLAY (5b), FULL DISPLAY (5c), SET EXPOSURE (5d), GREY DISPLAY (5e), TAKE PICTURE (5f), DISPLAY PIC (5g), and ENHANCE (5h).*

The subroutine GREY DISPLAY takes sensor pixels from 15 exposures and translates them into arrays of the smaller display pixels to represent intermediate brightnesses.

Figure continued

Figure 5 continued:

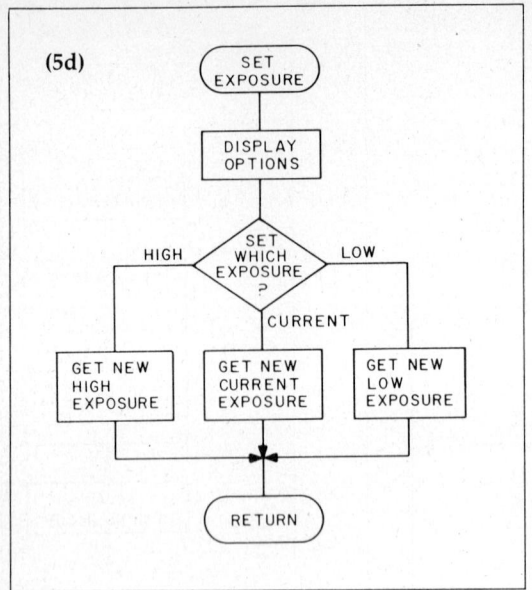

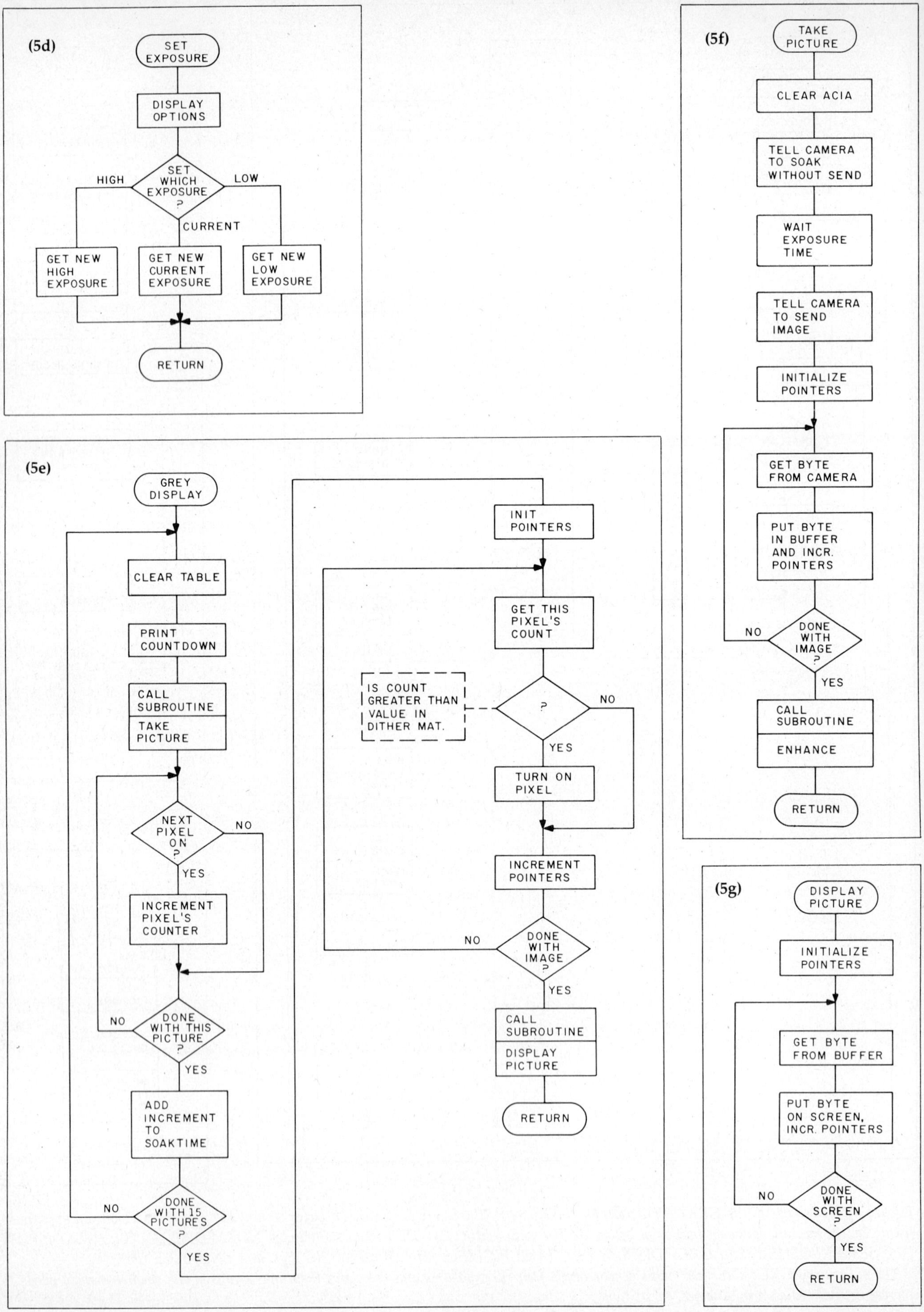

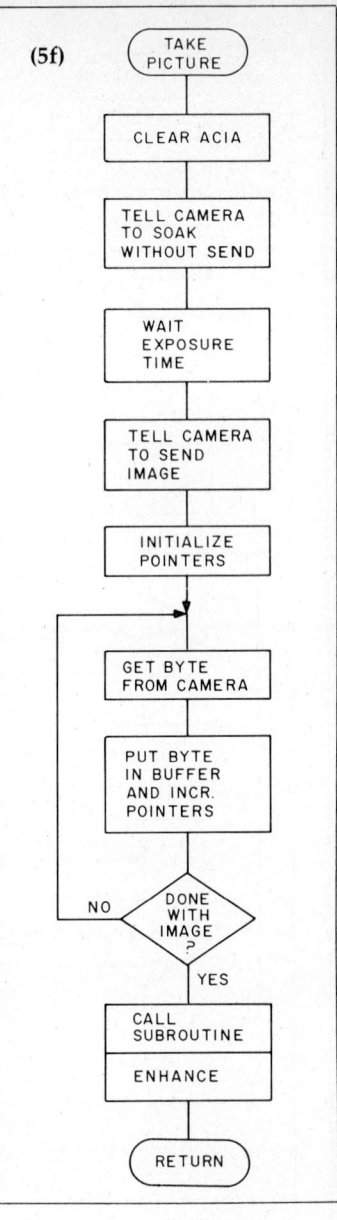

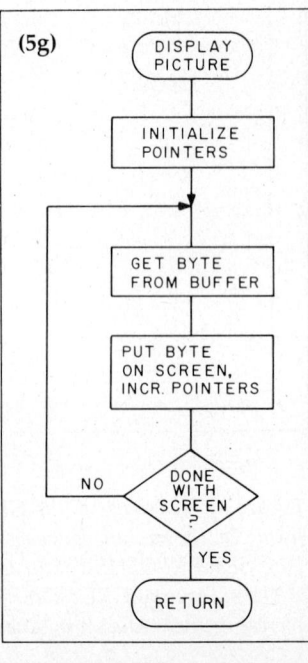

Figure continued

Figure 5 continued:

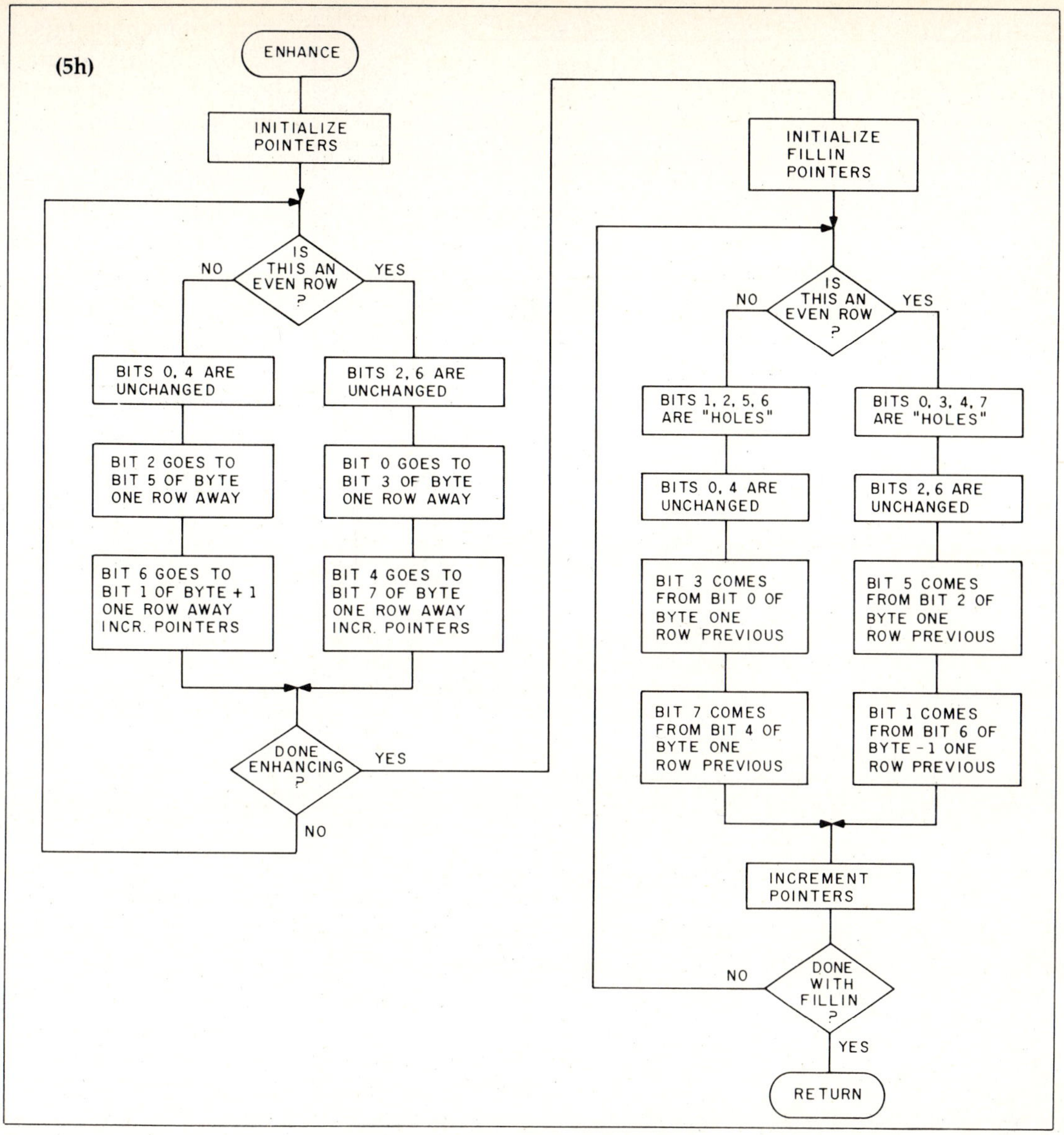

(5h) [Flowchart: ENHANCE → INITIALIZE POINTERS → IS THIS AN EVEN ROW? NO: BITS 0,4 ARE UNCHANGED; BIT 2 GOES TO BIT 5 OF BYTE ONE ROW AWAY; BIT 6 GOES TO BIT 1 OF BYTE +1 ONE ROW AWAY INCR. POINTERS. YES: BITS 2,6 ARE UNCHANGED; BIT 0 GOES TO BIT 3 OF BYTE ONE ROW AWAY; BIT 4 GOES TO BIT 7 OF BYTE ONE ROW AWAY INCR. POINTERS. → DONE ENHANCING? NO: loop; YES → INITIALIZE FILLIN POINTERS → IS THIS AN EVEN ROW? NO: BITS 1,2,5,6 ARE "HOLES"; BITS 0,4 ARE UNCHANGED; BIT 3 COMES FROM BIT 0 OF BYTE ONE ROW PREVIOUS; BIT 7 COMES FROM BIT 4 OF BYTE ONE ROW PREVIOUS. YES: BITS 0,3,4,7 ARE "HOLES"; BITS 2,6 ARE UNCHANGED; BIT 5 COMES FROM BIT 2 OF BYTE ONE ROW PREVIOUS; BIT 1 COMES FROM BIT 6 OF BYTE −1 ONE ROW PREVIOUS. → INCREMENT POINTERS → DONE WITH FILLIN? NO: loop; YES → RETURN]

will be completely white. These situations may be remedied by increasing or decreasing the exposure time or changing the lens aperture. You may need to focus, also. Eventually, a clear picture will appear on the video screen when the lens is properly adjusted.

The gray-scale portion of the program demonstrates what can be done with just a little bit of software enhancement, permitting you to create images with 14 intermediate levels of brightness (plus extreme dark and bright) and display them on the Apple's hi-res screen. The image of an automobile shown in photo 1 is an example.

The technique used to display the gray-scale pictures on the Apple II Plus and IIe computers is known as *ordered dithering*, in which half-tone values are constructed from multiple binary black or white images. The process requires the Micro D-Cam

Photo 1: *The Micro D-Cam was aimed at a car parked outside. The dithered digital gray-scale image shown here is displayed by an Apple II Plus.*

system to take 15 exposures of the same subject, each lasting a little longer than the previous one. (This normally takes only several seconds.) After each exposure is taken, every pixel in it is checked. If the pixel is on (showing a 1 value corresponding to brightness above that exposure's threshold), a counter location corresponding to that pixel is incremented. At the end of 15 passes, this process yields a table of values, each value describing the relative intensity of its corresponding pixel. For example, if a pixel's final value is 15, that pixel should be displayed maximally bright; if a pixel's value is 8, the pixel deserves a shade of gray halfway between the black and white extremes.

Once the pixel-intensity table has been constructed, a 4 by 4 dither

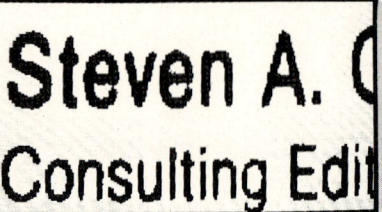

Photo 2: *A corner of the author's business card (2a) was easily reproduced by the Micro D-Cam (2b) because the printing represents only two levels of brightness. One potential use of the image camera is in optical character recognition.*

matrix is used to assign a display value (an array of binary pixels) to each screen position. In this software, the matrix is as follows:

```
 0   8   2  10
12   4  14   6
 3  11   1   9
15   7  13   5
```

Then, one pixel at a time, the values in the table of final image magnitudes are compared to the array-element values in the matrix. If the image value for the pixel is greater than the element's value, that array element is turned on. If the intensity value is 0, none of the matrix elements are displayed bright; if the value is 8, elements 0 through 8 are displayed bright; and if the value is 15, all the elements become bright. In this manner, 15 levels of luminance may be represented but at a certain loss of spatial definition. The process is repeated across the entire screen until each screen position has a value assigned to it.

It would definitely be possible to use different-size dithering matrices, with certain trade-offs. For example, a 2 by 2 matrix would yield only 5 levels of gray but would have much finer spatial definition, while an 8 by 8 matrix would yield 64 levels of gray but with much loss of spacial definition.

The GREY16 program overcomes many of the limitations associated with binary optical sensors. While black print on white paper (like my business card, shown in photo 2) is easily viewed by the Micro D-Cam with no enhancement, we don't live in a pure-black-and-white world, and three-dimensional objects need shading to be recognized on a two-dimensional video display.

This is most easily demonstrated with a series of photos of a pair of dice. Photo 3a shows the color and lighting conditions of our sample object. If we use the Micro D-Cam without gray scale, we obtain the binary picture in photo 3b. (This slightly vague yet quite representative picture of the dice would probably be usable in robotics or some recognition applications.)

For a more representative picture, we can invoke the G command in the GREY16 program to produce photo 3c. There is now no question of what the subject is or what value is shown on the dice. If the image were reproduced on a computer capable of displaying half-tones, it would look much more like photo 3a than this dithered Apple II Plus display.

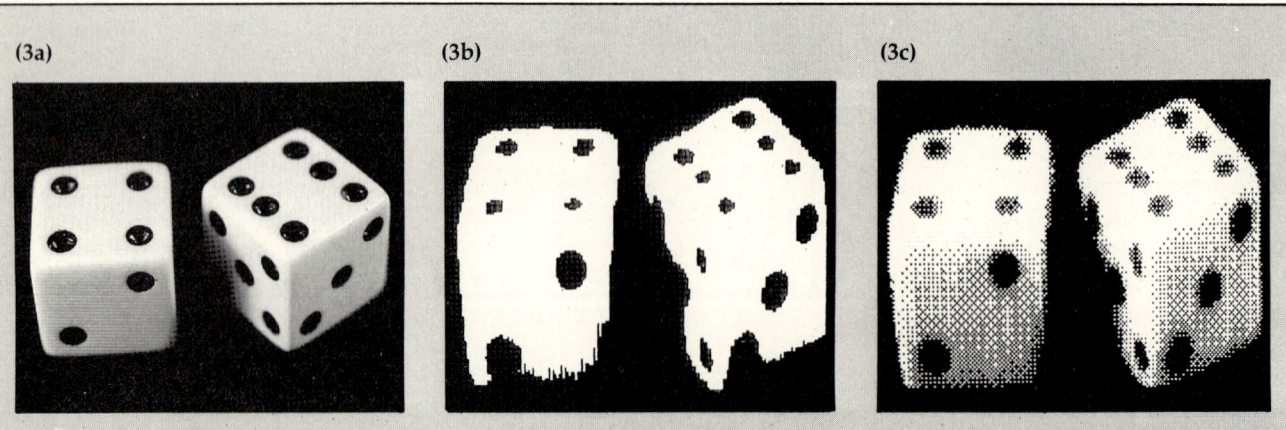

Photo 3: *A pair of dice (3a) was scanned by the Micro D-Cam. When only two levels of luminance are recorded and sent to the computer's display, the result is the output shown in photo 3b. When multiple gray-scale exposures and ordered dithering are invoked, the more easily recognizable output of photo 3c appears.*

Closing Observations

Two articles have not been enough to describe all the capabilities of the Micro D-Cam. If I had more time, I'd try some experiments using different lenses and filters. Theoretically, if three exposures were taken through red, green, and blue filters, we should be able to create a color image.

One interesting fact I did observe is that the IS32, like most silicon-based image sensors, is infrared-sensitive. My test was somewhat unscientific, and I have no precise data on the Optic RAM's spectral sensitivity. I merely lighted the subject with some infrared light-emitting diodes, but it was clearly seen by the Micro D-Cam even in visible-light darkness.

This mild success leads me to consider related experiments. Don't count on it, but in a few months you just might be reading about some sort of character-recognition wand I've built using an Optic RAM. In the meantime, if you find any other dynamic RAM chips that are suitable in this application or wish to show me a character-recognition program of your own, please write and let me know.

References

1. Ciarcia, Steve. "Analog Interfacing in the Real World." BYTE, January 1982, page 72.
2. Ciarcia, Steve. "Build the Micro D-Cam Solid-State Video Camera, Part 1: The IS32 Optic RAM and the Micro D-Cam Hardware." BYTE, September 1983, page 20.
3. Crow, Franklin C. "Three-Dimensional Computer Graphics." Part 1, BYTE, March 1981, page 54; Part 2, April 1981, page 290.
4. Grob, Bernard. *Basic Television: Principles and Servicing*, 4th ed. New York: McGraw-Hill, 1975.
5. Newman, William M. and Robert F. Sproull. *Principles of Interactive Computer Graphics*, 2nd ed. New York: McGraw-Hill, 1979.
6. Tomas, Joe. "Hardware Review: Dithertizer II." BYTE, February 1982, page 219.
7. Walker, Terry, Harry Garland, and Roger Melen. "Build Cyclops: First All Solid-State TV Camera for Experimenters." *Popular Electronics*, February 1975, page 27.
8. Williams, Thomas. "Digital Storage of Images." BYTE, November 1980, page 220.

The following items are available from

The Micromint, Inc.
25 Terrace Drive
Vernon, CT 06266
For orders: (800) 635-3355
For information: (203) 871-6170

1. *Complete Micro D-Cam unit including interface card, extension cable, IS32 Optic RAM, lens, remote housing, operator's manual, and utility software. Specify Apple II (II Plus or IIe) or IBM Personal Computer.*
   Assembled and tested............$295

2. *Same as item 1 except in kit form. Specify Apple II or IBM PC.*
   Complete kit....................$260

3. *IC32 Optic RAM sold separately.*
   Each...........................$42

4. *Serial-interface (RS-232C) Micro D-Cam for general use. Software listings for several different computers to be available soon. Call for price and delivery.*

Please add $4 shipping and insurance in continental United States, $20 overseas. New York residents, please include 7 percent sales tax.

Special thanks to Carl Baker and Jim Herrud of Micron Technology Inc. for their contributions to this project.

Build the H-Com Handicapped Communicator

During an engineering assignment a few years ago I went to meet a man we'll call Dave, the owner of a small development company and its chief designer. As I sat in the lobby waiting to see him, I couldn't help but notice the many plaques, patents, citations, and honors bestowed on the company. "Surely," I thought, "to possess such impressive credentials, the manager of this company must be a real dynamo." I pictured him barking orders and moving at a furious pace, carrying a memocorder in one hand and a wireless phone in the other, being pursued by a cadre of support personnel. How else could anyone accomplish so much?

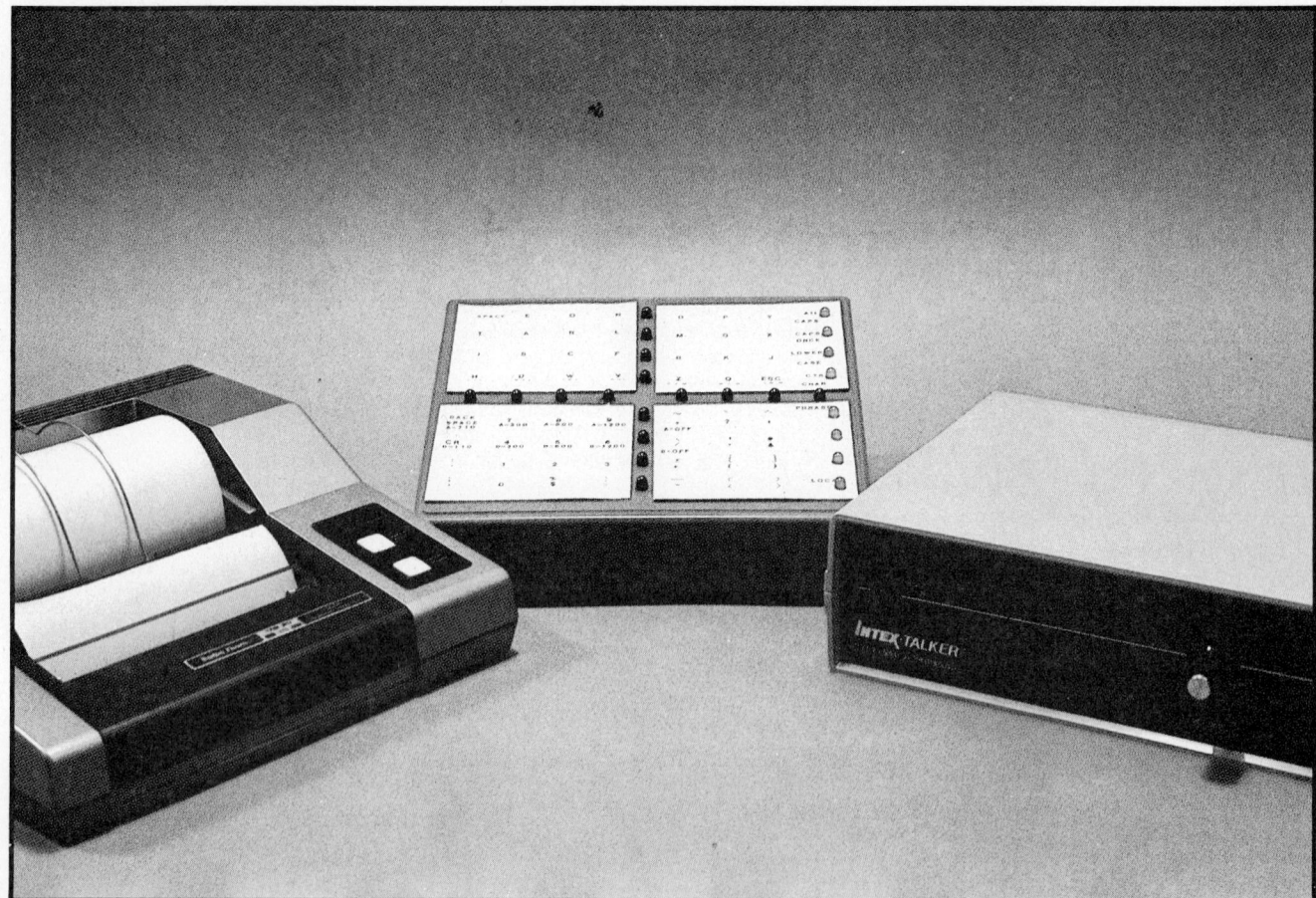

Photo 1: *The H-Com scanning communicator, a kind of keyboard simulator, can be used to send text directly to a printer, such as the Radio Shack CGP-115 shown here, or to a text-to-speech synthesizer, such as the Intex Talker, in this fully configured system. Using the serial-output commands and phrase mode, the H-Com can transmit words and sentences from a prestored vocabulary.*

The Intel 8748 self-contained microprocessor forms the heart of a scanning communicator

I don't now remember what we discussed at that first meeting. I only remember my shock at discovering that this super executive was a quadriplegic, suffering from a degenerative disease of the nervous system that left him with no fine motor control, virtually paralyzed.

During our meeting Dave used a one-switch scanning communicator, a sophisticated machine that enabled him to type on an electric typewriter. A scanning communicator presents a display of alphabetic, numeric, and punctuation characters. Under or beside each character is a lamp indicator. The device illuminates the lamp for one character (or group of characters) in a sequence. By biting down on a mouth switch at the right instant, Dave could cause the indicated character to be typed. The machine also stores a vocabulary of frequently used words and phrases. In later conversations with other staff members I learned that Dave often wrote entire design proposals using this technique.

Dave's body was frail, but he had one of the sharpest minds I've ever met. I've since given up dealing in stereotypes.

My purpose in relating this experience to you is not to solicit your sympathy but rather to inform you how technology has helped one man compensate for physical limitation. This meeting left me with a profound appreciation for the value of communication and the important role that electronics can play in aiding disabled people.

While it would be hard to duplicate the sophistication of the scanning communicator that Dave used, technology has advanced to a state where we can reproduce certain of its primary functions at minimal expense. In view of this, I decided to present a project that can serve both as an example of an application for the Intel 8748 single-chip microcomputer and as a demonstration of the potential benefits of technology.

Build the H-Com

This month's Circuit Cellar project is called H-Com, which stands for "handicapped communicator." It's intended to do the same job as a normal computer keyboard, but using only one "key," a single user-input point hereinafter referred to as the switch. Because there is only one switch in the H-Com, its user need control only one muscle to actuate it. Any kind of normally open momentary-closure switching contacts will work. An eye-blink detector would work, or the system could even use the biofeedback detector I wrote about in a previous Circuit Cellar article (see reference 4).

The H-Com has three outputs: two RS-232C ports and one audible horn. The RS-232C output ports can be turned on or off and the data rate set by user input. For serial communication, the full ASCII (American National Standard Code for Information Interchange) character set, including all control characters, can be generated. The horn can be used to beep out seven different patterns, intended principally for obtaining the attention of other people nearby.

The H-Com terminal has a prestored vocabulary of words and complete sentences that can be transmitted upon receipt of a single command. These canned transmissions can take the form of ASCII-encoded text sent to a voice synthesizer (such as the one discussed in reference 3) or control codes sent to an autodialing telephone (or modem) that directly links the user to help in an emergency. And the H-Com is designed with eventual expansion in mind. All of these design criteria require that the H-Com contain one of the devices we've used so often lately in high-performance electronic equipment—a microprocessor.

The microprocessors you're probably most familiar with are the general-purpose Z80, 6502, and 8088. But these chips are designed to be used in relatively large digital systems; other less well known microprocessors have been built to be easier and cheaper to use in simple control applications.

The Intel 8748

One of Intel Corporation's product lines is a set of VLSI (very large-scale integration) chips—containing processor, memory, and support-logic circuitry—of which the flagship product is the 8048. The 8048 features mask-programmed ROM (read-only memory), which is good for applications that require thousands of the chips to be installed in identical pieces of equipment, such as the keyboards of IBM Personal Computers. But small-scale experimentation can more practically use its cousin, the 8748, which sports on-chip EPROM (erasable programmable ROM). Figure 1 is a functional block diagram of the Intel 8748 single-chip 8-bit microcomputer, which is shown in photo 2.

The resident program memory in the 8048 consists of 1024 (1K) words 8 bits wide (in other words, the memory is 1K bytes), which are addressed in random-access fashion by the program counter. In the 8748 this memory consists of EPROM, which allows the processor's program to be loaded in the system designer's workshop rather than at the factory. To burn the program into the 8748's EPROM, external circuitry must activate the program mode, apply and latch an address, apply data, and pulse the chip's program line. Each word of memory is verified immediately after it has been burned. The entire EPROM contents can be erased by exposing the 8748 to ultraviolet light (see reference 2).

The 8748 contains 64 eight-bit registers, called the *resident data memory,*

Materials pertaining to the 8748 are reprinted courtesy of Intel Corporation.

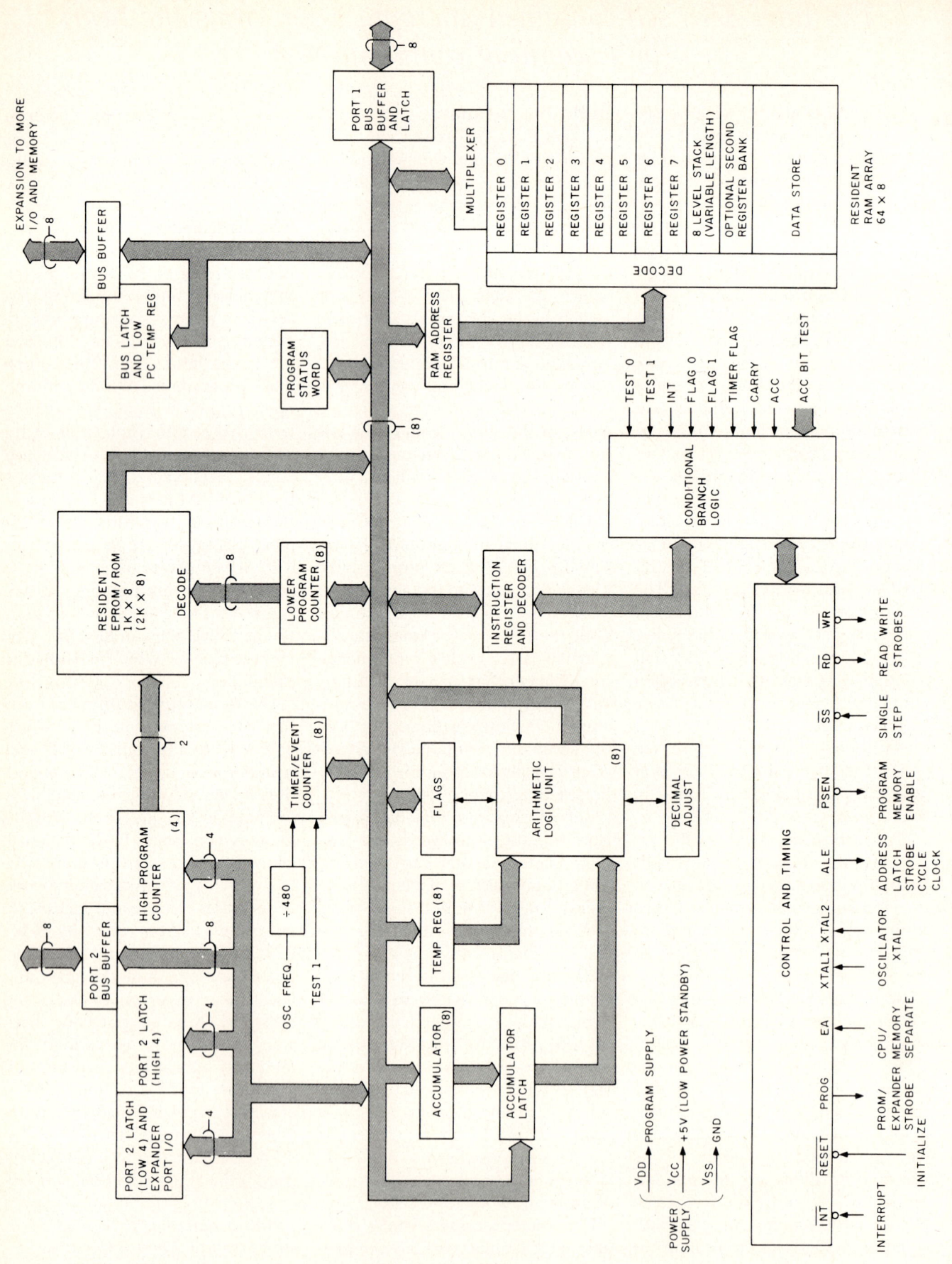

Figure 1: *A functional block diagram of the Intel 8748 self-contained microprocessor.*

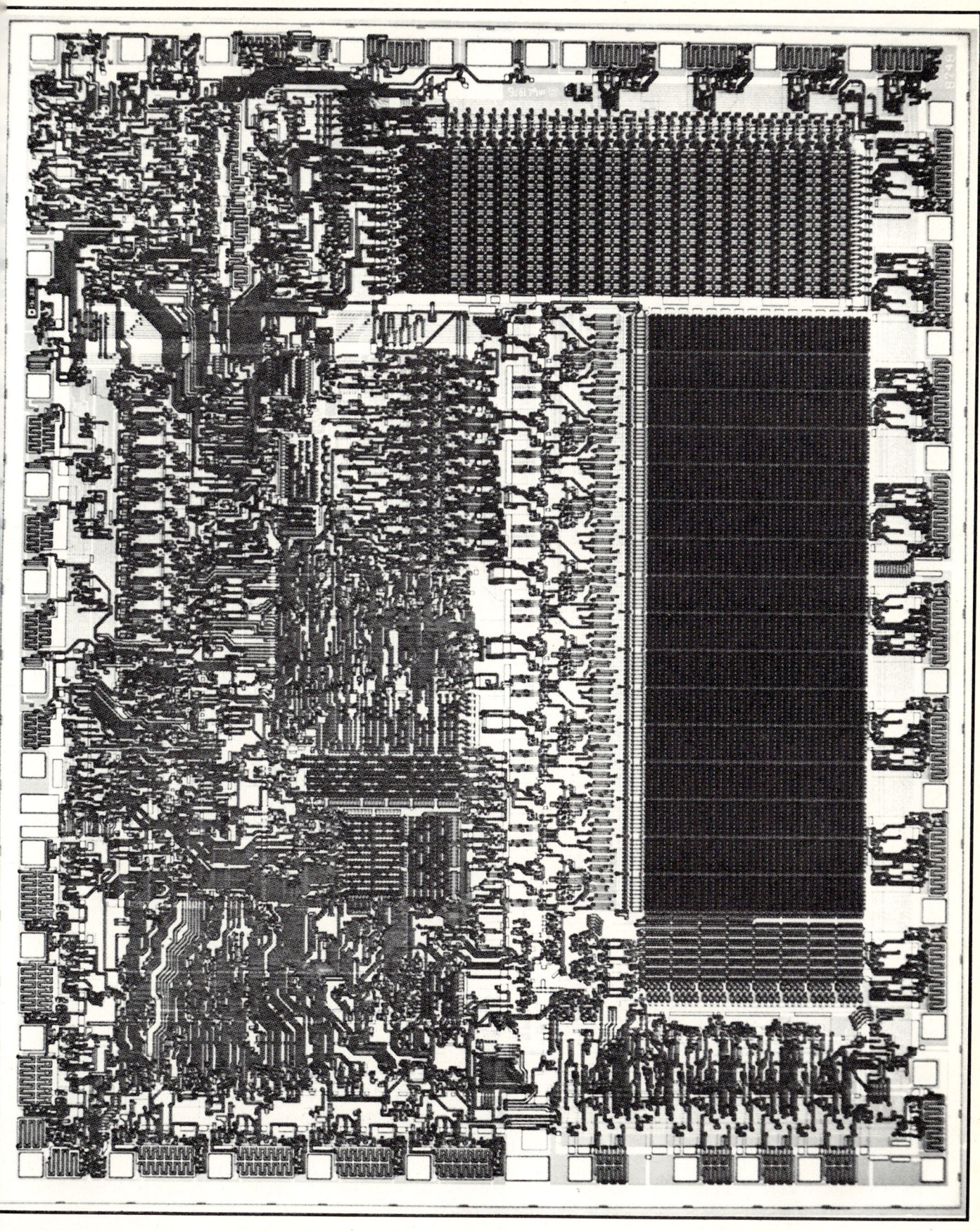

Photo 2: *Shown in this photomicrograph, Intel Corporation's 8748 microprocessor is largely self-sufficient, containing its own EPROM, scratchpad RAM, and I/O circuitry.*

> ## The Intel 8048/8748 Instruction Set
>
> *The processor contains the basic data-manipulation functions and can be divided into four major functional sections: the arithmetic/logic unit (ALU), the accumulator, the carry flag, and the instruction decoder.*
>
> *In a typical operation, data stored in the accumulator is combined in the ALU with data from another source on the internal bus (such as a register or I/O port), and the result is stored in the accumulator or another register. The ALU accepts 8-bit data words from one or two sources and generates an 8-bit result under control of the instruction decoder. The ALU can perform the following functions:*
>
> - *add with or without carry*
> - *AND, OR, exclusive OR*
> - *increment/decrement*
> - *bit complement*
> - *rotate left, right*
> - *swap nybbles in accumulator*
> - *decimal adjust accumulator (BCD)*
>
> *One machine instruction makes very efficient use of the working registers as program-loop counters: the DJNZ (decrement, jump if not zero) instruction allows the program to decrement and test the register in a single instruction.*

which can be used as scratchpad RAM (random-access read/write memory). The first eight locations in this array (numbered 0 through 7) are designated as special-purpose "working" registers and are directly addressed by several instructions. All 64 locations are indirectly addressable through either of the two RAM-pointer registers, registers 0 and 1. Because the first eight registers are more easily addressed, they are typically used to store frequently accessed data or intermediate results. The text box above discusses the 8748's instruction set.

The 8748 has 27 I/O (input/output) signal lines. Twenty-four of these lines are grouped into three I/O ports of eight lines each; these can be used for input, for output, or bidirectionally. The remaining three lines are single-bit "test" inputs, which can alter program flow when tested by conditional-jump instructions.

I/O ports 1 and 2 are each 8 bits wide and have identical characteristics. The lines of these ports are called *quasibidirectional* because they employ a special output-circuit structure that allows each line to serve as an input, an output, or both, even though the outputs are statically latched (that is, data written to these ports for output remains unchanged until new data is loaded into them). However, when used as input ports, these lines are nonlatching; this requires the external circuitry to keep the levels for each transferred byte valid until the 8748 reads the byte by an input instruction. The I/O ports are fully compatible with TTL (transistor-transistor logic); the outputs will drive one standard TTL load.

The third I/O port is called the *bus port*. It is also an 8-bit port, but it is truly bidirectional, having associated input and output strobe signals. If bidirectional operation is not needed, the bus port can serve as either a statically latched output port or a nonlatching input port. However, input and output lines on this port cannot be mixed. In some modes of operation, the bus port is used to address external memory.

In static-port operation, data is written and latched using the 8748's OUTL instruction; data is input using the INS instruction. The INS and OUTL instructions generate pulses on the corresponding $\overline{RD}$ and $\overline{WR}$ output strobe lines; however, in the static-port mode these signals are generally not used. In bidirectional-port operation, the MOVX instructions are used to read and write to the port. A write to the port generates a pulse on the $\overline{WR}$ output line, and output data becomes valid at the trailing edge of the pulse. Reading the port generates a pulse on the $\overline{RD}$ output line; input data must be valid at the trailing edge of the $\overline{RD}$ pulse. When not being written or read, the bus-port lines are in a high-impedance state.

The 8748 also contains a counter/timer register intended for use in enumerating external events and generating accurate time delays without placing an extra burden on the processor. This 8-bit binary up counter can be preset and read with two MOV processor instructions, which transfer the contents of the accumulator to the counter, and vice versa. The contents of the counter are not cleared by a processor reset; they can be initialized solely by the MOV instructions. Counting is stopped either by a processor reset or when a STOP TCNT instruction is executed. After counting has stopped, it can be restarted for use as a timer by a START T instruction or as an event counter by a START CNT instruction. Once started, the counter is continually incremented, overflowing to zero when its maximum value (hexadecimal FF) is reached but continuing its count until stopped by a STOP TCNT instruction or processor reset.

The 8748 contains all necessary circuitry for generating timing signals, with the exception that a frequency reference, which can be a crystal, inductor, or external clock pulse, must be connected. The on-board oscillator is a high-gain series-resonant circuit with a frequency range of 1 to 6 MHz. A crystal or inductor connected between the 8748's pinouts X1 and X2 provides the feedback and phase shift required for oscillation. A 6.144-MHz crystal allows easy derivation of all standard serial-communication frequencies.

Implementation of the H-Com

The H-Com consists of a small case with a character grid of 64 elements arranged into 8 horizontal rows and 8 vertical columns (see photo 3). Each element is the equivalent of a keyboard key.

The characters are arranged in the array such that the ones most frequently used are clustered in the upper left, the position reached most quickly during the scanning process. The least used characters (special punctuation) are placed at the end of the scan in the lower right. The rightmost (eighth) column is used to control the H-Com's operation rather than transmit characters. A practiced

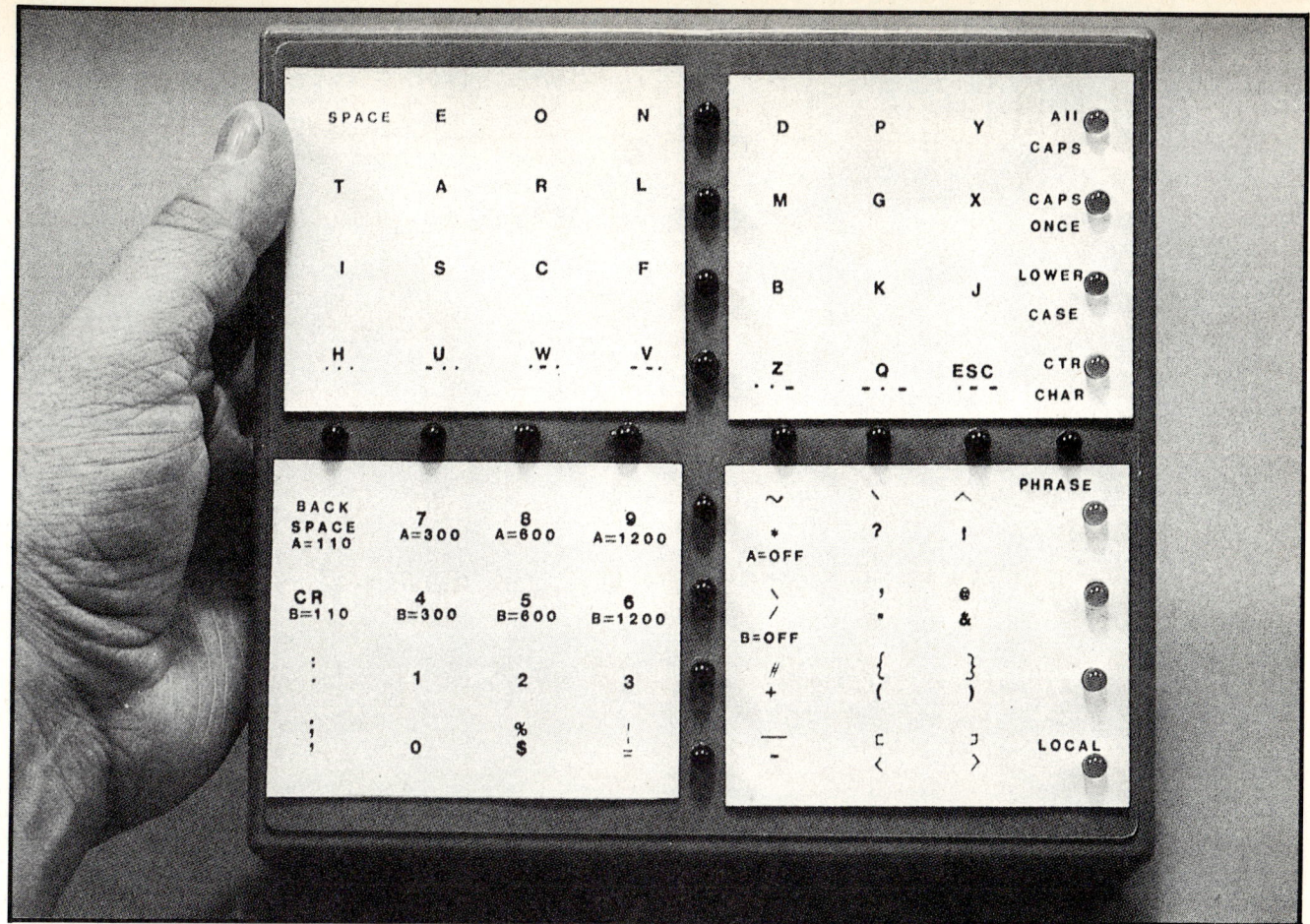

Photo 3: *The H-Com's character display contains 8 rows and 8 columns of characters and control functions, numbered from top to bottom and from left to right. The intersecting lines of red LEDs are used in scanning the row and column positions, while the yellow LEDs along the right edge indicate which mode is in use.*

user can select and transmit characters with relative ease and surprising speed.

Each of the 8 rows and 8 columns has a corresponding selection indicator, a total of 16 red LEDs (light-emitting diodes). The scanning operation proceeds as follows. The LEDs for the 8 rows are lighted individually in sequence from top to bottom: first row, second row, third row, and so on to the eighth row, then back to the first row and repeat. The row scan continues until the H-Com senses that the switch is closed, indicating that the user has made a selection of the row for which the LED is lit. The H-Com program stores the selected row number and proceeds to the column scan. In this second phase of selection, each of the column LEDs is lit in succession from left to right. Once again, the user closes the switch during the interval in which the LED is lit that corresponds to the column containing the desired character.

When both a row and a column have been selected, the microprocessor looks in a table to find the character associated with the row and column position (x and y coordinates, if you will). The character or function assigned to the position may vary according to the major mode of operation selected. If the character is in the printable set, the H-Com transmits it through either or both RS-232C ports.

H-Com Modes

The rightmost column, as I mentioned, is used for controlling the H-Com, mostly for shifting its six modes of operation. Beside each mode square is a yellow LED, which is lighted when the corresponding mode is in use. When the H-Com is powered up, it starts out in the All-Caps mode, in which it will transmit only the main character set consisting of uppercase A through Z, numerals 0 through 9, and commonly used punctuation. Separate modes generate lowercase characters, braces, ASCII control characters, and special functions.

For example, to send a Control-C, you first select the control-characters mode (by closing the switch first during the row-4 interval and then in the column-8 interval), and then select the particular character ("C") with the next row/column scan. Immediately after sending the Control-C character, the H-Com reverts to the All-Caps mode. One of the modes even lets you transmit lengthy prestored messages by selecting a two-character mnemonic key. Let's look at the six H-Com modes:

All Caps: This is the default mode. All characters are converted to uppercase (capital letters) before being sent.

Photo 4: *The prototype of the H-Com circuit, viewed from the rear to show the integrated circuits. The light-emitting diodes are mounted on the other side.*

One Cap: This mode, when selected, sends the first character after its invocation as uppercase, and then all subsequent characters as lowercase. This is useful for capitalizing words because normally only the first letter is uppercase.

Lowercase: In this mode, characters are sent out lowercase.

Control Characters: This mode is used to generate the control codes. It acts much like One Cap except that it converts the next character selected to its control equivalent for transmission. Because the Escape control code is treated as a normal character, you need not use the Control Characters mode to generate it. The control codes normally used for cursor control are accessed by Control-8, -4, -6, and -2. Also, seldom used punctuation is generated in this mode, not in one of the caps modes.

Phrase: This mode is used to generate sequences of many characters to form complete words, sentences, etc. The text strings are stored serially in a type-2716 EPROM, each phrase tagged with a mnemonic key. For the H-Com to transmit the sequence, you select the Phrase mode, the characters of the mnemonic key, and then the space character. When the H-Com has detected the scan selection of a space while in Phrase mode, the 8748 takes the key and looks through the EPROM until it finds the corresponding text string; it then sends the string exactly as if the letters were being selected one at a time. If there is no phrase associated with the entered key, the H-Com beeps the horn and returns the mode to All Caps or Lowercase, whichever was last selected. The internal storage format for the EPROM is shown in listing 1, a simplified example. Normally this listing would be several pages long and contain hundreds of words.

Local: This mode is used for tasks that don't involve sending characters. The first three rows of the character array do nothing in Local mode.

The fourth row in the array controls the horn. The dot and dash symbols in the squares indicate the beep patterns, which superficially resemble Morse code. To sound a pattern of three short honks, for example, you select Local mode, then the H key, which causes three short beeps to be emitted. Each letter of the fourth row beeps a different pattern.

In Local mode, the fifth row selects the operating parameters for serial port A. The first position in the row, labeled Backspace/A = 110, sets port A to communicate at 110 bps (bits per second). The second position,

7/A=300, sets port A to 300 bps, the third position to 600 bps, the fourth to 1200 bps, and the fifth (labeled */A=OFF) turns the port off. To turn port A on, you select the data rate desired (if you want it off, select Local and then */A=OFF). The sixth row controls port B in the same manner.

The seventh and eighth rows control the scanning rate of the row- and column-select LEDs. The seventh-row, first-column position sets the slowest rate, and each succeeding column sets a rate faster by a factor that increases geometrically.

H-Com Hardware

Shown in the schematic diagram of figure 2, the circuitry of the H-Com can be divided into seven sections: the power supply, the RS-232C drivers, the microprocessor, the LED decoder/drivers, the phrase-lookup EPROM, the horn-tone generator, and the input switch. The prototype circuit board is shown in photo 4.

The H-Com draws about 300 mA (milliamps) at 12 V (volts). Current could be drawn from a motorized wheelchair's battery, a separate battery pack, or a 110-V AC-powered supply. If a 12-V supply is chosen, the currently available Radio Shack CGP-115 printer can be used as a convenient portable display device. The +12-V potential is reduced to +5 V through a type-7805 voltage regulator to power the logic circuitry.

IC1, a type-556 dual-timer chip, serves two purposes. It produces an audio signal at pin 9 to sound the horn and generates a second AC signal used as input to a charge-pumping circuit to produce a −9-V supply for the RS-232C transmitter section.

The horn signal, the direct output of IC1, drives a loudspeaker, which generates a sound low enough in frequency and loud enough to be heard by someone in an adjacent room. (Solid-state piezoelectric transducers, while efficient and compatible with TTL circuits, are not loud enough or low enough.) A series resistor (about 100 ohms) keeps the volume at a comfortable yet noticeable level. Sounding of the horn is controlled by an output bit on the 8748.

User inputs to the H-Com are handled through the 8748's T1 test input. This line is one of three input pins (T0 and INT are the others) that allow conditional program branches without using I/O instructions of the type that load the accumulator from the input port. Because T1 is to be connected to a mechanical switch, a debouncing integrator (resistor/capacitor combination) and a Schmitt trigger (IC6) smooth out its transitions.

Control of the H-Com functions is handled through the three parallel ports. Four bits of port 1 are reserved for serial communication. (The four remaining bits could be programmed to provide more ports if necessary.) With the data rates and character framing generated by software, each

---

The only unconventional part of the circuitry is the phrase-memory section.

---

port transmits independently at data rates from 110 to 1200 bps. When the H-Com is first turned on, the program sets port 1 to 600 bps to be compatible with the CGP-115 printer. IC8 and IC9 are the familiar MC1488 and MC1489 RS-232C driver and receiver chips. The −9-V supply mentioned earlier is used in the 1488. These devices were chosen primarily for simplicity; they could be replaced with a couple of transistors if you wanted to reduce the number of integrated-circuit packages.

Port 2 drives the LED display. The high-order 4 bits of port 2 are connected to a 4- to 16-line decoder driver, IC2, which produces the row/column scanning action. Depending upon the 4-bit value appearing at IC2's input, one of the 16 LEDs will be lit. As the count is incremented, the next LED in the row or column lights up, and scanning takes place.

The low-order 3 bits of port 2 are connected to a 3- to 8-line decoder/driver, IC3. Functioning in a manner similar to IC2, this circuit drives the yellow LEDs that indicate what mode the H-Com is in. The remaining bit of port 2 controls the horn.

The program for the 8748 single-chip microcomputer, IC9, is stored in the on-chip 1K- by 8-bit EPROM.

The only unconventional part of the circuitry is the phrase-memory section. The signals to address this memory are not generated by the processor, as is commonly the case. Instead, they are generated by two 8-bit binary counters (IC5 and IC7).

Initially, the counters are cleared (reset) by a low-level signal on the $\overline{WR}$ (pin 10) line of the 8748 (IC9), under the direction of a bus-port write instruction. When the processor needs to look up a phrase from the memory, it reads the bus port. After each such read instruction, an active-low pulse appears on the $\overline{RD}$ line, increasing the value in the counters by 1. When you request transmission of a stored phrase, the 8748 clears these address counters and begins reading at the beginning of the 2716 EPROM. The 8748 keeps reading and incrementing the counters until it finds a match to the phrase key.

This circuit, although not commonly seen, requires few chips and uses a relatively simple searching algorithm. Also, because the counters produce 16 address bits, up to 64K bytes of text storage can be easily accommodated. In fact, simply changing the type-2716 EPROM to a type-27128 would add 14K characters. But even with as many as 64K characters of stored phrases, the search would take less than one second.

Words and phrases are stored in the EPROM as ASCII character strings preceded by one or more mnemonic key characters that identify the particular word or phrase. As you can tell from listing 1, the mnemonic key is stored first in the EPROM, followed by a space character (hexadecimal 20), followed by the word or phrase (any length), and concluded by a null character (hexadecimal 00). Phrase storage could also be used to remind you how to operate certain features, with a help message triggered simply by setting Phrase mode and then selecting H, P, and a space on successive scans.

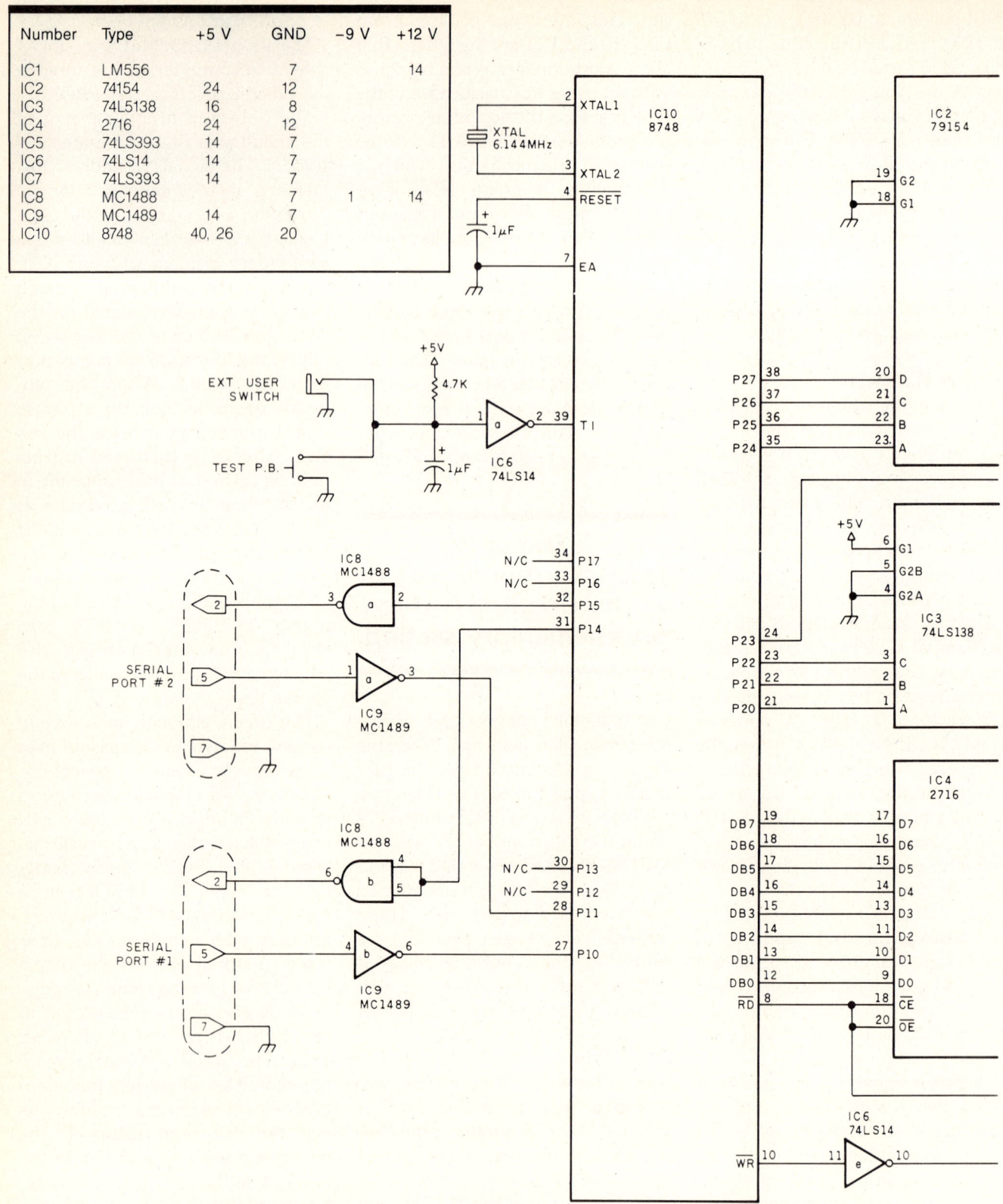

Figure 2: *The schematic diagram of the H-Com. The external EPROM (IC4, a 2716) is used for storage of mnemonically keyed phrases; addresses for the EPROM are generated by the two binary counters IC6 and IC7.*

H-Com Software

The source code of the control program stored in the 8748's memory is shown in listing 2. The program is structured to deal with one quirk of the 8748's instruction set, its eight-level fixed-size stack. When the stack pointer is incremented beyond 7, it "wraps around" to 0, reusing its memory area and subsequently limiting the programs to no more than eight levels of subroutine nesting.

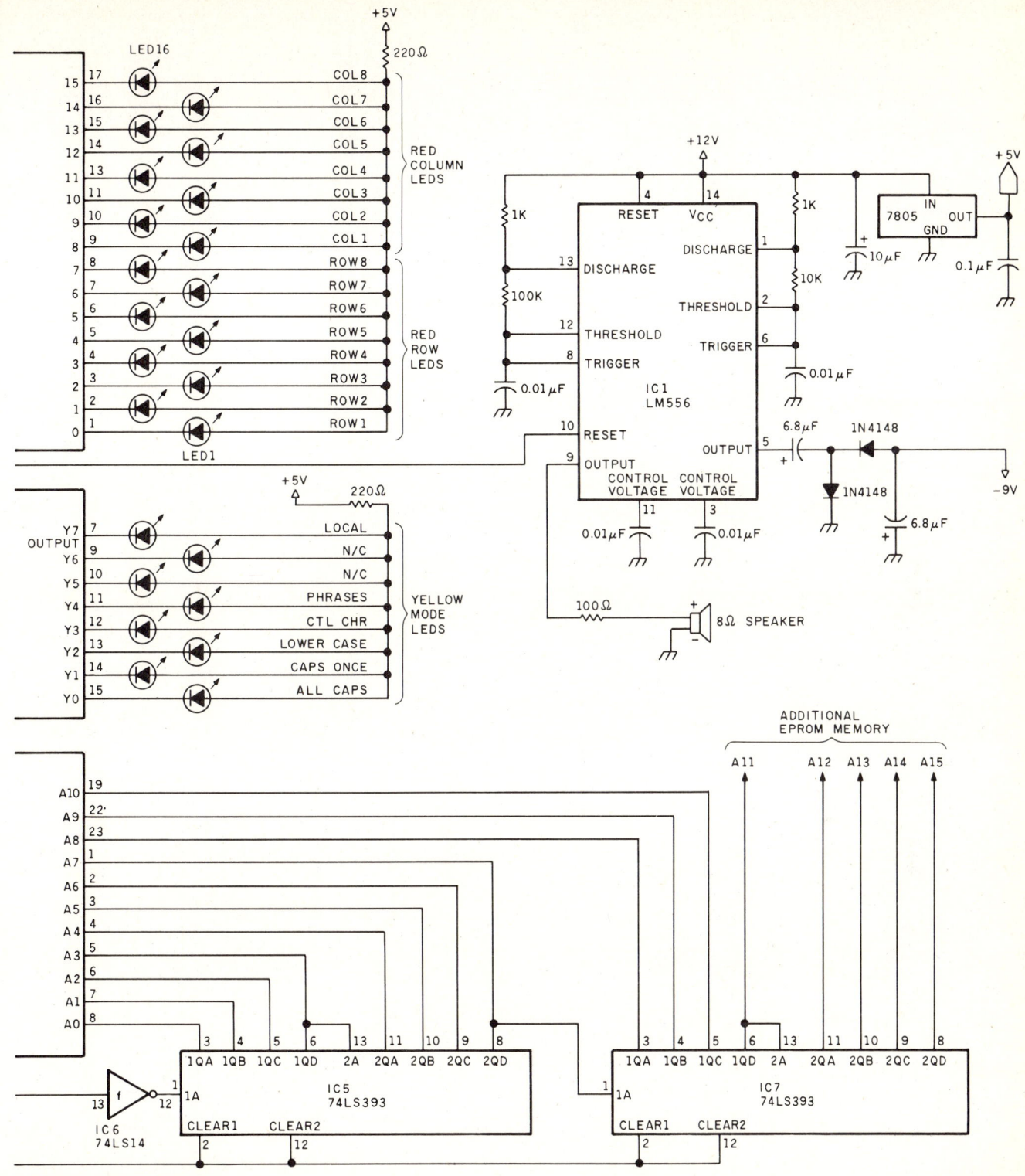

However, at any point in the program, control can branch to a second point without having to clean up the contents of the stack. The H-Com control program uses this feature.

But the jump (branching) procedure is odd, too. Conditional jumps are restricted to within the 256-byte page of memory containing the jump instruction. This characteristic is not particularly convenient, but it can be circumvented by condi-

Listing 1: *Part of the customized vocabulary of words, sentences, and phrases, as formatted for storage in the 2716 EPROM.*

```
                    ;EPROM text must be assembled with the following form:
                    ;phrase key
                    ;one space
                    ;text,which may include any ASCII char except a NUL (00H)
                    ;one NUL character (hex value of 0) to end the text

              CR    EQU     0DH

000D          
0000                ORG     0

0000 747420         DB      'tt '
0003 74686174       DB      'that ',0

0009 7720           DB      'w '
000B 77697468       DB      'with ',0

0011 747920         DB      'ty '
0014 74686579       DB      'they ',0

001A 747320         DB      'ts '
001D 74686973       DB      'this ',0

0023 6620           DB      'f '
0025 66726F6D       DB      'from ',0

002B 6820           DB      'h '
002D 68617665       DB      'have ',0

0033 777420         DB      'wt '
0036 77686174       DB      'what ',0

003C 786120         DB      'xa '
003F 49742077       DB      'It was so nice to get your letter.',CR,0

0062 786220         DB      'xb '
0065 506C6561       DB      'Please excuse my spelling - it never',CR
008A 68617320       DB      'has been one of my strong points.',CR,0

00AC 786320         DB      'xc '
00AF 4D79206E       DB      'My new address is 1234 Perth Avenue.',CR
00D4 49744269       DB      'It is a small cottage with pink walls',CR
00FA 616E6420       DB      'and a large oak tree in the front',CR
011C 79617264       DB      'yard.',CR,0

0122 6E20           DB      'n '
0124 4A6F686E       DB      'John Doe',CR
012D 31323334       DB      '1234 Perth Avenue',CR
013F 486F6D65       DB      'Homestead FL 33030',CR,CR,0

0154 687020         DB      'hp '
0157 48454C50       DB      'HELP  HELP  HELP  HELP',CR
016E 49206E65       DB      'I need medical help. Would anyone',CR
0190 72656164       DB      'reading this message please dial',CR
01B1 31323320       DB      '123 4567 and tell the person answering',CR
01D8 74686520       DB      'the phone that John Doe is having',CR
01FA 616E6F74       DB      'another seizure.',CR,0

020C FF             DB      0FFH    ;end of file marker

0000                END
```

Listing 2: *The control program for the H-Com, written for the Intel 8748, using an external EPROM for storage of canned phrases.*

```
                    ;This code is hereby placed in the public domain

                    ;test 0 and test 1 pins reserved for human interface
                    ;interrupts, internal timer, and alternate registers 0-3
                    ;also reserved for human interface

                    ;bus port reserved for text EPROM
                    ;writing to the port resets the serial address counter
                    ;reading from the port fetches the "next" byte, and
                    ;increments the serial address counter after the read

                    ;port 1 reserved for RS-232C I/O:
                    ;b0 - b3 (P10 - P13) input to 8748
                    ;b4 - b7 (P14 - P17) output from 8748
                    ;specific assignments made according to implementation

                    ;port 2 reserved for LED scan and beeper:
                    ;b0 - b2 (P20 - P22) select yellow "page" indicator LEDs
                    ;b3      (P23)       beeper control, high = beeper on
                    ;b4 - b7 (P24 - P27) select red row/column scan LEDs

                    ;main register set:
                    ;R0 - pointer for phrase match key
                    ;R1 -
                    ;R2 -
                    ;R3 -
                    ;R4 - LED scan rate
                    ;R5 - current key pressed, also data to send
                    ;R6 - shift counter for serial output stream
                    ;     also countdown storage for WSCAN
                    ;R7 - serial port parameters as follows:
                    ;     b0 - b1 baud rate port A
                    ;     b2 - A status 0 = active, 1 = inactive
                    ;     b3 - unused
                    ;     b4 - b5 baud rate port B
                    ;     b6 - B status 0 = active, 1 = inactive
                    ;     b7 - unused

                    ;RAM:
                    ;00H - 07H, main register set, see above
                    ;08H - 17H, 8 level stack area
                    ;18H - 1BH, alternate registers R0 - R3 reserved for human
                    ;           interface interrupt routine if necessary
                    ;1CH - 1FH, storage for phrase match key
                    ;20H - 3FH, high memory unused at present

                    ;ROM (8748 program memory)
                    ;000H - 0FFH main housekeeping program
                    ;100H - 1FFH serial ascii output routines
                    ;200H - 2FFH user interface routines (KBIN)
                    ;300H - 3FFH

000D                CR      EQU     0DH

0000 0409           START:  JMP     INIT            ;page 0

0003 83             EINT:   ORG     3
0003 83                     RET                     ;external interrupt vector

0007 83             TINT:   ORG     7
0007 83                     RET                     ;counter/timer interrupt vector
```

Listing 2 continued:

```
                        ORG    9            ;main program here
0009  BF22      INIT:   MOV    R7,#22H      ;set both ports hot, B = 600, A = 600
000B  BC32              MOV    R4,#50       ;led scan raate
000D  89FF              ORL    P1,#0FFH     ;port 1 all hi at start
000F  9A00              ANL    P2,#0        ;port 2 all lo at start

0011  85        ALLCAP: CLR    F0           ;caps flag set
0012  95                CPL    F0
0013  9AF8              ANL    P2,#0F8H     ;light all caps LED
0015  5400              CALL   KBIN         ;get key (row/col) in r5
0017  FD                MOV    A,R5         ;fetch keypress
0018  E3                MOVP3  A,@A         ;look up char
0019  AD                MOV    R5,A         ;put char in R5
001A  3400              CALL   SEND         ;send it to whichever ports are hot
001C  0411              JMP    ALLCAP       ;loop back to caps

001E  85        ONECAP: CLR    F0           ;set caps flag
001F  95                CPL    F0
0020  9AF8              ANL    P2,#0F8H     ;light caps once LED
0022  8A01              ORL    P2,#1
0024  5400              CALL   KBIN         ;get next key
0026  FD                MOV    A,R5         ;load key
0027  E3                MOVP3  A,@A         ;look up char
0028  AD                MOV    R5,A         ;put char in R5
0029  3400              CALL   SEND         ;send it, then drop into...

002B  85        LOCASE: CLR    F0           ;caps flag down
002C  9AF8              ANL    P2,#0F8H
002E  8A02              ORL    P2,#2        ;light lower case LED
0030  5400              CALL   KBIN         ;get next key
0032  FD                MOV    A,R5         ;load key
0033  E3                MOVP3  A,@A         ;look up char
0034  AD                MOV    R5,A         ;put char in R5
0035  3400              CALL   SEND         ;send it to whichever ports are hot
0037  042B              JMP    LOCASE       ;loop to lower case

;control sends one character and branches back to the
;previous mode - either all caps or lower case.  The
;character sent is not necessarily an ASCII control
;character, it's whatever is in the lookup table at
;address 380H.
;The previous mode branch is also used by other routines.

0039  9AF8      CTRL:   ANL    P2,#0F8H
003B  8A03              ORL    P2,#3        ;light ctrl LED
003D  5400              CALL   KBIN         ;get next key
003F  FD                MOV    A,R5         ;load key
0040  4340              ORL    A,#40H       ;set b6
0042  E3                MOVP3  A,@A         ;look up char
0043  AD                MOV    R5,A         ;put char in R5
0044  3400              CALL   SEND         ;send it to whichever ports are hot
0046  B611      MODE:   JF0    ALLCAP       ;loop to all caps if caps flag set
0048  042B              JMP    LOCASE       ;else loop to locase

;Phrase resets the serial address counter with the WR strobe
;and looks for the phrase requested.

004C  9AF8      PHRASE: ANL    P2,#0F8H
004E  8A04              ORL    P2,#4        ;light up the phrase LED
0050  B91C              MOV    R1,#1CH      ;R1 points to key storage

0052  5400      PHRA1:  CALL   KBIN         ;get next keypress
0054  FD                MOV    A,R5         ;load key
0055  E3                MOVP3  A,@A         ;look up char (note - it's lowercase)
0056  A1                MOV    @R1,A        ;put char in lookup key memory
0057  19                INC    R1           ;bump pointer
0058  03E0              ADD    A,#0E0H      ;NZ if nospace
005A  9650              JNZ    PHRA1        ;loop till space was loaded
005C  02                 
                        MOV    A,#0FFH      ;pattern to send to bus port
                        OUTL   BUS,A        ;actually all we use is the WR strobe
;The key, terminated with a space, is set up in mem at 1CH.
;The eprom is reset (equivalent of tape rewound).

005D  B91C      PHRA2:  MOV    R1,#1CH      ;reset key pointer (R1)

005F  08        PHRA3:  INS    A,BUS        ;read der prom data
0060  F27D              JB7    ERROR        ;end of prom - no phrase matched
0062  AD                MOV    R5,A         ;save eprom char
0063  37                CPL    A
0064  17                INC    A
0065  61                ADD    A,@R1        ;NZ if no match with key
0066  9675              JNZ    ADPROM       ;no match, advance eprom to next phrase
0068  FD                MOV    A,R5         ;match, see if itsa terminating space
0069  03E0              ADD    A,#0E0H
006B  967A              JNZ    BUMPR1       ;NZ if it was a nonspace
006D  08        PHRA4:  INS    A,BUS        ;if nonspace, we bump pointer & continue
006E  C646              JZ     MODE         ;we're at the phrase
0070  AD                MOV    R5,A         ;here's where we bail out
0071  3400              CALL   SEND         ;chuck it into R5 to send
0073  046D              JMP    PHRA4

0075  08        ADPROM: INS    A,BUS        ;advance eprom to just past next 00H
0076  9675              JNZ    ADPROM
0078  045D              JMP    PHRA2        ;reset R1 pointer and try next phrase

007A  19        BUMPR1: INC    R1           ;advance the key pointer and loop
007B  045F              JMP    PHRA3        ;we seem to be matching one up

007D  8A08      ERROR:  ORL    P2,#08H      ;raise honker control line
007F  5439              CALL   BEEPC
0081  9AF7              ANL    P2,#0F7H     ;lower honker
0083  0446              JMP    MODE

0085  8A08      LOCAL:  ORL    P2,#08H      ;light local LED
0087  5400              CALL   KBIN         ;get next key
0089  FD                MOV    A,R5         ;look at key pos (not char)
008A  77                RR     A
008B  77                RR     A
008C  77                RR     A            ;move row number to lo 3
008D  5307              ANL    A,#07H       ;strip to low 3
008F  43E8              ORL    A,#0E8H      ;address of branch table
0091  B3                JMPP   @A           ;indirect jump to command

0092  FD        NOISE:  MOV    A,R5         ;lo bit to accum
0093  67                RRC    A            ;save it
0094  AD                MOV    R5,A
0095  14A3              CALL   BEEPC        ;beep according to cy flag
0097  FD                MOV    A,R5         ;lo bit to accum
0098  67                RRC    A            ;save it
0099  AD                MOV    R5,A
009A  14A3              CALL   BEEPC        ;beep according to cy flag
009C  FD                MOV    A,R5         ;lo bit to accum
009D  67                RRC    A            ;save it
009E  AD                MOV    R5,A
009F  14A3              CALL   BEEPC        ;beep according to cy flag
00A1  0446              JMP    MODE

00A3  F6AE      BEEPL:  JC     BEEPL        ;CY = long beep
00A5  8A08      BEEPS:  ORL    P2,#08H      ;raise honker control line
00A7  5439              CALL   WSCAN
00A9  9AF7              ANL    P2,#0F7H     ;lower honker
00AB  5439              CALL   WSCAN
00AD  83                RET

00AE  8A08      BEEPL:  ORL    P2,#08H      ;raise honker control line
00B0  5439              CALL   WSCAN
00B2  5439              CALL   WSCAN
00B4  9AF7              ANL    P2,#0F7H     ;lower honker
00B6  5439              CALL   WSCAN
00B8  5439              CALL   WSCAN
```

Listing 2 continued

Listing 2 continued:

```
30BA 83        RET

30BB FD        SETA:   MOV   A,R5           ;look at key pressed
30BC 5307              ANL   A,#07H         ;low 3 bits have status
30BE 2F                XCH   A,R7           ;store for now
30BF 5370              ANL   A,#70H         ;save port b configuration
30C1 4F                ORL   A,R7           ;"or" in port a configuration
30C2 AF                MOV   R7,A           ;record both ports config
30C3 0446              JMP   MODE           ;and back

30C5 FD        SETB:   MOV   A,R5           ;look at key pressed
30C6 5307              ANL   A,#07H         ;low 3 bits have status
30C8 47                SWAP  A              ;swap nibbles (b config in hi nibble)
30C9 2F                XCH   A,R7           ;store new b, get old a
30CA 5307              ANL   A,#07H         ;save port a configuration
30CC 4F                ORL   A,R7           ;"or" in port b configuration
30CD AF                MOV   R7,A           ;record both ports config
30CE 0446              JMP   MODE           ;and back

;scan speed set - put new wait constant in R4

00D0 FD        SCANSP: MOV   A,R5           ;look at key pressed
00D1 43F0              ORL   A,#0F0H        ;point to high 16 bytes of page
00D3 A3                MOVP  A,@A           ;get new wait const
00D4 AC                MOV   R4,A           ;and store it
00D5 0446              JMP   MODE           ;and back

00E8           ORG   0E8H

00E8 46464692  DB    MODE,MODE,MODE,NOISE,SETA,SETB,SCANSP,SCANSP

00F0           ORG   0F0H                   ;geometric ratio, .83 common mult

00F0 FFD4B092  DB    255,212,176,146,121,100,83,83
00F8 45393028  DB    69,57,48,40,33,27,23,23

;RS232 OUTPUT - sends ASCII byte in r5 to hot ports

;Note: when the comments refer to "hi" or "lo" signals, ;they
;are referring to the TTL levels at port 1 of the
;microprocessor, NOT the levels on the RS232 connector.

;The RS232 levels are: 1 = negative, 0 = positive with respect
;to ground. An oscilloscope connected to the RS232 data output
;should show a normally negative (about minus 6 volts) DC level,
;with data appearing as bursts of positive voltage (about 10
;volts).

0100           ORG   100H

0100 B60B     SEND:   JF0   FOLD           ;if f0 set, fold all lowercase
0102 FF       SEND1:  MOV   A,R7           ;get bit storage
0103 37               CPL   A              ;invert bit 2 (ignore others)
0104 521B             JB2   SENDA          ;RS232 port A hot, ship byte
0106 FF               MOV   A,R7           ;get it again
0107 37               CPL   A              ;invert bit 6 (ignore others)
0108 D241     SEND2:  JB6   SENDB          ;B hot, send it
010A 83       SEND3:  RET

010B FD       FOLD:   MOV   A,R5           ;get character
010C 039F             ADD   A,#9FH         ;NC if 60H or lower
010E E602             JNC   SEND1          ;bye bye
0110 0385             ADD   A,R5           ;get char again
0112 F602             JC    SEND1          ;C if 7BH or higher
0114 FD               MOV   A,R5           ;get char again
0115 03E0             ADD   A,#0E0H        ;subtract 20H
0117 AD               MOV   R5,A           ;folded char in R5
0118 2402             JMP   SEND1          ;return to caller
```

Listing continues:

```
011B 15        SENDA:  DIS   I              ;disable interrupts
011C 09        SA1:    IN    A,P1           ;xxxx xxx1 if printer busy
011D 5681              JT1   SA1            ;so's we don't hang up forever
011F 121C              CALL  SA0            ;hi line = busy
0121 343D              MOV   R6,#7          ;send a start bit
0123 BE07              MOV   R6,#7          ;set up loop counter
0125 FD        SA2:    MOV   A,R5           ;get data
0126 67                RRC   A              ;Cy holds next bit
0127 AD                MOV   R5,A           ;store what's left
0128 3437              CALL  AOUTC          ;wiggle the port
012A EE25              DJNZ  R6,SA2
012C 3439              CALL  AOUTHI         ;two stop bits
012E 3439              CALL  AOUTHI
0130 FD                MOV   A,R5
0131 67                RRC   A              ;straighten out r5
0132 67                RRC   A
0133 AD                MOV   R5,A
0134 05                EN    I              ;re-enable interrupts
0135 2406              JMP   SEND2          ;maybe send to b,too

0137 E63D     AOUTC:   JNC   AOUTLO
0139 891F              ORL   P1,#1FH        ;raise bit4
013B 246D              JMP   WTBIT          ;and wait one bit time

013D 99EF     AOUTLO:  ANL   P1,#0EFH;lower bit4
013F 246D              JMP   WTBIT

0141 15       SENDB:   DIS   I              ;disable interrupts
0142 FF                SWAP  A
0143 47                SWAP  A              ;rotate nibbles for wtbit
0144 AF                MOV   R7,A

0145 09       SB1:     IN    A,P1           ;xxxx xx1x if printer busy
0146 5681              JT1   SB1            ;break the possible lockup w/button
0148 3245              CALL  SB0            ;hi line = busy
014A 3469              MOV   BOUTLO         ;send a start bit
014C BE07              MOV   R6,#7          ;set up loop counter
014E FD       SB2:     MOV   A,R5           ;get data
014F 67                RRC   A              ;Cy holds next bit
0150 AD                MOV   R5,A           ;store what's left
0151 3463              CALL  BOUTC          ;wiggle the port
0153 EE4E              DJNZ  R6,SB2
0155 3465              CALL  BOUTHI         ;two stop bits
0157 3465              CALL  BOUTHI
015A FD                MOV   A,R5
015B 67                RRC   A              ;straighten out r7
015C AD                MOV   R5,A

015D FF                MOV   A,R7
015E 47                SWAP  A              ;straighten out r7
015F AF                MOV   R7,A

0160 05                EN    I              ;re-enable interrupts
0161 240A              JMP   SEND3          ;b sent, back to caller

0163 E669    BOUTC:   JNC   BOUTLO
0165 892F    BOUTHI:  ORL   P1,#2FH         ;raise bit5
0167 246D             JMP   WTBIT           ;and wait one bit time

0169 99DF    BOUTLO:  ANL   P1,#0DFH;lower bit4
016B 246D             JMP   WTBIT

016D FF      WTBIT:   MOV   A,R7            ;get baud rate selection
016E 43FC             ORL   A,#0FCH         ;a points to one of top 4 addr
0170 A3              MOVP  A,@A            ;this page, get wait time in accum
0171 07      WTBIT2:  DEC   A
0172 00              NOP
0173 00              NOP                    ;it takes 15 machine cycles to
0174 00              NOP                    ;go through this inner loop
0175 00              NOP                    ;for a total of 225 clock cycles
                                            ;which is 36.62 microseconds if the
```

Listing 2 continued:

```
0176  00                    NOP                         ;clock is running at 6.144 MHz
0177  00                    NOP
0178  00                    NOP
0179  00                    NOP
017A  00                    NOP
017B  00                    NOP
017C  00                    NOP
017D  00                    NOP
017E  9671                  JNZ     WTBIT2
0180  83                    RET

0181  0446          UNHANG: JMP     MODE                ;back to the former loop

01FC                        ORG     1FCH                ;last 4 bytes this page
01FC  F2                    DB      242                 ;110 baud wait const
01FD  59                    DB      89                  ;300
01FE  2C                    DB      44                  ;600
01FF  16                    DB      22                  ;1200

;Kbin goes through the keyboard selection game and updates r5
;to the key position selected.
;b7 - b6 returned zero
;b5 - b3 designate rows 0-7 (numbered from top to bottom)
;b2 - b0 designate cols 0-7 (numbered from left to right)

0200                        ORG     200H
0200  5600          KBIN:   JT1     KBIN                ;hang if button pressed (active)
0202  9A0F                  ANL     P2,#0FH             ;set hi nibble = 0 (red LEDs)
0204  5439                  CALL    WSCAN               ;wait one scan interval
0206  560C                  JT1     COL                 ;this the one, skip out
0208  5445                  CALL    NEXROW              ;otherwise shine next row scan LED
020A  4404                  JMP     ROW

020C  560C          COL:    JT1     COL                 ;hang if button active
020E  0A                    IN      A,P2                ;get the LED phase in accum
020F  77                    RR      A                   ;bits 654 to 543
0210  5338                  ANL     A,#38H              ;strip to bits 543
0212  AD                    MOV     R5,A                ;store bits 5,4 & 3
0213  9A0F                  ANL     P2,#0FH             ;set hi nibble = 0
0215  8A80                  ORL     P2,#80H             ;set bit 7 for col scanning
0217  5439                  CALL    WSCAN               ;wait one scan interval
0219  561F          COL2:   JT1     GOT                 ;this the one, skip out
021B  544C                  CALL    NEXCOL              ;otherwise shine next col scan LED
021D  4417                  JMP     COL2

021F  561F          GOT:    JT1     GOT                 ;hang if button active
0221  0A                    IN      A,P2                ;get the LED
0222  47                    SWAP    A                   ;swap nibbles
0223  5307                  ANL     A,#07H              ;strip to bits 210
0225  4D                    ORL     A,R5                ;or in previous 3
0226  AD                    MOV     R5,A                ;store it in r5
0227  5439                  CALL    WSCAN               ;wait a tad, unless delete
0229  5600                  JT1     KBIN                ;oops, a keypress in the cancel
                                                        ;time window, start over
022B  FD                    MOV     A,R5                ;if A is xxxx xlll itsa mode shift
022C  43F8                  ORL     A,#0F8H             ;   "   1111 1lll   "  "  "  "
022E  37                    CPL     A                   ;   "   0000 0000   "  "  "  "
022F  9638                  JNZ     NOTMODE

0231  FD                    MOV     A,R5                ;A has 00nn nlll, where nnn is mode #
0232  77                    RR      A                   ;"  "  100n nnll
0233  77                    RR      A                   ;"  "  1110 nnnl
0234  77                    RR      A                   ;"  "  1110 0nnn
0235  43F8                  ORL     A,#0F8H             ;"  "  1111 lnnn
0237  B3                    JMPP    @A                  ;jump indirect via table at 2F8H
                                                        ;which in turn points to the vectors
0238  83        NOTMODE:    RET                         ;on this page - if you understand this
0238                                                    ;sequence, give youself a jellybean, and
0238                                                    ;thank Intel for a nonmarchable stack

0239  FC        WSCAN:      MOV     A,R4                ;get current scan rate
023A  AE                    MOV     R6,A                ;256 gives 2.4 sec per row/col
023B  27                    CLR     A
023C  3471                  CALL    WTBIT2              ;9.375 milliseconds
023E  5644          WSCAN2: JT1     WSCAN3              ;return if switch on
0240  CE                    DEC     R6                  ;bump scan count
0241  FE                    MOV     A,R6
0242  963B                  JNZ     WSCAN2              ;look at it
0244  83        WSCAN3:     RET

0245  0A        NEXROW:     IN      A,P2                ;get current phase
0246  0310                  ADD     A,#10H              ;bump hi nibble
0248  537F                  ANL     A,#7FH              ;make sure b7 is lo
024A  3A                    OUTL    P2,A
024B  83                    RET

024C  0A        NEXCOL:     IN      A,P2                ;get current phase
024D  0310                  ADD     A,#10H              ;bump hi nibble
024F  4380                  ORL     A,#80H              ;make sure b7 is hi
0251  3A                    OUTL    P2,A
0252  83                    RET

;This here is the vector table which gets you off of this page.
0253  0411          V0:     JMP     ALLCAP
0255  041E          V1:     JMP     ONECAP
0257  042B          V2:     JMP     LOCASE
0259  0439          V3:     JMP     CTRL
025B  044A          V4:     JMP     PHRASE
025D  0411          V5:     JMP     ALLCAP
025F  0411          V6:     JMP     ALLCAP
0261  0485          V7:     JMP     LOCAL

;This here is the indirect page jump address lookup table, used
;by the JMPP A,@A instruction. The indirect jump in an 8748
;chip cannot make it across a page boundary (multiple of 256
;bytes). The doubly confusing double jump used here is a way
;around this limitation.

02F8                        ORG     2F8H
02F8  53555759              DB      V0,V1,V2,V3,V4,V5,V6,V7

;page 3, lookup table for keyboard layout

0300                        ORG     300H
0300  20656F6E              DB      'eondpy '
0308  74617226              DB      'tarlmgx'
0310  69736366              DB      'iscfbkj'
0318  68757776              DB      'huwvzq',1BH,20H
0320  08373839              DB      08H,'789*?!'
0328  0D343536              DB      0DH,'456'/"&
0330  2E313233              DB      '.123+()'
0338  2C302430              DB      ',0$=-<>'

0340  14011200              DB      14H,01H,12H,0CH,0DH,07H,18H,20H
0348  14011200              DB      09H,13H,03H,06H,02H,0AH,0AH,20H
0358  08151716              DB      08H,15H,17H,16H,11H,1BH,1BH,20H

0360  7F371E39              DB      7FH,37H,1EH,339H,7EH,60H,5EH,20H
0368  001C351D              DB      00H,1CH,35H,1DH,5CH,27H,40H,20H
0370  3A311F33              DB      3AH,31H,1FH,33H,23H,7BH,7DH,20H
0378  3B30257C              DB      3BH,30H,25H,7CH,5FH,5DH,20H

0000                        END
```

69

tionally jumping to an unconditional-jump instruction (which is not so restricted). Unconditional jumps include normal direct jumps to any place in program memory and several types of indirect jumps within the page. (For an extreme example of this technique, look at location hexadecimal 0237 in listing 2, where the mode switching occurs. Here, subroutine KBIN is called, with the calling routine expecting control to return with the character-selection code held in register 5. But if you actuate the switch to select a mode, such as the Phrase mode, the subroutine calculates where to continue execution and simply jumps there. Structured programming hasn't made much progress on the 8748.)

The H-Com program is arranged in three sections, plus a lookup table. These four modules fit conveniently in the 8748's four pages of program memory. The first page (page 0) is where the code for all the various modes of operation reside; each code section considers itself the main routine and calls the other sections as subroutines. The first section of code sets up the major modes (All Caps, Phrase, etc.)

The second section (page 1) is the text-transmission section. It sends the contents of register 5 out to one or both RS-232C channels, according to which are active at the time. It sends the data at the most recently selected data rate or at the default data rate set up by code in the first page. If the H-Com "hangs up" waiting for a device-ready status that never comes, you can resume the active scanning mode by pressing the switch.

The third section (page 2) is the scanning subroutine. As we've seen, it scans the rows and then columns until you make a selection. When in the column scan, you can return to the row scan by pressing the switch twice instead of once. If any position in the first seven columns is selected, this subroutine returns to the calling routine with the element *position* (not an ASCII value) in register R5. The calling routine must either convert this into a character or take some appropriate action (e.g., beeping the horn). If a position in the eighth (mode-select) column of the array is selected, this subroutine disregards the normal subroutine return and jumps to the appropriate mode routine.

The first half of page 3 of program memory is the character-lookup table. Its layout corresponds to the character-display arrangement, which serves to minimize access time. If you would prefer some other "keyboard" layout, merely change this table.

The H-Com program does not make use of the 8748's interrupts, interval timer, or alternate registers R0 through R3. These have been reserved for customization of the system to an individual user. The alternate register set R4 through R7 is used for phrase-key storage, and keys longer than three characters use the high end of scratchpad memory. Other than this, the memory above the alternate registers is unused.

The software for this project was written by Ralph McElroy. To encourage use and further development of the H-Com and similar devices, we are placing the software in the public domain.

Parting Thoughts

This project has been on my mind for some time. Its subject matter was suggested by my meeting with Dave, but I'm doing it now because of the recent increase in the number of letters I've received describing how disabled individuals are being helped by the speech synthesizers I've presented in these articles.

I can guarantee that I'll continue to investigate speech-related topics, but specific projects like this one will require some reader feedback and suggestions. I'd like to hear your comments and suggestions. If there is sufficient interest in the H-Com, I may make arrangements for it to be manufactured commercially. For information on its availability, contact Intex Micro Systems Corporation, 725 South Adams Rd., Suite L8, Birmingham, MI 48011, telephone (313) 540-7601.

If you want to see how a research group at Tufts University approached the same problem, you can read an article in the September 1982 issue of BYTE (reference 5); that issue also contained a number of articles on computer applications to help disabled people.

Special thanks to Ralph McElroy for his contributions to this project.

References

1. Baker, Bruce. "Minspeak." BYTE, September 1982, page 186.
2. Ciarcia, Steve. "Build an Intelligent EPROM Programmer." BYTE, October 1981, page 36.
3. Ciarcia, Steve. "Build the Microvox Text-to-Speech Synthesizer." Part 1, BYTE, September 1982, page 64. Part 2, BYTE, October 1982, page 40.
4. Ciarcia, Steve. "Mind Over Matter: Add Biofeedback Input to Your Computer." BYTE, June 1979, page 49.
5. Demasco, Patrick, and Richard Foulds. "A New Horizon for Nonvocal Communication Devices." BYTE, September 1982, page 166.
6. Schwejda, Paul, and Gregg Vanderheiden. "Adaptive-Firmware Card for the Apple II." BYTE, September 1982, page 276.

Ciarcia's Circuit Cellar

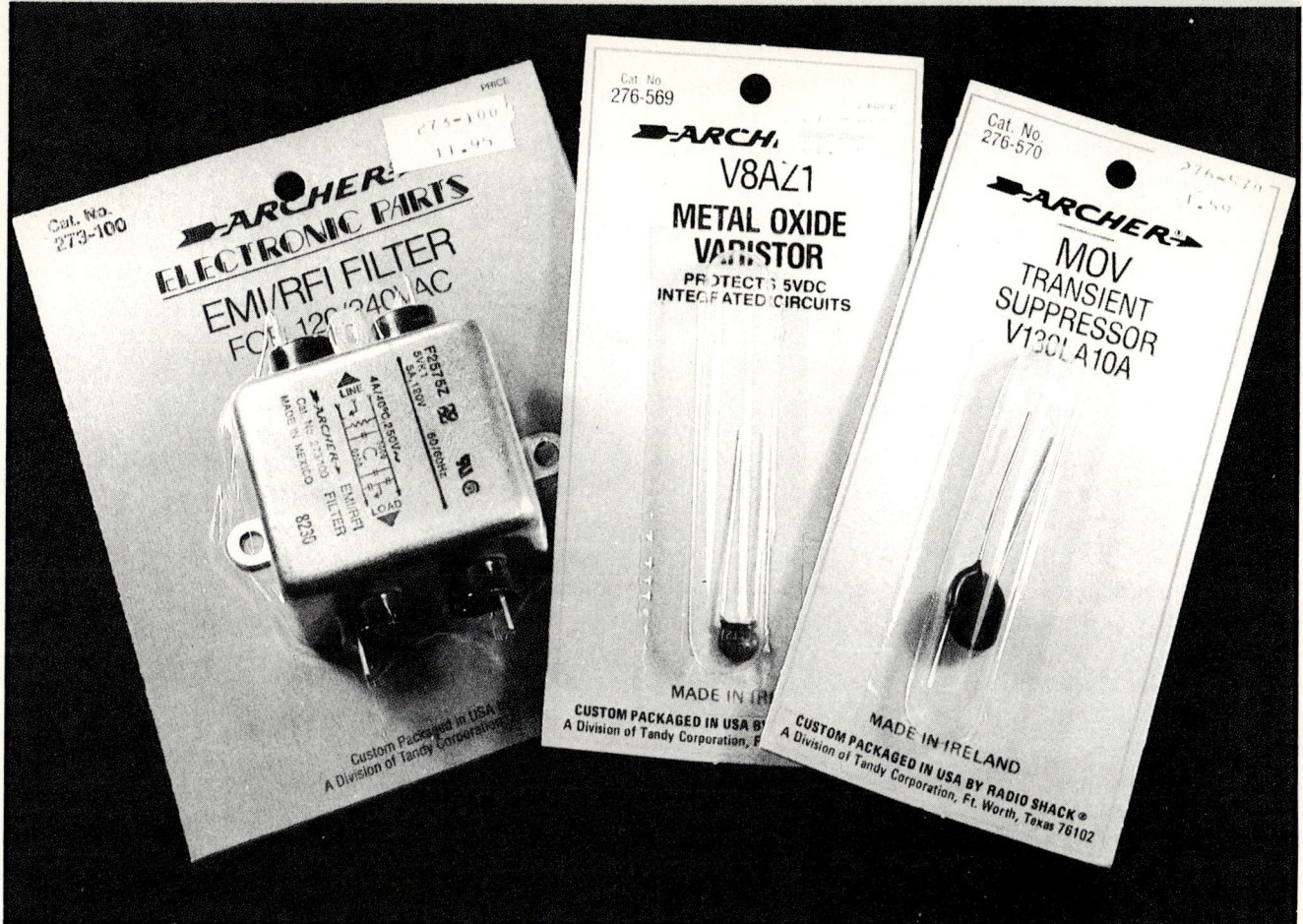

Photo 1: *Your computers and other electronic equipment are vulnerable to disturbances transmitted through the power line. Most of the components necessary for transient and noise suppression can be purchased from Radio Shack. Shown here are a commercial EMI/RFI line filter made by Corcom and two General Electric metal-oxide varistors.*

Keep Power-Line Pollution Out of Your Computer

A visitor once called the Circuit Cellar my mountaintop wilderness retreat. Since he lived in the center of Manhattan, the few oak and birch trees around my house seemed to him like a forest, and because he could view scenery further away than a block or two he must have felt like he was on Mount Whitney. Well, my area is one of the higher points in Connecticut, but that isn't very high. It's barely a prairie-dog mound to someone from Montana.

Life in a rural location has its special pleasures. I get to plow the snow from my own driveway, trim back the ever-encroaching foliage and rake the leaves, pile four cords of wood for the stove each winter (see reference 1), fight off the local animal population, and spend large sums of money repairing damage done to my electronic equipment by electrical disturbances.

This last item is the only one that really annoys me. Every year I can count on experiencing some equipment failure attributable to an external electrical impulse, usually coming in through the power line. For three years, just like clockwork, the first thunderstorm in June wiped out a DECwriter II terminal connected to one of my computers. After the first two times of spending a few hours replacing blown chips, I got smart and installed sockets. (Now I even know in advance which chips will be blown.) Last summer I kept the printer unconnected when I wasn't using it.

But the elements were not to be denied. During an August thunderstorm, lightning struck my house. I can't say for sure where the bolt ac-

An expensive lesson produces the cheapest Circuit Cellar project yet

tually hit (there were no burn marks or other visible clues), but I suspect the point of entry was the power line. I remember seeing an indistinct flash of light, hearing a tremendous crash, and then standing in darkness. My assistant Jeanette saw a bright blue glow behind one of the computers.

Such a tremendous power surge is not kind to semiconductor-based equipment. The casualties included one computer, one video camera, two video monitors, a microwave receiver, and probably several other assorted items I haven't found yet. The damage did not include the DECwriter (safely unplugged since May), but it was over $3000.

In December, thunderstorms are not an immediate threat, but as I write this in early September the memory is still fresh and I still have a month of potentially violent weather to contend with. I am forced to consider some defensive measures. Perhaps by relating my experiences I can save you from a similar fate.

Of course, lightning isn't the sole cause of electrical disturbances; you don't have to wait for a thunderstorm to be a victim. Many kinds of trouble can be ducted into your computer through the power line.

In the January 1981 Circuit Cellar article (reference 2), I wrote about electromagnetic interference (EMI) and radio-frequency interference (RFI). This month I'd like to pick up the saga by describing other forms of electrical pollution that occur on power lines. Afterward, I'll describe a few simple, inexpensive means of dealing with them.

The Power Line: A Hostile Environment

The lines leaving your local utility company's generating plant carry electrical power that in most respects is pure, smooth, and constant. However, as the power is routed through the distribution network, it comes under the increasing aberrant influence of external forces and the connection or shedding of electrical loads.

Your susceptibility to these aberrations depends on your location in the distribution system. If you are close to the power plant, you should have relatively few, with the low source-impedance of the generator and short distance of the transmission line limiting the influence of external forces. But rural customers at the end of the line usually experience the full effect. While the utilities try to distribute power evenly, the presence of a large-scale user of electrical power along the line between the generator and you can greatly affect the quality and quantity of the power you get.

If you own a personal computer, you should be concerned about the quality of the power you feed it. Power-line irregularities cause problems for computers and other digital equipment because certain kinds of extraneous electrical pulses can be interpreted as data or instructions, causing errors in operation. You face hazards every time you plug in a piece of electronic equipment, but there are certain precautionary measures that can protect your computer.

The degree of sensitivity depends somewhat on the type of equipment and the type of disturbances. As the operating speed of digital equipment increases, its tolerance to power-line pollution lessens. High-speed processors and memory components are susceptible to fast transients. (Dynamic memories, which must be periodically refreshed, are particularly susceptible.) Disk drives and displays, on the other hand, are more affected by lasting surges and sags in operating voltage.

Common Sources of Woe

Electrical power-line disturbances can come from either natural or man-made sources. Of the many ways the power line can be disturbed, the several varieties of voltage fluctuation most often cause problems with computer equipment. These fluctuations can be categorized by source and severity, as follows:

Blackouts. A blackout is a total power outage—the voltage goes to zero. Obviously if no alternate source of power is available as a backup, computer equipment will be severely affected, and data will be lost. Blackouts generally affect only a small number of utility customers (fewer than 5 percent) during a year and generally last less than 10 seconds.

Brownouts. A brownout is typically a corrective action taken by the utility when power demand exceeds generating capacity. The utility reduces the output voltage from a nominal 120 V (volts) by 5 to 15 percent. When the voltage is thus reduced, the resistive load presented to the generators by the distribution network consumes less power.

Generally speaking, most consumer and industrial equipment designed for use in North America functions properly when supplied with current within the range from 105 to 130 V. But when operating at either extreme, the equipment is more vulnerable to disruption from some other power-line anomaly. Fortunately, power companies rarely reduce the voltage by more than 7 percent.

Voltage transients. The phenomena of voltage transients include surges of voltage above the specified normal, voltage sags below, and instantaneous voltage spikes that leap far above the nominal levels.

Surges and sags are long-duration events occurring at some point in the distribution network when electrical equipment is routinely turned on or off nearby. The magnitude of the surge or sag depends upon the size of the load being removed from or placed on the network.

Sags are often produced by the turning on of electric motors, which have high starting currents. (You've probably noticed lights dimming

The important element in lightning protection is the lightning rod, a pointed shaft of copper to which a half-inch copper cable is fastened. The cable in run down the side of the building, where it is clamped to an 8½-foot copper-plated steel rod driven into the earth. The rod system pictured here costs $150.

How It Starts

As the electric charge builds up in a cloud, the electric field in the vicinity of the charge center increases to the point where the air starts to ionize. A column of ionized air, called a **pilot streamer**, begins to extend toward the earth at a velocity of about 100 miles per hour. After the pilot streamer has moved perhaps 100 feet to 150 feet, a more intense discharge called the **stepped leader** occurs. This discharge inserts additional negative charge into the region around the pilot streamer and allows the pilot streamer to advance for another 100 to 150 feet, after which the cycle repeats. As its name indicates, the stepped leader progresses toward the earth in a series of steps, with a time interval between steps on the order of 50 microseconds.

In a cloud-to-ground flash, the pilot steamer does not move in a direct line toward the earth but instead follows the path through the atmosphere where the air ionizes most readily. Although the general direction is toward the earth, the specific angle of departure taken by each succeeding pilot streamer from the tip of the previous streamer is unpredictable. Therefore, each 100- to 150-foot segment of the stroke will likely approach the earth at a different angle. This changing angle of approach gives the overall flash its characteristic zig-zag appearance.

As a highly ionized column, the stepped leader is at essentially the same potential as the charged area from which it originates. Thus, as the stepped leader approaches the earth, the voltage gradient between the earth and the tip of the leader increases. The increasing voltage further encourages the air dielectric between the two regions to break down.

Attracting Lightning

Objects extending above their surroundings are likely to be struck by lightning. Thin metallic structures, such as flag poles, lighting towers, antennas, and overhead wires, offer a very small cross-sectional area relative to the surrounding terrain, but ample evidence exists to show that such objects apparently attract lightning.

The ability of tall structures or objects to attract lightning serves to protect shorter objects and structures nearby. In effect, a tall object establishes a protected zone around it; within this zone, other structures and objects are protected against direct lightning strikes. As the height differen-

How Lightning Strikes

A lightning flash is characterized by one or more strokes with typical peak currents of 20 kA (kiloamperes) or higher. In the immediate vicinity of the stroke's impact on the earth, hazardous voltage gradients exist. It is difficult to establish a definite grounding-conductance value necessary to protect equipment and personnel. The current in a lightning strike is so high that even 1 ohm of resistance can theoretically produce hazardous potentials.

When lightning strikes a building unprotected by a lightning rod, the stroke seeks out the lowest-impedance path to earth (most likely through the electric wiring or water pipes).

tial between the shorter surrounding objects and the tall one decreases, the protection provided to the shorter objects decreases. Likewise, as the horizontal distance between the tall and short structures increases, the protection afforded by the tall structure decreases.

Lightning Rods

A protective device that makes use of this phenomenon is the lightning rod, shown in photo. Generally just a sharp copper spike, the lightning rod is attached to the highest point on the structure to be protected. When lightning strikes, the current is shunted directly through a heavy copper wire from the rod to a grounding electrode buried in the earth.

Although the duration of a strike is typically less than 2 microseconds, the voltage generated is high enough to cause flashover strikes to conducting objects located as much as 14 inches away from the conducting path. For this reason, metallic objects in close proximity to down conductors should be electrically bonded to the conductors.

But circuits not in direct contact with the lightning discharge path can experience damage, even in the absence of overt coupling by flashover. Because the high current associated with a discharge builds up so fast, large inductively produced voltages are formed on nearby conductors. Experimental and analytical evidence shows that the surges thus induced can easily exceed the tolerance level of many components, particularly solid-state devices. Inductive surges can be induced by lightning current flowing in a down conductor or structural member, by a stroke to earth in the vicinity of buried cables, or by cloud-to-cloud discharges occurring parallel to long cable runs, either above ground or buried.

The Moral

The objective of all lightning-protection systems is to direct the high currents away from susceptible elements or limit the voltage gradients developed by the high current to safe levels. In a given area, certain structures or objects are more likely to be struck by lightning than others; however, no object, whether man-made or natural, should be assumed to be immune from lightning. The voltages that could be induced by such discharges present a definite threat to signal and control equipment, particularly equipment employing semiconductor components.

Power-Line Conditioner Sources

Cuesta Systems Inc.
3440 Roberto Court
San Luis Obispo, CA 93401
(805) 541-4160

Dymarc Industries Inc.
21 Governor's Court
Baltimore, MD 21207
(800) 638-9098
(301) 298-2629

Electronic Protection Devices
Division CNS Electronics Corp.
5-9 Central Ave.
Waltham, MA 02154
(800) 343-1813

Electronic Specialists Inc.
171 South Main St.
Natick, MA 07160
(800) 225-4876 (orders)
(617) 655-1532

Isoreg Corporation
410 Great Rd.
Littleton, MA 01460
(617) 486-9483

RKS Industries
4865 Scotts Valley Dr.
Scotts Valley, CA 95066
(800) 892-1342
(408) 438-5760

Sun Research Inc.
POB 210
Old Bay Rd.
New Durham, NH 03855
(603) 859-7110

Power-Line Filter Sources

Cornell-Dubilier Electronics
Box B-967
New Bedford, MA 02741
(617) 996-8561

Corcom Inc.
1600 Winchester Rd.
Libertyville, IL 60048
(312) 680-7400

Curtis Industries Inc.
8300 North Tower Ave.
Milwaukee, WI 53223
(414) 354-1500

Genisco Technology Corporation
18435 Susana Rd.
Rancho Dominguez, CA 90221
(213) 537-4750

Hopkins Engineering Company
12900 Foothill Blvd.
San Fernando, CA 91342
(213) 361-8691

The Potter Company
Division of Varian
POB 337
Wesson, MI 39191
(601) 643-2215

Siemens Corporation
8700 East Thomas Rd.
Scottsdale, AZ 85252
(602) 941-6366

Sprague Electric Company
87 Marshall St.
North Adams, MA 01247
(413) 664-4411

Stanford Applied Engineering
3520 De La Cruz Blvd.
Santa Clara, CA 95050-1997
(408) 988-0700

---

when an air conditioner comes on.) Surges are generally the result of network switching by the utility or of a sudden reduction in demand for power in the network; during the period necessary for the utility's electromechanical compensation system to function, an overvoltage transient condition can exist.

The most damaging power-line disturbance is the high-speed, high-energy *voltage spike*. People speaking loosely about "power-line transients" are probably talking about this type of event. Lasting usually less than 100 microseconds, spikes can be up to 6000 volts. Such high-energy transients are produced by the switching off of inductive loads by the opening of switch contacts, short circuits, or blown fuses; severe network load changes; or lightning. Inductive-load switching accounts for the majority of spikes.

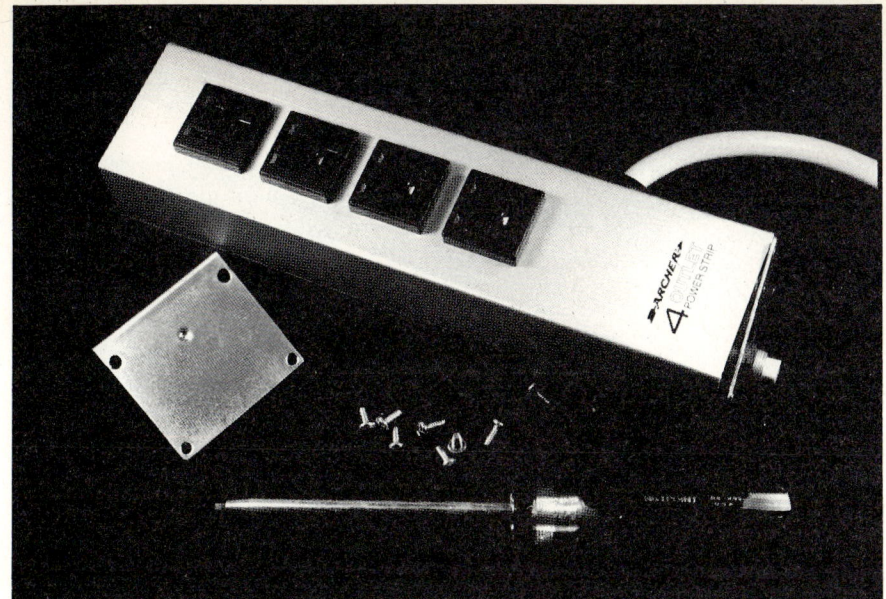

Photo 2: *You can save approximately $40 on the price of a transient-protected power strip by adding the protection yourself, as demonstrated on the Radio Shack Archer 61-2620 unit. First, unscrew the end plates.*

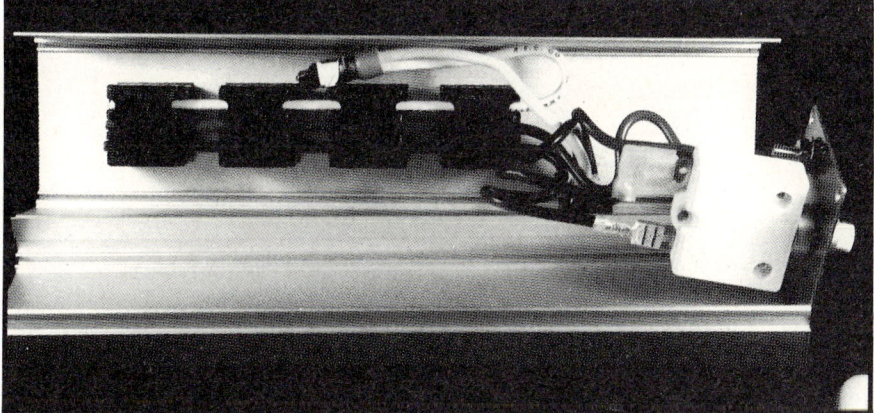

Photo 3: *Open the strip case, exposing the four receptacles and the white circuit-breaker block. The three wires conducting power run the length of the strip: black is the hot side, white is the neutral return, and the green wire is earth ground.*

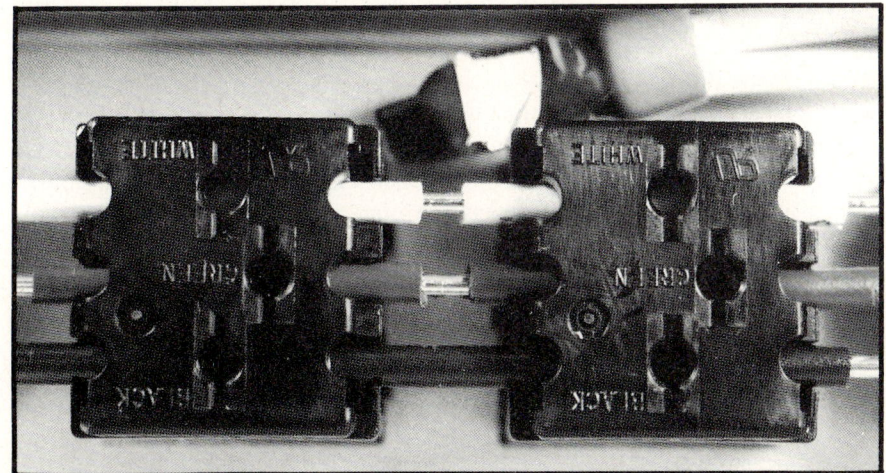

Photo 4: *Using an X-acto knife or similar tool, strip insulation from the wires between the receptacles (which I number 1 through 4, from left to right) according to the following system: between 1 and 2, strip the green and black; between 2 and 3, strip the green and white; between 3 and 4, strip the black and white.*

When the coil of an inductive load such as a transformer or motor is suddenly deenergized, the collapsing magnetic field must dissipate its energy, and it does this by placing a large voltage back into the circuit that energized it. Let's examine the process in detail.

As the circuit through the inductor is broken, current in the inductor continues to flow, charging the distributed capacitance in the windings. At some point, the charge voltage becomes sufficient to leap across the switch gap as a spark. This sudden shorting action discharges the winding's capacitive charge back into the circuit until the spark ceases. This process repeats in a cycle until there is too little energy left in the coil to create an arc across the contacts. The waveform of inductance-generated transients is oscillatory. For example, a contact opening while conducting 100 mA (milliamperes) in a 1-H (henry) inductance will produce a 3000-V spike, assuming about a 0.001-μF (microfarad) stray winding capacitance.

Whenever you plug in a vacuum cleaner, hair drier, or other appliance (even your computer), you could be creating some potentially serious transient disruptions for other equipment on the same power line. The equipment need not even be on the same wiring circuit. The capacitance of household wiring is often sufficient to couple a transient from one wire to another (differential mode) or from the wire to the ground (common mode).

Lightning is the most violent and most destructive source of transient energy. A direct lightning hit is catastrophic, but direct hits seldom occur. A more frequent danger is that a lightning strike on a power line miles away may result in a thousand-volt spike rushing throughout your home. Such hits happen frequently enough to cause much grief. (Because lightning is such a significant source of transients, I've explained it in detail in the text box "How Lightning Strikes." A secondary, and more widespread, effect of a lightning hit on a power line is a voltage sag over a large part of the network as the power company's safety circuits compensate for the spike.)

Electrical noise. Miscellaneous electrical noise is the final source of power-line disturbances. It is best understood as high-voltage high-frequency interference. Noise in the range from 10 kHz (kilohertz) to 50 MHz (megahertz) is the most common cause of computer failures. Because of its frequency, noise can be either broadcast through free space from its source or conducted directly through the power lines. Digital electronic equipment is a prime source of high-frequency noise.

Power-Line Protection

I'm not trying to make you afraid to plug your computer into the wall outlet. There are remedies for virtually all the problems I've mentioned, although some are more practical for some computer users than others.

If surges or sags are a constant problem for you, you can try having the power company change the tap on your local step-down transformer or installing a constant-voltage transformer on your premises. These measures, although expensive, are effective. If you are plagued by blackouts or have equipment that should never be shut down, I suggest that you consider obtaining an *uninterruptible power supply*, abbreviated UPS. Using a UPS gives you confidence in the quality of your power and effectively isolates your computer from damaging perturbations. However, a UPS is also quite costly.

In the case of electrical noise and EMI, there are filters and construction techniques that can be employed to reduce interference, but a better answer is to find the pollution at the source and eliminate it. My article in the January 1981 BYTE outlined most methods of filtration and preventive design. While I'll try not to belabor the point, a power-line filter is an important noise- and transient-suppression device.

The best answer to transients is to suppress their voltages to a harmless level, either with filters or a special category of components called *transient suppressors*.

Photo 5: *You can now solder a varistor between each of the stripped wire pairs, mounting it flat against the back face of the receptacle so that the case will fit together again.*

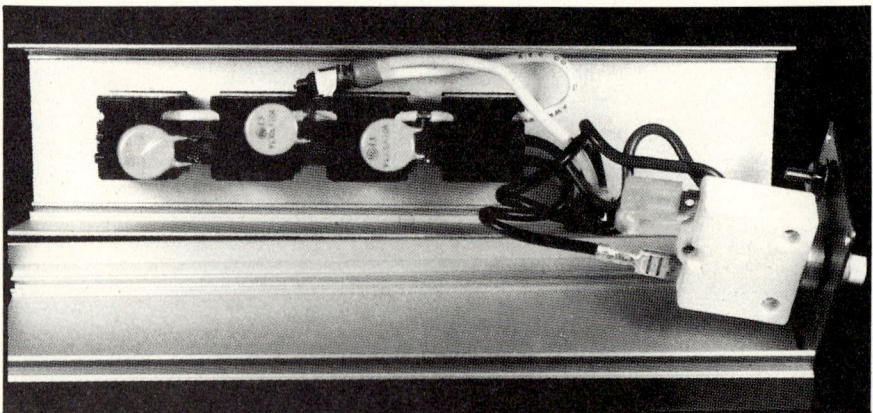

Photo 6: *The outlet strip with three MOVs installed provides both common-mode and differential-mode transient suppression. After you have finished soldering, carefully reassemble the power strip's enclosure and screw it back together.*

Power-Line Filters

A power-line interference filter is an electronic circuit used to control RFI and EMI conducted into and out of equipment. The filter is intended to provide unwanted interference signals with a high series impedance (into the vulnerable equipment) and low shunt impedance (to ground). It generally consists of a set of passive components that act as a mismatching network for high-frequency signals—a low-pass filter. The network attenuates RF energy above 10 kHz, while passing the 60-Hz power.

The simplest possible filter is a single capacitor wired in parallel or a single coil wired in series with the power line. More typically, several capacitors and/or coils are used together, connected into different configurations variously called L, π, and T filters.

Though containing only a few components, such passive bilateral networks have complex transfer characteristics that are extremely dependent upon the impedances of the source and load. Because you can't predict these impedances for all applications, it is not possible to unequivocally state that a specific filter configuration will work the same way in two different environments. But to allow

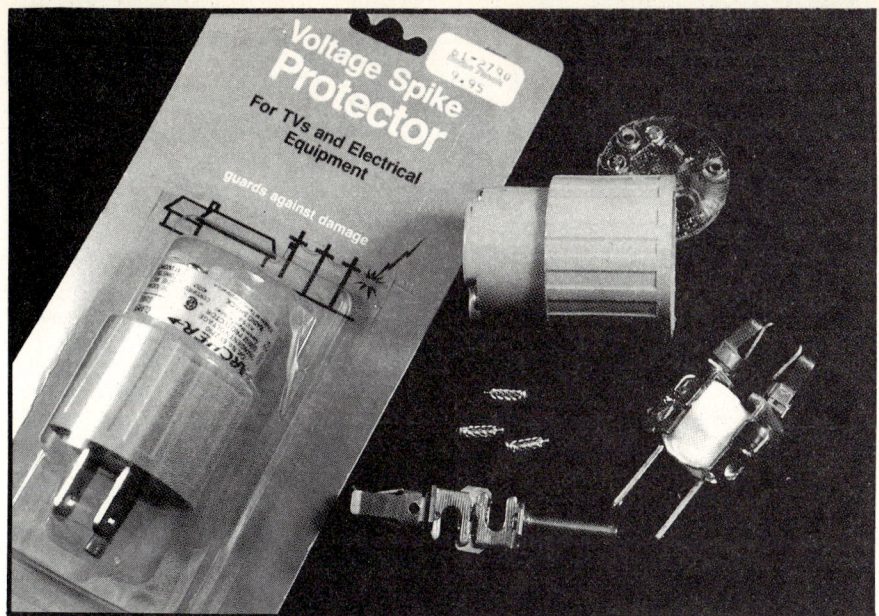

Photo 7: *For quicker and easier, though incomplete, protection, you can plug your computer into a simple voltage-spike protector such as the Radio Shack 61-2790. As you can see from the disassembled unit, the metal-oxide varistor (wrapped in fiberglass tape) is connected between only the hot and neutral lines (black and white). It has no varistor connection to the ground lead and therefore does not protect against common-mode transients.*

signed according to theory using a 50-ohm assumed impedance probably won't work as well as one empirically derived using the actual equipment and power line.

Transient Suppressors

Protection from the various kinds of line transients is obtained by suppressing or diverting them. The three types of circuits most often used for this are filters, crowbars, and voltage-clampers.

As I previously alluded, *filters* comprising inductances and capacitances are widely used for interference protection, including transients. Since most transient signals are high frequency, the suppression by a filter is often effective, provided it can withstand the associated high voltages.

Crowbar circuits use a switching action, such as turning on a thyristor or arcing across a spark gap, to divert transients. But crowbars that incorporate SCRs (silicon-controlled rectifiers) and triacs are much too slow to effectively suppress 100-μs (microsecond) transients. Most often they are incorporated in low-voltage DC power-supply output circuits where overvoltage conditions occur at more manageable speeds (milliseconds). Spark-gap devices, which include carbon blocks and gas tubes, are fast

electrical specifications to be minimally compared, however, resistive source and load impedances of 50 ohms each are generally used.

Two similar power-line filters, even built with the same circuit topology and component values, may not perform identically; the mounting and wiring of the filter can be critical influences on its performance. A power-line filter is best installed at the point in your equipment where the power line comes inside the case rather than at the far end of a long cord. The filter's purpose is to attenuate high-frequency signals: this purpose is defeated if these parasitic signals can gain access to the equipment by capacitively coupling to the power cord at a point behind the filter.

It's not always possible to disassemble your computer to add a line filter, but the best location for a power-line filter is bolted to the chassis of the electronic equipment it protects, or at least in the immediate vicinity, such as at the power receptacle.

While you could construct a line filter using the formulas and designs from a magazine article, I heartily recommend that you buy a packaged unit instead. The selection is easier and much more controlled using commercial line filters (see the text box on page 74). So much depends upon component selection and layout that the only way to make sure power-line interference has been eliminated is to actually test the filter in your equipment. A circuit de-

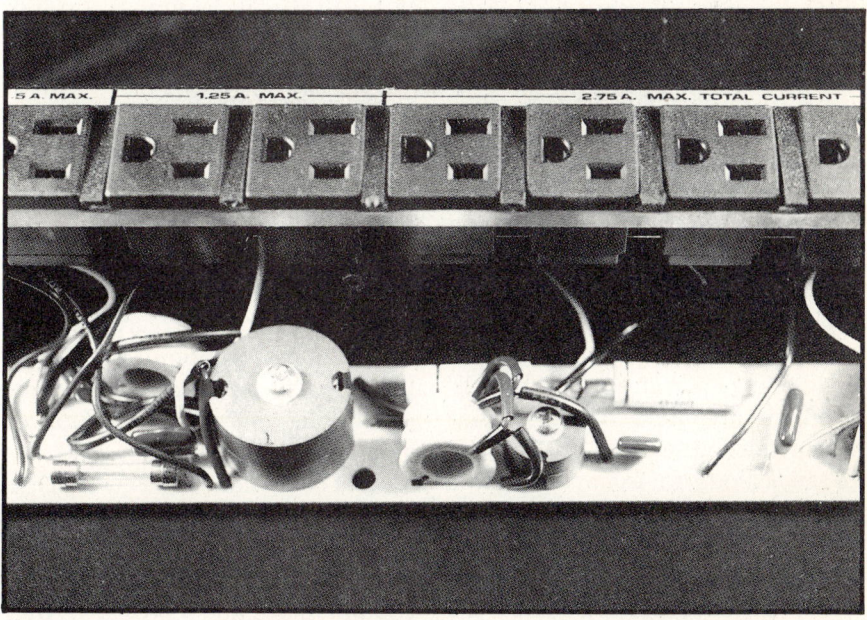

Photo 8: *Some line filters are made to work in specific circumstances. This Radio Shack power-line-filter strip (stock number 26-1451) was devised to cure interference problems with the TRS-80 Model I computer; it contains two separate LC (inductance/capacitance) interference filters but no varistors. If you have this strip, I suggest you install some MOVs.*

and effective, but they trigger at relatively high voltages, making them unsuitable as the sole protection for semiconductor circuitry.

Voltage-clamping devices, on the other hand, have impedances that vary as a function of either the voltage across or the current through them. The circuit being protected is unaffected by the presence of the clamping device unless the incoming supply voltage exceeds the clamping level, as would be the case when a transient hits. The various kinds of high-speed voltage-clamping devices include selenium cells, zener diodes, silicon-carbide varistors, and metal-oxide varistors. Of these, the *metal-oxide varistors*, or MOVs, hold a significant price/performance advantage and are highly applicable in personal computing applications.

MOVs to the Rescue

Metal-oxide varistors are voltage-dependent nonlinear devices that behave somewhat like a back-biased zener diode. When a voltage lower than its conduction threshold is applied across it, the MOV appears as a nonconducting open circuit. But if the applied voltage becomes greater than this set point (when a transient hits), the MOV begins to conduct, clamping the input voltage to a safe level. In effect, the MOV absorbs the transient and dissipates the energy as heat.

An MOV is made of zinc oxide combined with small amounts of bismuth, cobalt, and manganese. The individual zinc-oxide grains form many p/n (positive-doped/negative-doped) junctions that combine in a multitude of series and parallel arrangements. This diversity of microstructure causes its nonlinear semiconducting characteristics. An MOV is inherently more rugged than a single-junction semiconductor device (a zener diode, for example) because energy is uniformly absorbed throughout the bulk of the component.

The physical dimensions of the MOV determine its characteristics, its conduction-threshold voltage varying as a function of thickness, and its energy-dissipating capacity varying according to volume. MOVs are available in operating voltages from 6 to 2800 V, with peak current capacities of up to 50,000 A (amperes). MOVs respond to transients in only a few nanoseconds and are relatively inexpensive. The chief producer of MOVs is the General Electric Company.

Protect Your Computer

Large companies sometimes solve power-line problems by producing their own power. In the home or small office, it's more practical to protect your computer and peripherals through comprehensive application of filtering and transient suppression.

Most of the commercially available filtered power strips contain MOVs as their primary suppression device. Even those costing $50 or $75 rarely contain more than $5 worth of transient protection. By purchasing the suppression components separately and installing them yourself, you can save a lot of money.

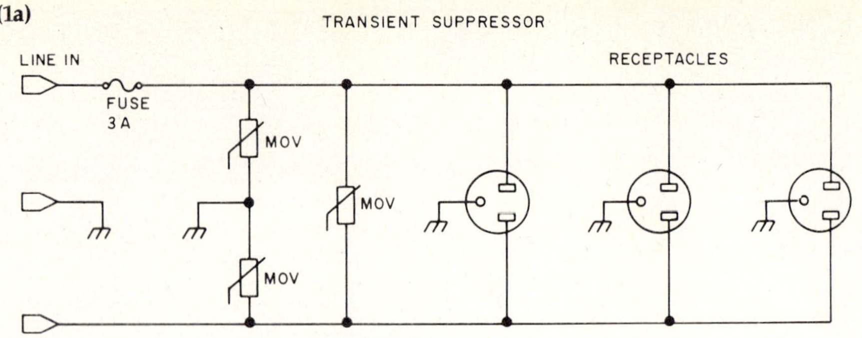

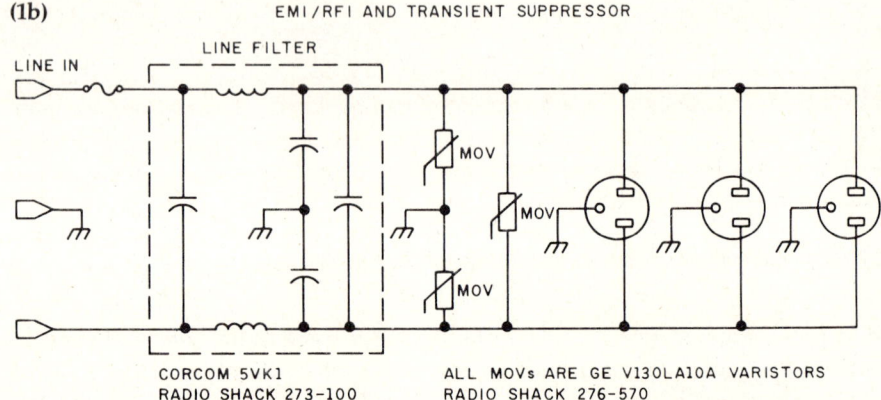

Figure 1a: *The Radio Shack four-outlet power strip can be easily modified to protect equipment from high-energy power-line transients. Three General Electric V130LA10A metal-oxide varistors (MOVs—Radio Shack number 276-570) are connected between the hot, neutral, and ground wires of the power line.*

Figure 1b: *For added protection against low-energy electromagnetic and radio-frequency interference, the Corcom 5VK1 line filter (Radio Shack 273-100) can be installed in the circuit.*

The majority of the projects I've presented in Circuit Cellar articles can be built for $50 to $2000, but the project this month wins hands down for economy. For the most part, line filters and MOVs are available off the shelf, and adequate transient suppression for your computer might cost as little as $1.59!

You can take two approaches in installing suppression. If you are interested in protecting only a few items of equipment, MOVs can be wired across the AC line where it enters the enclosures. You can find the General Electric V130LA10A MOV component at Radio Shack for $1.59 (stock number 276-570). This device is ideally suited to 120-VAC applications. It has an energy rating of 38 joules (watt-seconds) and will clamp to 340 V at 50 A within 35 ns (nanoseconds). Its peak-current rating is 4500 A. (For heavier duties, you'll need to use V130LA20A or V130PA20A MOVs.)

(As a rule, if you are going to be

working inside the equipment you should also install line filters. You can buy Corcom type-5VK1 5-A RFI power-line filters at Radio Shack for $11.95 (stock number 273-100). These units, like the one shown in photo 1, are adequate for most consumer applications and fit in very nicely with existing equipment.)

The easier alternative is to modify a regular power strip to include transient suppression. Radio Shack's 4-outlet strip (number 61-2620, costing $15.95) is perfect for this application. Merely open it up and install three MOVs, as demonstrated in the series of photos 2 through 6, connected as shown in figure 1. One MOV is installed directly between the black (hot) and white (neutral) leads, the second MOV is connected from the black lead to the green (ground) wire, and the third from the white to the green. While you might squeak through by installing one MOV across the line, complete common-mode and differential-mode suppression requires three MOVs. (Photo 7 shows a commercial adaptation of the simplified scheme.) The price for all the parts of the protected power strip is $20.72. If you were to buy a larger power strip or build your own distribution box, you could also add a power-line filter. And if you have a filter strip already on your computer, you might want to check its degree of transient protection (see photo 8).

An Ounce of Prevention...

This project may not seem very exciting. I didn't find the idea very exciting, either, until the flash and subsequent smoke coming out of my favorite article-writing computer provided all the excitement I'll need for months. Most of the $3000 worth of damage I had was for equipment plugged into a single circuit, some of it on the same power strip. I had always known the protective value of MOVs, but I thought it wouldn't happen to me. A few dollars' worth of parts could have saved a lot of aggravation.

Voltage spikes and power-line disturbances aren't always the result of storm activity. Transient-caused equipment failures can happen any-time. The events I've described just served as a catalyst for presenting the subject. And even if lightning never hits you, you should know that many of the new computers I have been evaluating this year have shown an increased sensitivity to external interferance, including power-line glitches. You wouldn't want to find your new computer rebooting suddenly at a critical point or discover the memory to be scrambled after you plug in a printer on the same outlet. Transient suppression constitutes an ounce of prevention. You can spend thousands for the cure.

References

1. Ciarcia, Steve. "A Computer-Controlled Wood Stove." February 1980 BYTE, page 32.
2. Ciarcia, Steve. "Electromagnetic Interference." January 1981 BYTE, page 48.
3. Roberts, Steven K. *Industrial Design with Microcomputers*. Englewood Cliffs, NJ: Prentice-Hall, 1982.

Build the Circuit Cellar Term-Mite ST Smart Terminal
Part 1: Hardware

National Semiconductor's NS455A Terminal-Management Processor permits an easy, economical terminal design

Did the the personal computer revolution begin in 1975 with the MITS Altair 8800 microcomputer? Most people think so, but I believe that the first personal computer product appeared two years earlier under the unassuming name of TV Typewriter. This construction project, described by Don Lancaster in *Radio Electronics* magazine (see reference 6) was a simple video-display terminal, a basic building block for those of us who were dreaming about building a computer. The circuit logic was a wiring nightmare of controlled race conditions, but it worked. Of course, its uppercase-only 16-line by 64-column display and total ignorance of control codes (it didn't even scroll) seem primitive today.

A few months after Don's article was published, the Mark 8 computer project appeared in *Radio Electronics* (see reference 10). The Mark 8, based on the Intel 8008 microprocessor, was the first real microcomputer (though the word had not yet been coined) and was the trigger that launched me and many others into the microcomputer hobby. Many of us who had built Don's terminal might not have been otherwise able to comprehend and use the Mark 8 as quickly as we did. (I built something different from the Mark 8, but the first article I ever wrote was for the Mark 8 constructor's newsletter. Coincidentally, my first BYTE article described how to build a vector-graphics display for an 8008-based system—see reference 2.)

Advancing Display Technology

Wiring up the TV Typewriter was a monumental job, but the basic circuit was really not unlike a commercial video-display terminal of the same period. If you ever opened the case of a video terminal from the early 1970s, you were probably amazed at the complexity. There were usually several large printed-circuit boards (each containing 70 to 100 integrated circuits), a large power supply, the keyboard, wires and diodes, and of course the cathode-ray tube (CRT) replete with high-voltage wires and "Danger—Do Not Touch" signs.

These early terminals were basically "glass Teletypes," performing only simple functions and displaying only uppercase characters. The lack of sophistication matched the level of integrated-circuit (IC) technology available at the time.

Many discrete logic circuits were needed to detect even the simplest functions such as linefeed and cursor-home. Each command was treated independently by the hardware: it was necessary to have separate circuitry to detect each control character and cause the appropriate function to occur. For instance, for the terminal to respond correctly to the ASCII (American National Standard Code for Information Interchange) Return character, the terminal-control logic had to be able to detect when a hexadecimal 0D value was received from the host computer or typed on the keyboard, to change the current cursor position to the beginning of the line, and possibly to scroll the screen up one line (if automatic linefeed is on) and blank the new line. Connecting sets of NAND and NOR gates to accomplish this is a considerable task.

For a long time, advances in IC technology were met by demands for increased performance in terminals. The first real simplifying breakthrough came with the microprocessor. Using the power of this new development, designers could implement features that had been prohibitively expensive and could freely add new functions to terminals. Off-line editing with character insertion and deletion, function keys with multiple-character transmission, and multiple-page display memories were just a few of the features that found their way into the terminal marketplace. And as microprocessors became more advanced, terminals incorporating the latest silicon intelligence could no longer be called "dumb." The watchword in the terminal trade became *smart*.

Somewhat surprisingly, the first microprocessor-based intelligent terminals were no less complicated inside than the dumb variety. Computer circuitry had replaced much of the discrete logic, but the expanded functions had also necessitated increased complexity in the low-level display-driver circuitry. An integrated

Photo 1: *The prototype board of the Term-Mite ST intelligent video-display terminal.*

solution to discrete video circuitry was needed.

The second technological achievement resulting in lower circuit complexity was the development of integrated CRT-controller (or video-controller) chips, such as the National Semiconductor DP8350 and the Intel 8275. Usually used in combination with a microprocessor, programmable CRT controllers incorporate many of the discrete counters, registers, and character-attribute circuits needed in a modern terminal. (See references 5 and 9.)

The new controller chips made it easy to do tricks with character attributes: blinking, blanked, or underlined characters; half-intensity or reverse video; and expansion to double height, double width, or both. Also, a terminal manufacturer could now easily make a whole family of terminals just by changing the control firmware. Either a simple "glass Teletype" model or a sophisticated editing terminal with write-protected fields and multiple display pages could be built with only minor hardware differences simply by changing the programs controlling the microprocessors. Figure 1 on pages 83 and 84 should give you an idea of how the combination of the microprocessor and the video-controller chip served to make hardware design much simpler while again increasing the terminals' sophistication.

For a long time now I've wanted to present a smart-terminal project from the Circuit Cellar, but even with the reduced circuit complexity afforded by a CRT controller I've never been

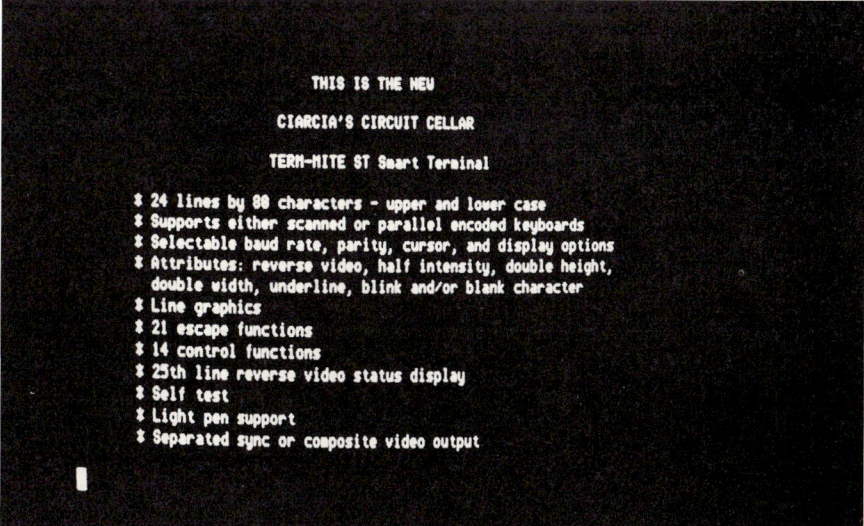

Photo 2: *Unretouched photo of the Term-Mite's screen.*

Photo 3: *Video terminals designed only a few years ago had to use over 100 discrete-logic circuits to obtain even rudimentary control functions.*

able to devise a reasonable design containing fewer than about 30 IC packages. (Yes, the MPX-16 computer I began in November 1982 does contain 121 chips, but my battle scars from that project still pain me at times.) While I was still deliberating, developments in technology caught up with me.

Semiconductor makers had provided both crucial elements: the microprocessor and the CRT controller. The next logical step was to incorporate their functions into a single IC package. National Semiconductor Corporation has done just that, and more, with the NS455A *Terminal-Management Processor* (TMP). Incorporating most of the processor, video, and communication functions in a single 48-pin dual-inline package, the NS455A allows the design engineer to reduce a terminal's chip count while maintaining a high level of performance. Just six chips can perform the basic operations.

In two articles, this month and next, we'll look at the NS455A's characteristics and see how to build an intelligent video terminal, called the Term-Mite ST, which is equal to many on the market costing $1000 or more. Its 21-chip design provides the most-needed features, as shown in table 1, such as 24 lines of 80 characters each, uppercase and lowercase; a full set of character attributes; and line (block) graphics. A block diagram of the Term-Mite ST terminal appears in figure 2.

Inside the NS455A TMP

Integrated into the NS455A TMP are all the system-control functions except the video RAM (the random-access read/write memory used to store the display data) and I/O (input/output) buffers. The TMP replaces the separate microprocessor, program ROM (read-only memory), CRT controller, DMA (direct memory access) logic, character generator, UART (universal asynchronous receiver/transmitter), and data-rate generator typically used in other terminal designs. In place of these, the TMP provides a control processor, display-timing control circuitry, and direct interface logic for the keyboard, monitor, memory, and serial communication.

A complete listing of the NS455A's capabilities is shown in table 2, while figure 3 on page 86 shows a pinout diagram (3a) and a block diagram (3b). The architecture and instruction set for the TMP are derived, with some differences, from that of the Intel 8048-series of microprocessors. Extra instructions have been added and the architecture tailored to allow the NS455A to serve more efficiently as a terminal controller. Within the TMP are three distinct functional sections: processor, I/O, and display

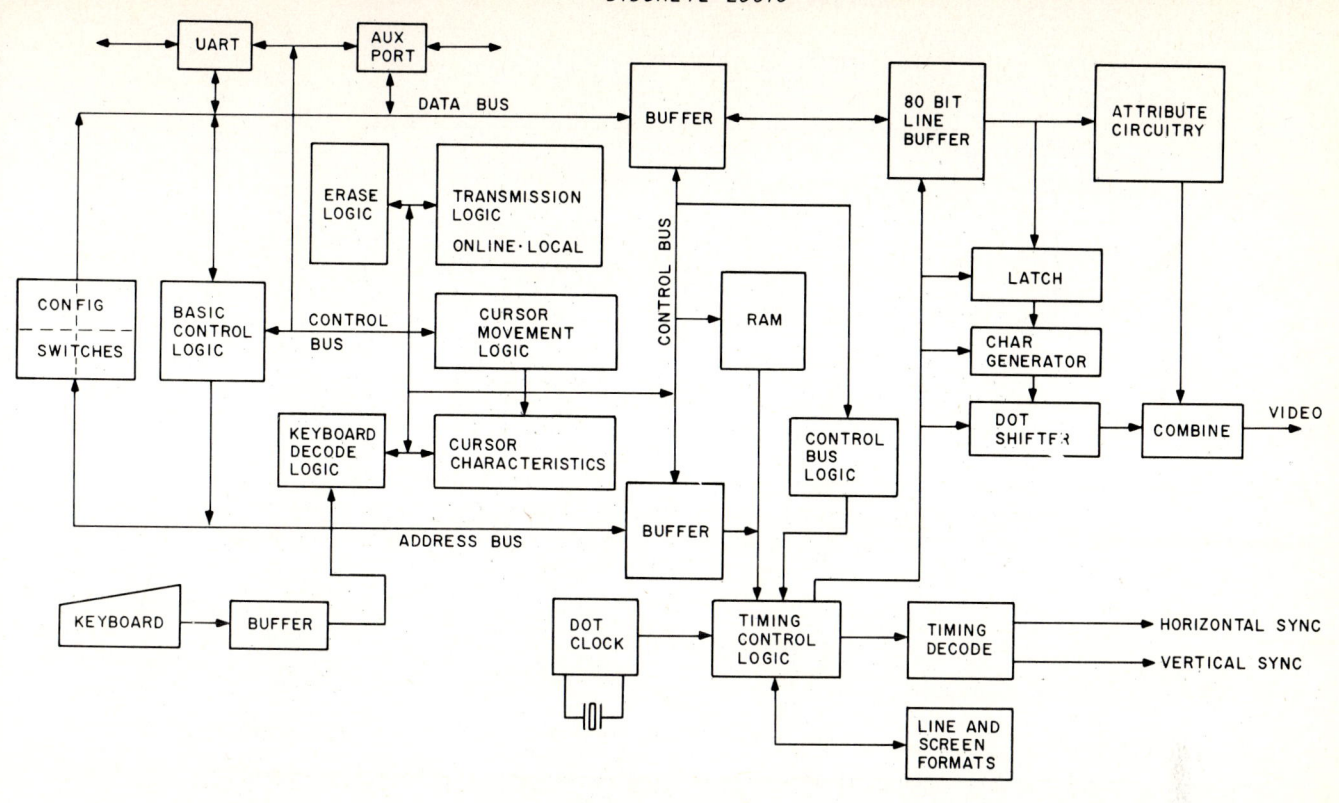

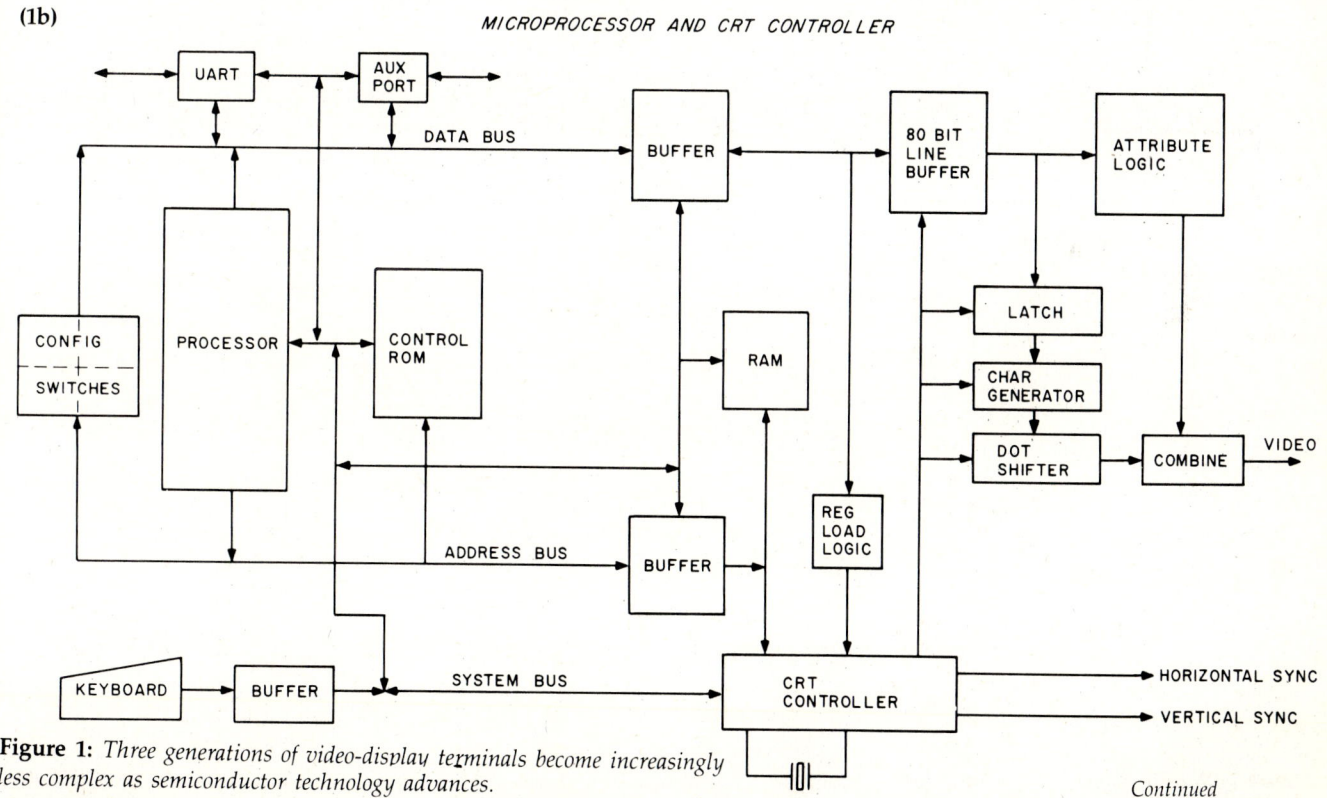

Figure 1: Three generations of video-display terminals become increasingly less complex as semiconductor technology advances.

Continued

driver. Let's look at each of these in turn.

Processor and Memory

Since the processor in the TMP is a modified implementation of the Intel 8048 architecture, I'll review the 8048 to make comparison easier.

The basic 8048 was designed as a self-contained microprocessor; it includes ROM for its factory-set permanent program as well as temporary storage in the form of scratch-pad RAM. It operates on 8-bit data but has an 11-bit program-counter (PC) register, so that it can address up to 2K bytes of program. The standard 8048 has an eight-entry fixed-size

Figure 1 continued

(1c) INTEGRATED MICROPROCESSOR AND VIDEO CONTROLLER

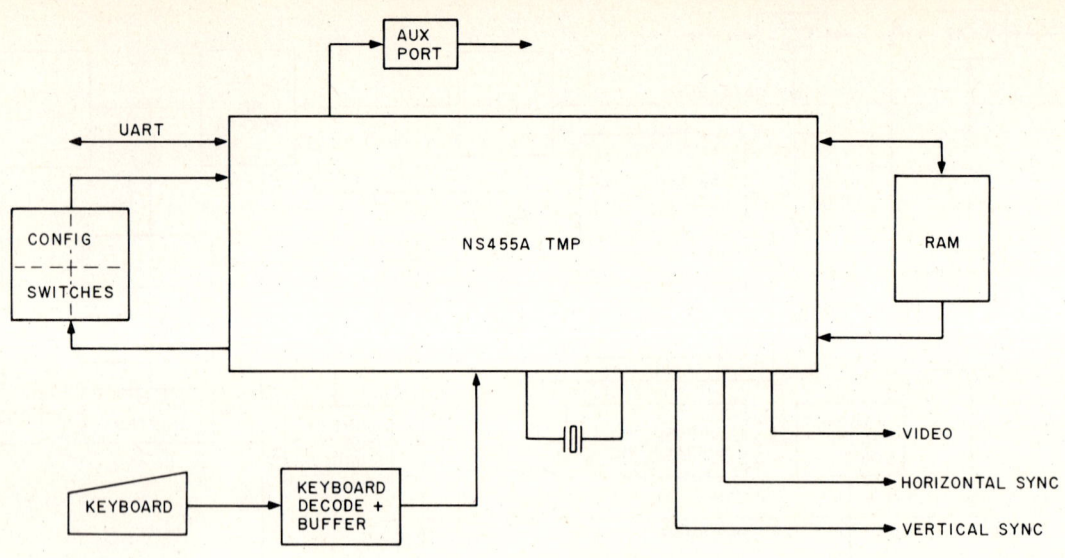

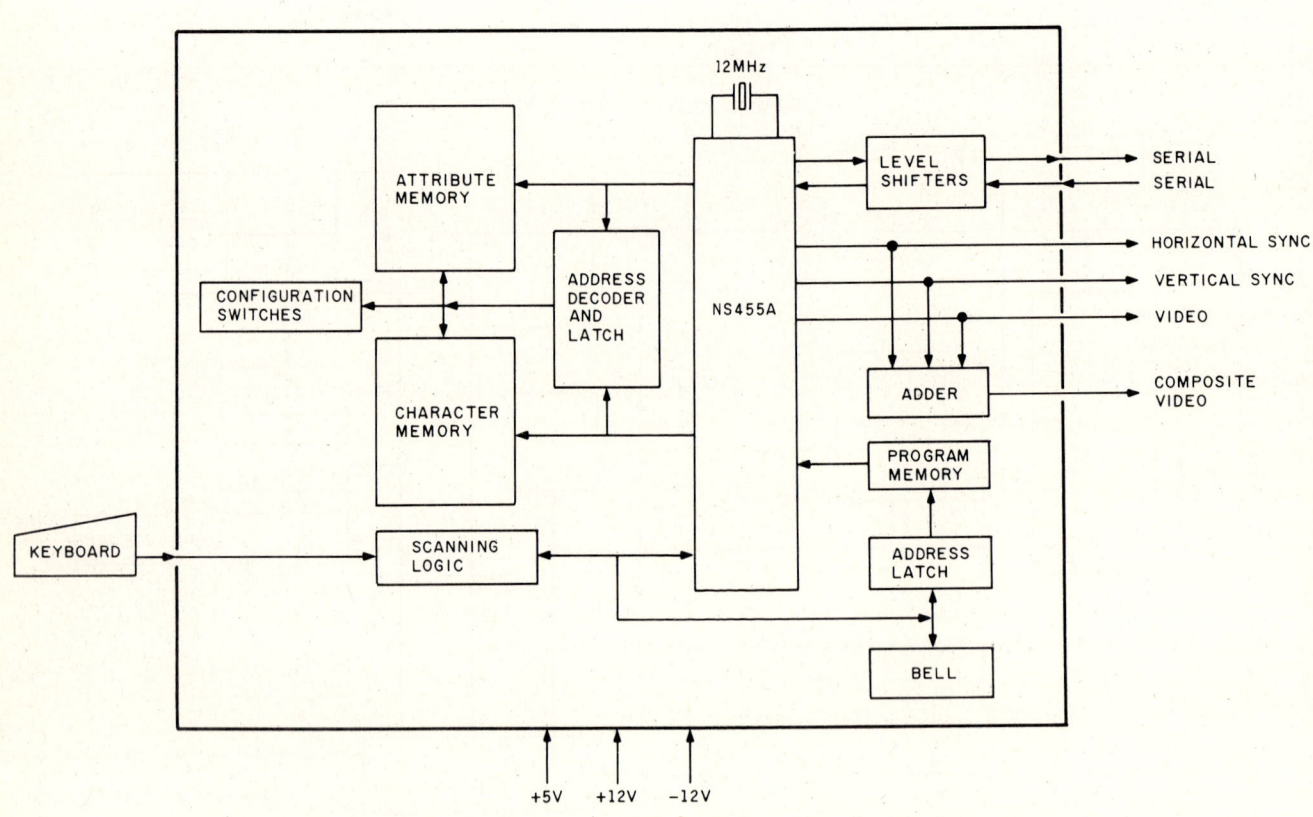

Figure 2: *A block diagram of the Term-Mite ST terminal circuit board.*

stack used to store return addresses during subroutine calls and interrupt handling; the stack pointer consists of 3 reserved bits in the processor status word (PSW). A set of general registers in the RAM, R0 through R7, can be used for fast-access storage. In addition, an alternate register bank, also located in the on-chip RAM area, can be selected and used just like the primary set. Finally, the standard 8048 has three parallel I/O ports, which can be used to communicate with external peripherals or to address additional memory. (I used the 8748, a cousin of the 8048, in my recent H-Com project; see reference 3.)

When National Semiconductor en-

1. 24 lines by 80 characters, uppercase and lowercase
2. supports either scanned or parallel-encoded keyboards
3. selectable data-rate, parity, cursor, and display options
4. attributes: reverse video, half intensity, double height, double width, underlined, blinking, and/or blank character
5. 21 Escape functions
6. 14 Control functions
7. line (block) graphics
8. twenty-fifth-line reverse-video status display
9. self-test
10. separate-sync or composite-video output

Table 1: *Features of the Term-Mite ST terminal.*

1. enhanced 8048 instruction set and architecture
2. on-board ROM, 2K by 8-bit; up to 8K by 8-bit external
3. on-board RAM, 64 by 8-bit
4. programmable display format
5. 16-bit display-memory bus (direct-video and attribute-RAM interface)
6. built-in timer
7. real-time clock (may be programmed for 1 Hz)
8. video-control signals
9. eight independent character attributes
10. pixel graphics
11. programmable cursor
12. CRT refresh at 50 or 60 Hz
13. light-pen support
14. on-board UART, programmable data rate up to 19,200 bps
15. character generator (128 characters in 7- by 11-dot character area)
16. single +5-V (volt) supply
17. interface compatibility with popular 8- and 16-bit microprocessors
18. up to 18-MHz clock frequency
19. 48-pin package
20. 8-bit parallel port (multiplexed with external ROM)

Table 2: *Features of the National Semiconductor NS455A Terminal-Management Processor. This integrated circuit normally comes with a terminal-control program mask-programmed into the 2K bytes of on-board ROM.*

hanced the architecture for use in the Terminal-Management Processor, instructions that manipulate 16-bit data were added, so that the TMP could manage up to 64K bytes of display memory. To support these instructions, an 8-bit high-order extension, called the *high accumulator*, was added to the existing 8-bit accumulator. In addition, 8-bit high-order extensions were added to the R0 and R1 registers to allow them to point to the full 64K bytes of display RAM. The original roster of three I/O ports was trimmed to a single bidirectional port, and 2 more bits were added to the program counter to accommodate up to 8K bytes of program storage.

All the other changes from the 8048 processor were directly associated with the additional tasks of video-display driving and I/O required to do the job. Display-management registers were added to help with the screen-refresh chores. A cursor register serves to load characters into the display RAM as well as to mark the current cursor position. A whole set of video-timing-chain registers is used to set the display configuration for the screen: how many characters per line, how many lines per frame, horizontal and vertical synchronization timing, etc.

While the characters to be displayed are stored in RAM, the program that drives the NS455A's processor resides in ROM. The NS455A may operate with either internal or external ROM; external ROM may function all by itself (disregarding the internal ROM), or it may supplement the internal program. Although address space is provided for up to 8K bytes of program, the standard on-chip ROM size is only 2K bytes. The off-the-shelf NS455A TMP comes with a standard program, masked into the 2K-byte ROM, which is intended to illustrate the capabilities of the chip and serve as a tutorial example of terminal programming.

I/O-Port Section

The single I/O port is an 8-bit bidirectional parallel type, with data transferred on pins RE0 through RE7. It is written into and read from using the processor instructions OUT PORT and IN PORT. In the Term-Mite ST, the encoded keyboard is read and the RS-232C handshaking-protocol signals (Data Terminal Ready, Clear to Send, and Ready to Send) are transmitted through this port. The keyboard uses only 7 bits, so a signal to sound the terminal's "bell" beeper is sent out on the extra bit.

The serial input and output functions are handled by the on-chip UART through the serial-in line and serial-out line. The UART also contains the data-rate generator, which can be set up by software for virtually any data rate. The standard program, however, contains only the set of 12 most-used rates from 110 to 19,200 bps (bits per second). The serial I/O lines need only to be buffered by level-shifting devices (the MC1488 and MC1489) to give you a complete RS-232C data path to a host computer or other data-communicating device.

Display-Driver Section

The third section is the video-display driver and control section. It is made up of the character generator, the CRT-refresh logic, the character-attribute logic, and many now-integrated, formerly discrete functions. Because the NS455A provides all these capabilities by itself, the only additional ICs needed are display

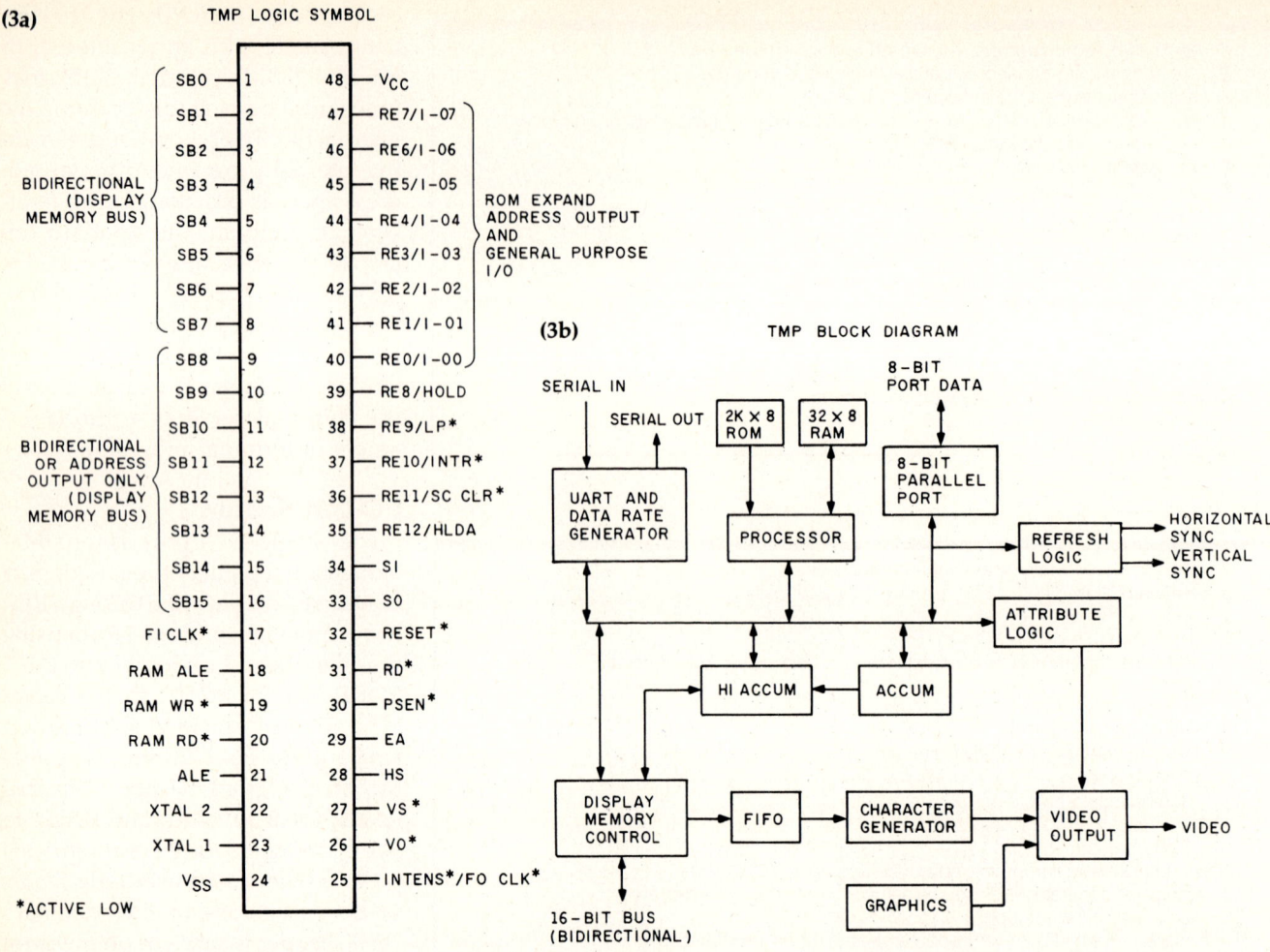

Figure 3: *A pinout diagram (3a) and a block diagram (3b) of the National Semiconductor NS455A TMP.*

memory, an RS-232C buffer and a driver, and a bus-interface latch or two. In fact, an absolutely minimal video terminal could be constructed from 6 chips, and a relatively high-performance unit could be built from only 15.

You may not be as familiar with the complex workings of the CRT driver and control section as with the other two preceding sections. Indeed, because the screen image must be refreshed 60 times every second, the video circuitry stays very busy. The best way to see what happens is to follow a character from display memory until it appears on the screen.

First, the character is read by the display-access logic from the display RAM into the FIFO buffer (a small first-in, first-out storage area). The FIFO buffer contains only four entries, but it is needed so that the processor and display-refresh logic, which both require continual access to the RAM, will not contend with each other for that access.

(The refresh logic has to push characters through the analog video section at a constant rate; if the refreshing is delayed, the display will flicker where characters or parts of the image are being missed. In older designs using CRT-controller chips, the usual technique was to allow the processor to run only during the horizontal and vertical retrace times of the electron beam—barely 40 percent of the time. Such approaches, once considered necessary, can severely limit throughput; you can recognize the terminals that employ this scheme as the ones whose displays flicker when they are updated.)

A FIFO buffer solves the problem. The refresh logic can fill the buffer faster than the video logic can empty it, so when the buffer is sufficiently full, the processor is allowed to grab and use a display-memory cycle. The processor has to be held off only intermittently, and then not for very long.

As the character code leaves the FIFO buffer, the proper pattern of display dots (picture elements, or pixels) in a bit-mapped dot matrix is selected by the character generator according to the character's shape. What dot value leaves the generator at a given instant depends not only on what the character is but also the current scan line (vertical position) of the video raster. In the case of the Term-Mite ST, the character information is retrieved from the character generator's ROM, where the dot patterns are stored, using the ASCII code value as a high-order address; the scan-line position is used to set up the low-order address bits. The dots for all the characters in a scan

Strobe Number	A15	A14	A13	A12	
0	0	0	0	x	page 0 memory
1	0	0	1	x	page 1 memory
2	0	1	0	x	page 2 memory
3	0	1	1	x	page 3 memory
4	1	0	0	x	terminal-characteristics switch
5	1	0	1	x	miscellaneous-status switch
6	1	1	0	x	UART-configuration switch
7	1	1	1	0	scanned keyboard-input port
8	1	1	1	1	auxiliary/printer-output port

Table 3: *Memory-mapped I/O addressing as used in the Term-Mite ST. The final hardware of the project does not support the printer port.*

Bit	Attribute
0	reverse video
1	half intensity
2	blinking
3	double height
4	double width
5	underlined
6	blanked
7	graphics

Table 4: *Character attributes supported by the NS455A in the Term-Mite ST.*

line are assembled into a parallel field containing dark and bright dots. These dots are then serialized and shifted out to the screen, one at a time.

Information on other character attributes may be sent along with the basic black- or white-dot data. Specialized logic was incorporated in the 455A to modify the dot output according to the display attributes selected, but, generally speaking, the logical process for accomplishing this has not changed from that previously implemented by many external logic gates.

Eight special character attributes are provided by the 455A: blinking, double height, double width, graphics, half intensity, reverse video, underlined, and blanked character. A terminal built around the 455A can specify the attributes in two ways: internally or externally.

Internal attributes make use of two attribute latches (AL0 and AL1) inside the TMP chip. These latches can be read by external circuitry. Their states are determined by the most significant bit (MSB) in the character-code byte (the character set is therefore limited to the 128 standard ASCII 7-bit characters). If the MSB is a 0, AL0 is activated; if the MSB is a 1, AL1 is activated. The latch status and the incoming character dot values completely specify the final appearance of the displayed character.

When the TMP is configured for *external attributes*, a 16-bit-wide display memory is used. The lower 8 bits specify the character, and the upper 8 bits are used to handle the attribute data. Although this arrangement doubles the memory needed in the video RAM section, it gives you the freedom to use any possible combination of attributes for any character.

The remaining video logic involves the horizontal sync (synchronization) pulse and the vertical sync pulse. The horizontal sync pulse is generated by the video circuitry in the TMP, but its timing is under program control. It is the terminal designer's responsibility to decide the crystal frequency

> A terminal built around the 455A can specify attributes in two ways: internally or externally.

used to drive the TMP, the number of characters per line, the size of the character cell, and the tolerances of the CRT's driver circuits. In most cases, all of these variables must yield something close to 15,750 Hz (hertz) for the frequency of the horizontal sync signal; the frequency must allow the display of 80 characters per line plus enough time to permit the electron beam to retrace to the beginning of the next scan line. The vertical sync pulse will occur at either 50 or 60 Hz depending on where in the world the terminal is to be used (in North America 60 Hz would be used). The necessary programming has already been done in the standard NS455A and in my Term-Mite ST project.

Memory-Mapped I/O

Because there is only one parallel port and one serial port, I/O by conventional methods is limited in the TMP. But, fortunately, by mapping I/O registers into memory addresses, you can greatly expand the TMP's I/O capabilities. (See reference 4.) For example, a printer port could be placed at address hexadecimal F000. The processor would just act like it is putting data into a memory location at that address, but the data would actually be sent to the printer.

Technically, there are 16 bits of memory-address space, which could in theory define 64K (65,536) I/O locations. However, it's only necessary to put in circuits to decode only the 4 high-order bits to designate nine specific memory-address allocations. These are listed in table 3. As with all construction projects, I had to observe practical limits when I froze the design. Even though the NS455A can support the printer port, the Term-Mite ST as presented in this article does not incorporate it, and only eight of the nine I/O addresses are available.

Term-Mite ST Design Details

The Term-Mite ST is an intelligent video-display terminal built from 21 integrated circuits (or 18 if you use a parallel-encoded keyboard). Shown in the schematic diagram of figure 4 on pp. 89-90, circuit is intended to operate with the same ordinary 2K-byte terminal-control program mask-programmed into the generic NS455A but contained instead in a 2K-byte type-2716 EPROM (erasable

Number	Type	+5 V	GND	+12 V	−12 V	Number	Type	+5 V	GND	+12 V	−12 V
IC1	6116	24	12			IC12	74LS373	20	10		
IC2	6116	24	12			IC13	74LS373	20	10		
IC3	74LS373	20	10			IC14	2716	24	12		
IC4	74LS373	20	10			IC15	74LS374	20	10		
IC5	74LS138	16	8			IC16	74LS154	24	12		
IC6	74LS240	20	10			IC17	74LS123	16	8		
IC7	74LS240	20	10			IC18	74LS244	20	10		
IC8	74LS240	20	10			IC19	MC1488		7	14	1
IC9	74LS240	20	10			IC20	MC1489	14	7		
IC10	74LS240	20	10			IC21	74LS86	14	7		
IC11	NS405	48	24								

Table 5: *Power connections for integrated circuits in figure 4 (page 46). Another EPROM could be substituted for the 2716 in some applications. Note: The Term-Mite actually uses a NS405 chip which is a specialized version of the generic NS455.*

programmable ROM). The Term-Mite ST could potentially handle up to 8K bytes of external program memory (a type-2764 EPROM), which would allow the control program to be enhanced—perhaps to include more features or to emulate the display protocols of popular commercially sold terminals.

The display format is 24 lines of 80 characters, with a 25th reverse-video status line. The particular TMP version I have chosen uses a 12-MHz crystal and displays characters in a 5- by 7-dot matrix in a 7- by 10-dot character area. The masked program automatically configures the correct horizontal and vertical frequencies.

In figure 4, IC1 through IC5 constitute the video-display memory section. Two type-6116 static RAM chips (IC1, IC2) form a 2K-word display memory of 16-bit words. The low-byte chip (IC2) contains the ASCII character codes, while the high-byte chip (IC1) holds the screen attributes. The attributes supported are listed in table 4, with their relation to bits in IC1 shown.

IC3 and IC4 are the type-74LS373 address latches for the display memory. When any access to external display memory occurs, the address of the location is set on lines SB0 through SB15 and loaded into these two latches on the occurrence of the address-latch-enable strobe (RAM ALE) signal. Address bits A12 through A15 are decoded through a 74LS138 (IC5) to provide eight enable lines for the memory-mapped I/O as described previously. Depending upon whether the instruction is a memory-read or memory-write operation, either the active-low $\overline{\text{RAM RD}}$ or $\overline{\text{RAM WR}}$ line will be logic 0 (active).

Not all the decoded address-strobe signals are used in the Term-Mite ST, and only eight of the nine possible are implemented. To cut down on the number of chips, I decided to limit the display memory to a single 4K-byte page and not include a printer-output port. The software as supplied still supports four pages and the printer, so you may expand on

> As with all construction projects I had to observe practical limits when I froze the design.

the basic design if you feel resourceful and feel like wiring a few more chips.

The remaining memory-mapped I/O devices are three configuration switches (buffered through IC6, 7, and 8—three 74LS240 chips) and 12 bits of scanned keyboard data (buffered through IC9 and 10). Each device is addressed and its data gated onto the bus during the $\overline{\text{RAM RD}}$ pulse. Next month I'll describe these switches and the scanning logic in greater detail.

On the right-hand side of the TMP (IC11) is the program-memory (EPROM) and user-I/O circuitry. In a process similar to that described for the display memory, an address is loaded into the latches IC12 and IC13 during the active state of the ROM ALE; the EPROM data is read during the $\overline{\text{PSEN}}$ pulse. Type-2716, -2732, or -2764 EPROMs may be used (with proper jumper selection).

A parallel keyboard, instead of a scanned keyboard, may be connected to the Term-Mite ST through the type-74LS244 buffer IC18. Of the 8-bit input port, 7 bits are used to transfer ASCII data, while the eighth bit is borrowed for RS-232C handshaking. The keyboard strobe (active-low) connects directly to the RE10 TMP line and generates an interrupt when active.

Serial communication is handled directly by the TMP through an on-board UART. The data rate and protocol are set via configuration switches, and full handshaking is supported. The MC1488 buffer and MC1489 driver are connected directly to the TMP.

The TMP also has a direct output line for a bell signal trigger. The pulse is generated whenever a Control-G code is output or whenever the cursor reaches column 72 on the display screen. The trigger pulse is only a few microseconds long, so that a monostable multivibrator, or one-shot (IC17), is needed to stretch the pulse and drive a self-contained piezoelectric transducer.

Video output from the TMP is in the form of separate horizontal sync, vertical sync, and luminance signals. IC21, in combination with some discrete components, merges these to generate a composite-video signal. Because of the wide bandwidth re-

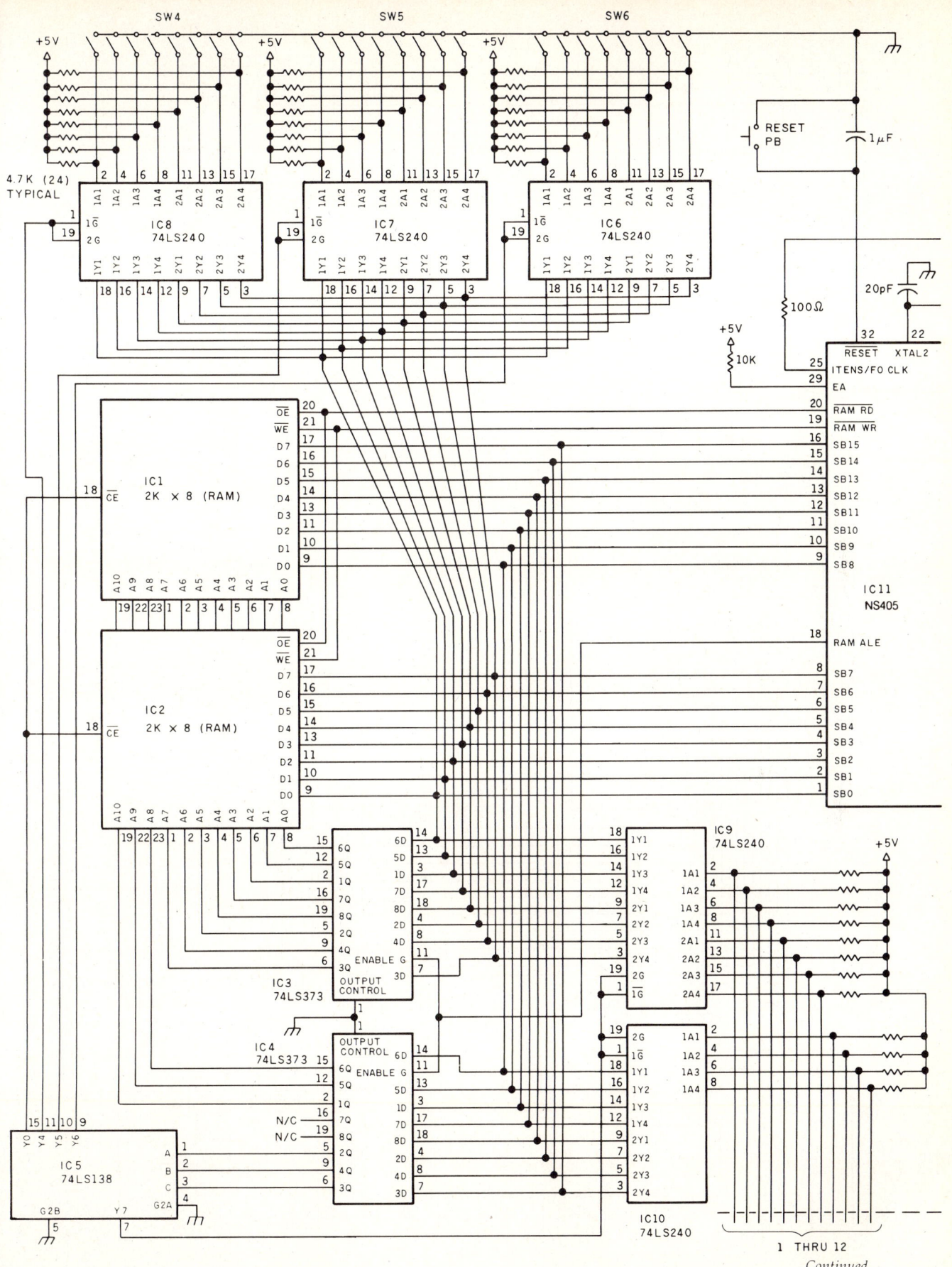

Figure 4: *Schematic diagram of the Term-Mite ST. The NS455A's usual stock terminal-control program is here contained in a 2K-byte type-2716 EPROM (erasable programmable ROM).*

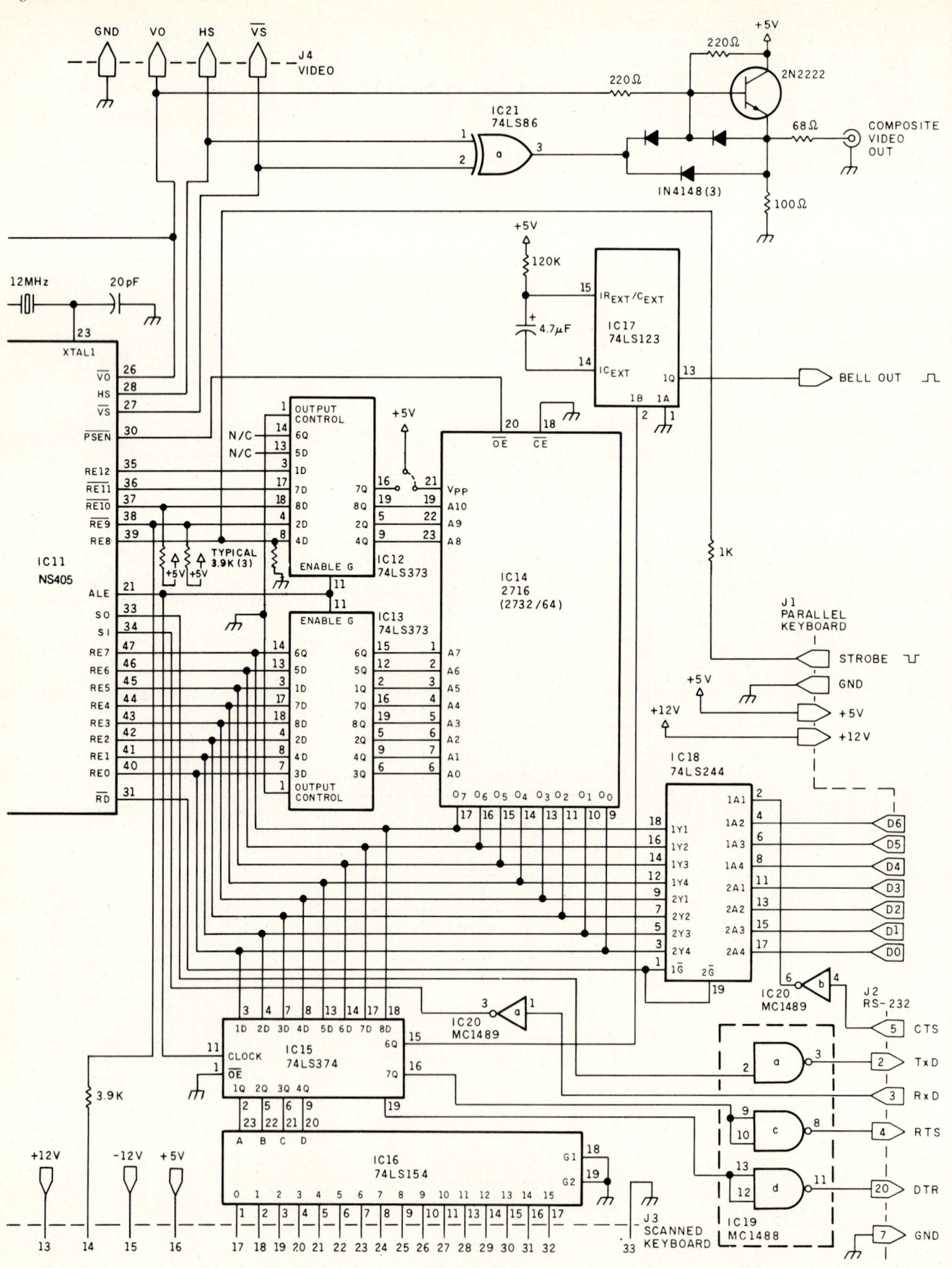

Figure 4 continued

quired for the 80-column display, you would probably have limited success using an RF (radio-frequency) modulator and TV set. I recommend that you use a high-quality CRT monitor for best results.

Although the current 2K-byte control-software release does not support it, the hardware provides a light-pen input in the form of an interrupt to the TMP. With proper programming, you could get the TMP to remember where the electron beam was scanning when the interrupt happened and subsequently return a value giving the location to the program. And with the proper software running in the host computer, all sorts of menu-driven tasks could be handled in this way.

Special thanks to Bob Harbrecht of National Semiconductor Corporation for his help with this project.

References

1. Cayton, Brian, and Mort Herman. "The CRT 9007 Video Processor and Controller." BYTE, April 1983, page 96.
2. Ciarcia, Steve. "Make Your Next Peripheral a Real Eye Opener." BYTE, November 1976, page 78.
3. Ciarcia, Steve. "Build the H-Com Handicapped Communicator." BYTE, November 1983, page 36.
4. Ciarcia, Steve. "Memory-Mapped I/O." BYTE, November 1977, page 10.
5. Haas, Bob. "Single Chip Video Controller." BYTE, May 1979, page 52.
6. Lancaster, Don. "TV Typewriter." *Radio Electronics*, September 1973, page 43.
7. Lancaster, Don. *TV Typewriter Cookbook*. Indianapolis: Howard W. Sams & Company, 1976.
8. Roberts, H. Edward, and William Yates. "Altair 8800 Minicomputer." Part 1, *Popular Electronics*, January 1975, page 33. Part 2, *Popular Electronics*, February 1975, page 56.
9. Tennant, Chris. "The Intel 8275 CRT Controller." BYTE, May 1979, page 130.
10. Titus, Jonathan. "Build the Mark 8 Minicomputer." *Radio Electronics*, July 1974, page 29.
11. Wierenga, Theron. "Construction of a Fourth-Generation Video Terminal." Part 1, BYTE, August 1980, page 210. Part 2, BYTE, September 1980, page 126.

The following items are available from:

The Micromint, Inc.
25 Terrace Drive
Vernon, CT 06266
For orders: (800) 635-3355
For information: (203) 871-6170

1. Complete Term-Mite ST video-display-terminal kit including NS455A, printed-circuit board, IC sockets, DB-25S serial connector, and all other components but without keyboard or CRT monitor. Board size is approximately 4½ by 6½ inches with 0.156-inch 44-pin edge connector.
   Price $239

2. Complete assembled and tested Term-Mite ST terminal circuit board.
   Price $279

Please add $5 for shipping in the continental United States and $25 for all other countries. New York residents please add 7 percent sales tax. Allow four to six weeks for delivery.

Build the Circuit Cellar Term-Mite ST Smart Terminal

Part 2: Programming and Use

The supplied standard control software supports several character attributes and various configuration options

Last month, in Part 1, I introduced you to the Term-Mite ST intelligent video-display terminal, shown in photo 1. It is designed around the new National Semiconductor NS455 Terminal-Management Processor (TMP). This self-contained terminal-controller chip permits the essential features (see table 1) to be provided by only 21 integrated circuits.

This month I'd like to pick up the story beginning with a discussion of the firmware, that is, the program logic inscribed in ROM (read-only memory) inside the NS455A that runs on the internal microprocessor to control all the terminal's functions. We'll look at the functions of the firmware, the configuration-switch settings, the Escape sequences and control functions, and, finally, demonstrate putting a few characters on the screen.

Factory-Supplied Control Program

The control program has four main sections: the initialization code, the main keyboard-scanning and wait loop, the display-processing routine, and the interrupt-processing routines. (Since keyboard scanning may be a new technique to some of you, the logic flow of this code section will be described in detail.) The program is constructed in a modular fashion; a general flowchart is shown in figure 1, while a memory map is shown in figure 2.

The initialization routine is executed when the terminal is first turned on or when it is reset. The routine first checks all the display memory and the serial I/O (input/output) circuits; then it reads the configuration switches and loads the NS455A's registers with their initial values. Many values must be loaded into registers before the terminal can work: among them are the timing-chain constants that specify character-cell parameters; values for cursor definition, horizontal, and vertical timing; values for the character attributes and other display controls; values to determine character positions; and values that set the data rate, parity, full- or half-duplex transmission mode, etc.

After everything has been initialized, the program enters the character-processing loop, which is often referred to as the *main wait loop*. If a scanned keyboard is being used, scanning occurs every 10 milliseconds (ms) during this loop. When the hardware detects a character, a branching instruction passes control to the display-processing routine. (A more definitive explanation of keyboard scanning follows.)

Characters typed on a parallel-encoded keyboard are handled somewhat differently, as are characters received from the host computer or another terminal. When a character arrives from one of these sources, the hardware of the Term-Mite ST generates an interrupt, and control automatically passes into one of the interrupt-handling routines. In the case of the parallel keyboard, its data-strobe signal generates an external interrupt (an interrupt relying on hardware outside the processor); the interrupt handler causes the TMP to read the typed character from the I/O port, queue it for display, transmit it (if necessary), and return to wait for the next character.

When the hardware receives data on the serial input line, it generates an internal interrupt. Control branches to several routines that determine what type of interrupt occurred and take appropriate action. The first test checks for received characters in the buffer of the UART (universal asynchronous receiver/transmitter). If a character is found in the buffer, it is queued for display (i.e., placed in the holding area for data to be shown on the screen). If no character is found, the register contents are tested for a match with the special value that indicates an empty-transmit-buffer condition, in which case the program tries to fill the transmit buffer. The final test is for the internal timer; timer interrupts happen every 10 ms and are used to trigger the keyboard scanning. If the Term-Mite ST is not set up to use a scanned keyboard, the internal-timer interrupt produces no activity.

Because of the operating differences between scanned and encoded keyboards, the control program contains two routines for reading a keyboard character. The encoded keyboards are handled by the external-interrupt routine. Very few processor instructions are required to fetch the character, since the data comes in from the I/O port already encoded as values in the ubiquitous ASCII (American National Standard Code for Information Interchange) character set established by ANSI (the American National Standards Institute). In contrast, getting a character from the scanned keyboard takes a fairly lengthy subprogram, which must examine the switch matrix of the keyboard for closures and convert that data into a meaningful ASCII character. The two keyboard routines, however, merge at the point where the character has been identified; a common section of code is used to display, transmit, and queue the character.

The display-processing subroutine is entered from the main wait loop when the program finds that the input-character buffer is not empty. If the character appearing in the buffer is part of an Escape sequence, the processing routine decides whether to wait for additional information (more characters) or to take immedi-

Photo 1: *A prototype of the Term-Mite ST circuit board, measuring only 4½ by 6½ inches. The design incorporates the National Semiconductor NS455A Terminal Management Processor.*

1. 24 lines of 80 characters each, uppercase and lowercase
2. supports either scanned or parallel-encoded keyboards
3. selectable data rate, parity, cursor, and display options
4. attributes: reverse-video, half-intensity, double-height, double-width, underlined, blinking and/or blanked character
5. line (block) graphics
6. 21 Escape functions
7. 14 control functions
8. twenty-fifth-line reverse-video status display
9. self-test
10. separated-sync or composite-video output

Table 1: *The features of the Term-Mite ST intelligent video-display terminal.*

ate action. ASCII control codes are processed immediately. If the received character is an ordinary displayable character, it is simply displayed and possibly transmitted through the serial port.

Keyboard-Scanning Logic

The scanned keyboard is fundamentally a set of push-button switches arranged in a set of rows and columns and wired together with diodes. Every 10 milliseconds, triggered by the internal-timer interrupt, the Term-Mite ST looks at each row and column in turn to find out if any of the switches have been closed.

The basic scanning algorithm is shown in figure 3. The first thing the routine does is check to see if the keyboard is currently enabled (it is possible to turn the keyboard off). The scanning loop is initialized for 16 columns of key switches. The wire along the first column is driven to the voltage that represents logic 1 while the row lines are monitored. If the logic-1 voltage appears on the output of any row, the terminal knows that the key at the intersection of that row and column is being pressed. Whenever the scanning loop detects a hit, program control momentarily leaves the loop while the row and column coordinates are used to look up the appropriate ASCII value in a code-conversion table. The lookup routine also notes the current status of the Control, Shift, and Caps-Lock functions.

The key value is compared to the value found during the last scan; if they are the same, the routine assumes that the key simply has not yet been released from the previous stroke and ignores the key-pressed condition. (When typing, most people hold down each key long enough for many scans to occur. Except when the Repeat key is in use, the terminal assumes that continued closure of the key switch should not produce further output.) If the scanning routine finds that the most recently read key value is indeed a new character, it stores the value and resumes scanning. If two key switches are found simultaneously closed in one column, the two characters are processed in turn before the scan is restarted.

After all the columns have been scanned, the routine checks the character-value storage to see if any keys were pressed. If the number of "hits" found is greater than four, it exits with no output. Valid characters, produced by one to four key presses per scan, are queued in the keyboard buffer for display and output. The routine also checks the Repeat key; if it is being held down, the program initially delays 1 second and then begins to queue the same character again at intervals of 0.1 second. If no keys were detected during the scan, the program cleans out the keyboard-buffer storage area, resets the interrupt mask, does some housekeeping, and returns from the interrupt.

(Note: The keyboard-scanning rou-

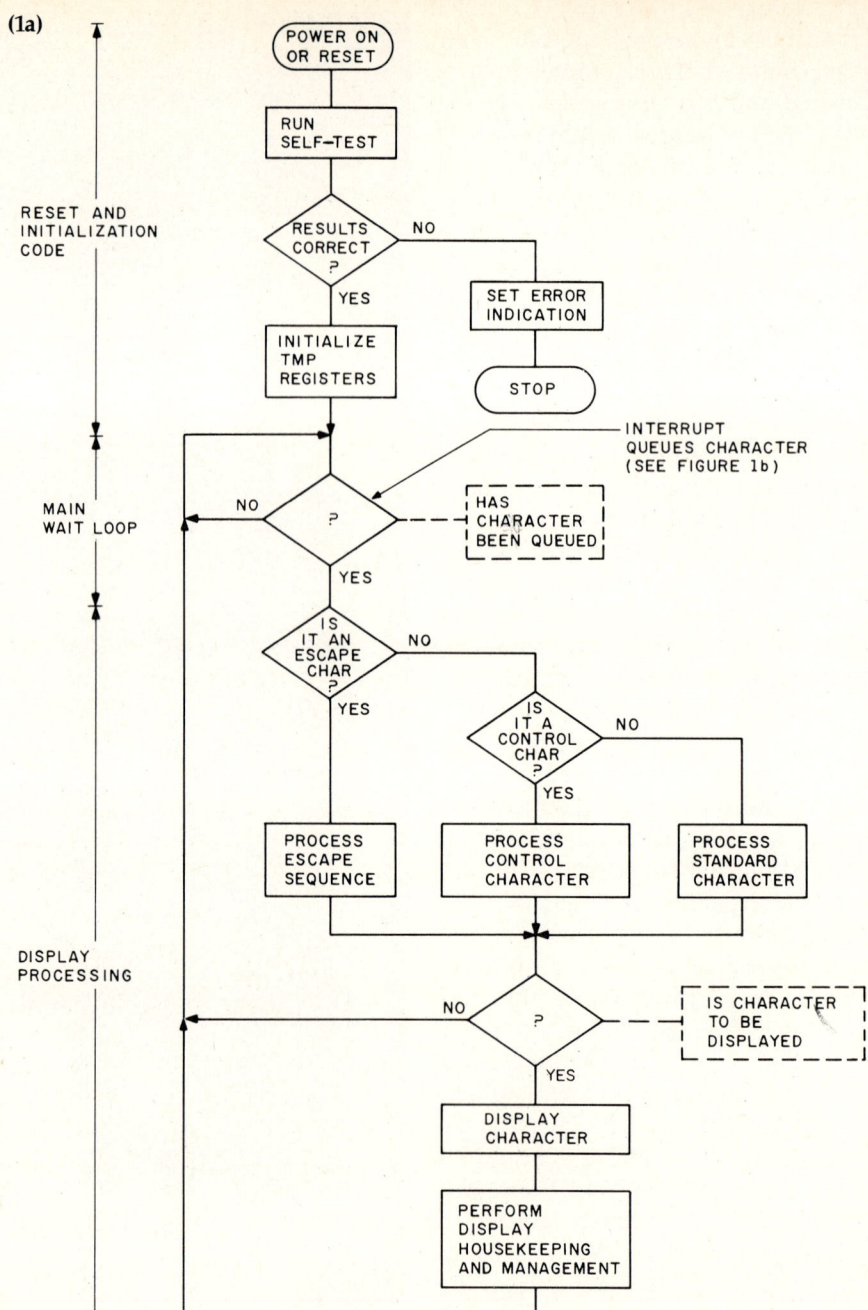

Figure 1: *A flowchart of the control program supplied standard by National Semiconductor for the NS455A. The main code is shown in 1a while the interrupt sections are in 1b.*

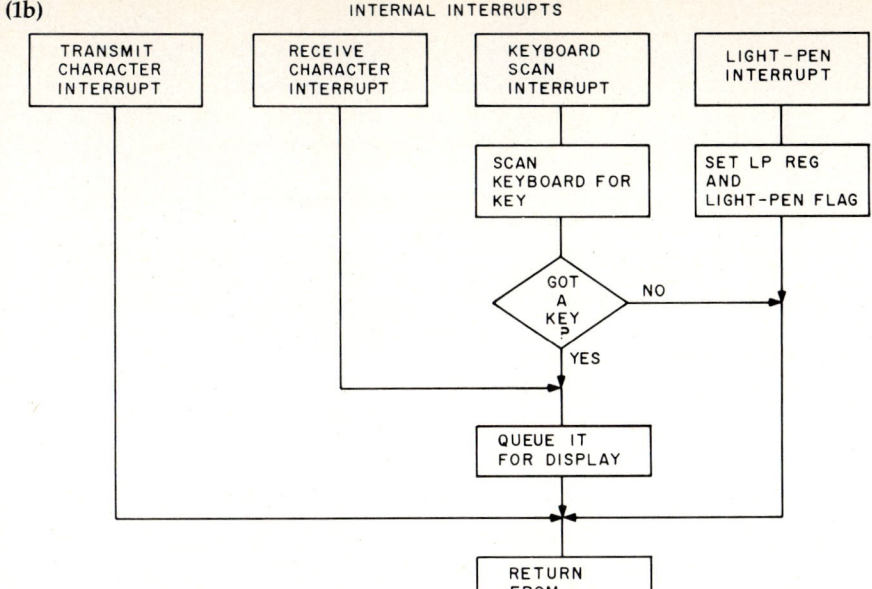

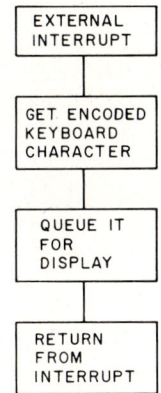

tine is somewhat hardware-dependent. A numeric keypad and separate input lines for Control, Shift, Repeat, and Caps-Lock keys may not be available or work the same way in all keyboards. The standard program is written to operate with an Oak full-travel membrane (FTM) keyboard. Other keyboards, including the one shown in photo 2, can be made to work with the standard NS455A, but the Oak keyboard can be used with a minimum of trouble.)

Configuring for Use

The biggest annoyance of today's advanced intelligent terminals is that they can be complicated to use. When you unpack a new unit you can expect to spend at least an hour trying to set the switches for all the proper optional modes and functions. At such times I fondly recall first plugging in my completed TV Typewriter in late 1973 (see references 2 and 4). There were no confusing configuration switches or Escape sequences. I just typed.

The Term-Mite ST, though small, was designed to be powerful, so it has to include some options that you set before use. I have tried to keep them as simple as is compatible with flexibility. These options are both hardware- and software-configurable.

In the Term-Mite ST, three eight-position DIP (dual-inline pin) switches let you set up the unit to assume certain operating conditions and parameters when you first turn it on. Once the terminal has been turned on, most of the switch-preset parameters can be changed by on-line commands received either from the keyboard or through the RS-232C port. In this same manner, some additional parameters that don't have switches can be set up. The commands are sequences of ASCII characters, either single nonprinting control codes or ordinary characters preceded by an Escape character (Escape sequences).

Control codes, as in most ASCII-encoded applications, are generated at the keyboard by simultaneously holding down the Control key and one other alphabetic or character key. The binary value emitted is within the special low range of ASCII codes designated for the control of devices. These codes do not normally cause any symbol to be printed or displayed, so they are referred to as *nonprinting*. A list of the control codes as used by the Term-Mite ST (running with the standard firmware) is shown in table 2 on page 98. An ASCII control code is often abbreviated by the corresponding printing character preceded by a caret or an up-arrow; thus "^G" stands for Control-G.

Escape sequences are more complex. These consist of characters that are mostly in the range of regular ASCII values, but the normally printable codes are transmitted following the special ASCII Escape character (decimal 27). This Escape character is so named because the characters that follow it "escape" from their normal meanings. (In the context of an Escape sequence, almost any meaning is possible for any character, although some Escape sequences are widely used, and one set has achieved the status of an ANSI standard equal to ASCII itself—see reference 1.) In the Term-Mite ST, an Escape sequence consists of at least two keystrokes: the Escape key followed by an uppercase letter (in the set A through Z, with some unused). The Escape sequences activate various functions of the Term-Mite ST. Only direct cursor addressing and the set-attribute-value function re-

Photo 2: *The Term-Mite ST can be connected to the stack-pole keyboard shown here, but the NS455A can be used more easily with an Oak full-travel membrane (FTM) keyboard from Oak Switch Systems Inc., POB 517, Crystal Lake, IL 60014. An enclosure from Pac Tec (Enterprise and Executive Aves., Philadelphia, PA 19153) enhances the terminal's appearance.*

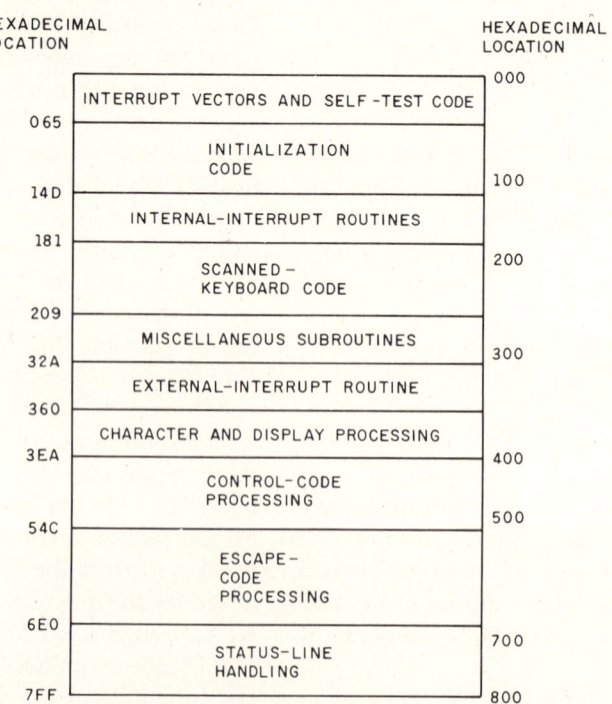

Figure 2: *Memory use by the NS455A control program.*

quire more than one character following Escape.

When power to the Term-Mite ST is first turned on, the three groups of *configuration switches* are read and their values stored in appropriate registers in the TMP. The switches appear to the processor as memory-mapped I/O devices; logic 1 is considered to be the on or closed position. (National Semiconductor's software refers to the groups as switches 4, 5, and 6, so I have used the same designations in the Term-Mite schematic.) Their configurations and various settings are shown in tables 4, 5, and 6 and in figures 4, 5, and 6.

Programming the Term-Mite

Your use of the Term-Mite ST can be simple or complex, depending on how you write your host software: to use the control commands and Escape sequences extensively or not

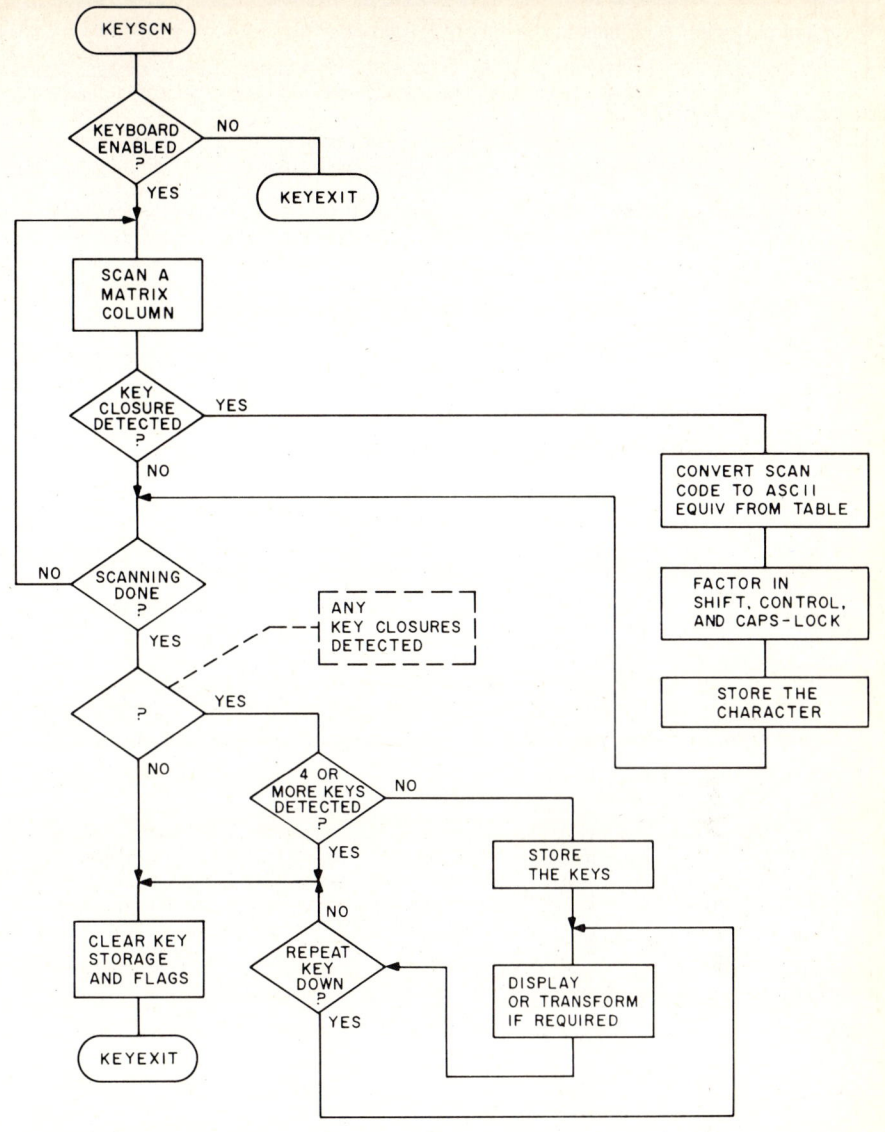

Figure 3: *A flowchart of the keyboard-scanning routine.*

at all. If you just want to write BASIC programs, you'll probably never have to do more than use the default switch settings. If, however, you want to use the terminal as part of a screen-template inquiry-transaction system, you'll want to use all the character attributes and graphics options.

Some forms of the Escape sequences cannot be generated through the keyboard (ones with leading-zero attribute values, for example); these must be supplied by the host computer. One of the easiest ways to do this is to use the CHR$ function in BASIC, giving the decimal values of the required characters. I used a Radio Shack TRS-80 Model 100 portable computer to generate the displays depicted in the accompanying photos. With these few simple examples, I'll try to give you a feel for the use of the control functions. The communication protocol is peculiar to the Model 100, but the basic approach and much of the code can be used on other machines.

The first example is shown in listing 1 on page 102, a demonstration of displaying blinking characters. Lines 20 and 25 in this BASIC program clear the screen with a Control-L, then position the cursor at the sixteenth line with a sequence of the type ESC, M, x, y. An ESC, I, 251 sequence sets the blinking attribute for the single word "BLINKING" (attribute bit 2 set

97

Control Code	Effect
Control-G	*Ring the bell.* The ^G code causes a strobe pulse to be sent out bit 5 of the I/O port to trigger a 100-ms one-shot multivibrator, which can be connected to a piezoelectric transducer.
Control-H	*Backspace.* The destructive backspace moves the cursor to the left; the new position is blanked. Wrap-around occurs from line to line, including a screen wrap from the home (upper left) to last position.
Control-I	*Horizontal tab—fixed every eight locations.* The tab function is handled by the Tab key or by the ^I code. Each line is divided up into fixed eight-character tab zones. Each Tab received causes the cursor to jump from its current position to the start of the next tab zone, proceeding to the right. Screen scrolling occurs at the bottom of the screen.
Control-J	*Linefeed.* ^J is the standard ASCII Linefeed character.
Control-K	*Vertical tab—fixed every eight lines.* ^K moves the cursor down the screen eight rows. If the cursor is at the bottom of the screen, the display scrolls by eight lines. Intervening lines are always blanked. The cursor column position always remains the same after ^K; no carriage return is performed.
Control-L	*Clear screen and home the cursor.*
Control-M	*Standard ASCII Return character.* ^M moves the cursor to the leftmost column.
Control-N	*Cursor up.* The cursor is moved up one line by the ^N code. The movement is nondestructive, and the cursor will wrap around the screen from top to bottom when the top line is reached.
Control-O	*Cursor down.* Similarly, the cursor moves down one line, nondestructively, when ^O is detected. Again, wrap-around from bottom to top occurs.

Table 2: *The functions of ASCII control codes in the Term-Mite.*

Escape Sequence	Effect
ESC, A	*Auxiliary (printer) port on.* An ESC, A sequence will turn on the auxiliary printer port, if it exists (the port hardware is not implemented in Term-Mite). Everything displayed on the screen will go out the auxiliary port as well. Note that Return and Linefeed characters are not sent to the display routine unless the terminal is in Control mode. (All control characters are then put on the display graphically.) The "AUX ON" message is displayed on the status line when this Control mode is active.
ESC, B	*Display switch-register status in status line.* An ESC, B causes the UART configuration switch and data-rate code to be displayed on the status line.
ESC, C	*Control mode on.* An ESC, C sequence causes the unit to enter the Control mode. This mode of operation permits you to see all the normally nondisplayable ASCII characters (e.g., Return, Linefeed) on the display screen. The control characters are displayed as reverse-video, half-intensity uppercase letters. "A" through "Z" plus some punctuation are used. In addition, the message "CTL MODE" is displayed on the status line.
ESC, D	*Toggle on-line/local mode.* You can set the terminal to on-line or to local mode from the keyboard using this Escape sequence. It is a toggle function; i.e., each use causes the terminal to change from the current state to the other one. The status line also displays the current state.
ESC, E	*Toggle full-/half-duplex mode.* This is also a toggle function (see ESC, D) except that you can go from full-duplex (FDX) to half-duplex (HDX) communication and back. The terminal's current state is displayed on the status line.
ESC, F	*Control mode off.* ESC, F turns the Control mode off. The ASCII control characters now resume their normal operation and function. The message disappears from the status line.
ESC, G	*Set graphics mode on.* The ESC, G sequence flips the status line to normal video and enables the graphics attribute for characters specified in the AL1 register. This is the mode to use when doing forms drawing with the supplied terminal software. This state remains in effect until turned off by another Escape sequence (ESC, H).
ESC, H	*Set graphics mode off.* To turn off the graphics or line-drawing mode, the ESC, H sequence is used.
ESC, I, v	*Set Attribute Value.* You can set attributes to any combination by using the ESC, I, v sequence. If internal attributes are being used, the contents of AL1 are replaced with the 8-bit binary value v. When using external attributes, the value v is loaded into the current-attribute-value register. All subsequent characters will have this value loaded into their external attribute memory unless it is changed by another ESC, I, v. Bits are added together to obtain v for each combination of attributes as shown below. | Attribute Bit	Attribute
bit 7	graphics
bit 6	blanked
bit 5	underlined
bit 4	double width
bit 3	double height
bit 2	blinking
bit 1	half intensity
bit 0	reverse video	

Table 3: *The effects of Escape sequences in the Term-Mite ST.*

Control Code	Effect
Control-P	*Cursor left.* ^P moves the cursor left one column, nondestructively. Screen wrap-around from right to left will cause the cursor to move up a line each time the wrap-around occurs.
Control-Q	*Cursor right.* The fourth cursor-movement code is ^Q. The cursor moves right by one column, nondestructively, and screen wrap-around happens as above. Wrap-around moves down a line until the end of the screen is reached, and then the cursor moves back to the top of the screen. No scrolling is done.
Control-R	*Cursor home.* The ^R code moves the cursor to the home (uppermost left) position. Nothing else happens.
Control-S	*Send a "break" signal on serial line.* ^S generates a "break" signal (300 ms of "spacing" condition) on the RS-232C line.
Control-T	*Change and display data rate.* The data rate for the RS-232C lines can be changed via the ^T code. Each time ^T is entered, the data rate is displayed (in 00 to 15 code form) after being bumped to the next higher rate. Both receive and transmit rates are affected. They are as follows: 00 = 110 bps 08 = 3600 bps 01 = 134.5 bps 09 = 4800 bps 02 = 150 bps 10 = 7200 bps 03 = 300 bps 11 = 9600 bps 04 = 600 bps 12 = 19,200 bps 05 = 1200 bps 13 = 19,200 bps 06 = 1800 bps 14 = 19,200 bps 07 = 2400 bps 15 = 4800 bps

Escape Sequence	Effect
	A logic 0 in a bit position enables the corresponding attribute.
ESC, K	*Keyboard enable (X-on).* The keyboard can be selectively enabled or disabled. ESC, K performs the X-on or enable function.
ESC, L	*Return light-pen value.* An ESC, L sequence causes the currently latched values in the horizontal light-pen register (HPEN) and the vertical light-pen register (VPEN) to be transmitted back to the host system via the main RS-232C port. HPEN is sent (one binary character) followed by VPEN (also one binary character). Term-Mite does not support the light-pen hardware.
ESC, M, x, y	*Load cursor position (x,y).* The cursor position is dynamically alterable by means of this Escape sequence. The two parameters following the basic Escape code are used to set the *x* and *y* (respectively) positions of the cursor. The *x* value is the column position. Up to 79 columns are allowed. The *y* value specifies the row or line number, 0 through 23 being valid. The origin point is the home position on the screen (upper-left corner) and all the values are calculated as offsets from that point. The actual parameter values begin with the displayable ASCII character set, that is, blank through lowercase "o." To specify cursor position (5,6), for example, the parameters would be the two characters "%&" (hexadecimal 25 and 26). The 80 *x* values would run from blank (hexadecimal 20) through "o" (6F) and 24 *y* values from blank (hexadecimal 20) through "7" (37).
ESC, O	*Keyboard disable (X-off).* The ESC, O sequence disables the keyboard from further operation. This is essentially an X-off function. It must be reenabled by an ESC, K sent from the host system or entered from the keyboard itself.
ESC, P	*Print screen contents.* You can dump the entire displayed contents of the screen to the auxiliary or printer port by typing the ESC, P key sequence.
ESC, Q	*Run self-test diagnostic and reset.* ESC, Q causes the system to rerun the self-test and initialization routines. All current machine status conditions will be replaced by the power-on defaults. The current screen contents are lost.
ESC, R	*Block send the current row.* This is one of the Block Send commands. An ESC, R sequence causes the current line (from left margin to the cursor position) to be transmitted character by character to the host system. If the cursor is on the left margin (i.e., there's nothing on the line yet), nothing is transmitted.
ESC, S	*Block send the current screen.* The other Block Send command is ESC, S. This command transmits the data on the screen, character by character, from the home position to the current cursor location over the RS-232C link to the host system.
ESC, T	*Erase to end of line.* An ESC, T erases the current line from the cursor position to the right margin. It includes the cursor position in the erase operation.
ESC, W	*Wipe switch-register status from status line.* The switch-register and data-rate information can be wiped from the status line by entering ESC, W.
ESC, X	*Auxiliary port (printer) off.* This reverses the effect of ESC, A.
ESC, Y	*Erase to end of page.* You can erase the entire screen from the present cursor location to the end of the screen by typing ESC, Y. The present cursor location is included in the erase function.

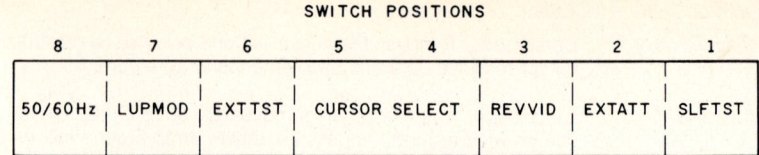

Figure 4: *The assignments of the switch positions in configuration switch 4. (Switches 1, 2, and 3 do not exist.)*

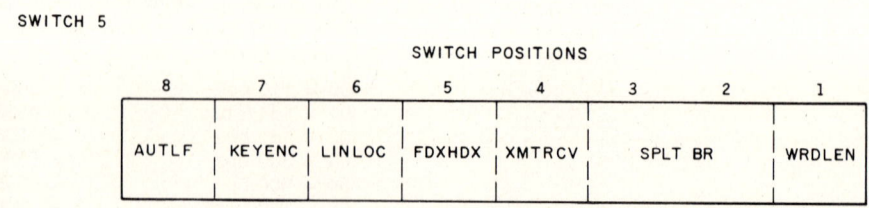

Figure 5: *The assignments of the switch positions in configuration switch 5.*

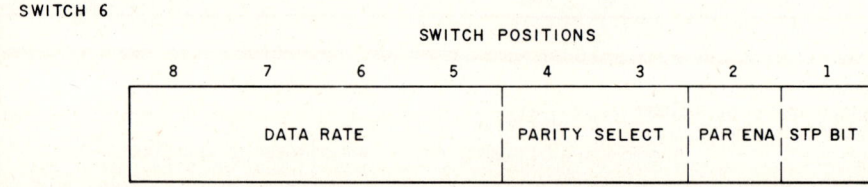

Figure 6: *The assignments of the switch positions in configuration switch 6.*

to 0). An ESC, I, 255 restores the normal character display for the final word.

Listing 2 functions in a similar manner to show off the reverse-video attribute. In this instance, the reverse-video attribute is activated by an ESC, I, 254 sequence (attribute bit 0 set to 0). The result is shown in photo 3a on page 105.

Listing 3 on page 103 combines four attributes. It starts by clearing the screen, positioning the cursor at column 21, row 6 (21,6), and printing "WE CAN PRINT THE REGULAR WAY"; next, the cursor is repositioned to (0,10) and the sequence ESC, I, 230 is sent. The value of 230 (bits 0, 3, and 4 set to 0) sets the double-width, double-height, and reverse-video attributes. Photo 3b on page 105 shows the appearance of the screen.

A few peculiarities do apply to the use of the double-width and double-height characters. The ASCII value of any character to be displayed in either or both of these attributes must be written in *all* the character positions that the expanded character will occupy. For example, if you wish to print "TEST" in double width, the BASIC PRINT statement should actually send the string "TTEESSTT", since the wide characters will occupy two regular character positions. In double height, the duplicate characters should be written one above the other. The other catch to using double height is that the second row of characters must also have the *blanked* attribute set. In the example of listing 3, with the attributes set for double height, double width, reverse video, and (on the second display line) blanked, the string "OR THIS WAY FOR SPECIAL EMPHASIS" is printed in large letters. After the pyrotechnics, another Control-I is used to reset the attributes.

One final demonstration program, shown in listing 4, shows many of the characteristics already mentioned with the addition of line graphics.

Switched Signal	Effect
50/60 Hz	Because the NS455A is designed for worldwide distribution, provision has been made for handling operation at either 50 Hz or 60 Hz. A logic 1 specifies 60 Hz, while a logic 0 sets 50-Hz operation.
LUPMOD	This bit affects the loop-back mode of the serial I/O line. When it is 0, the serial output line is logically connected to the serial input line inside the chip. A logic 1 sets the serial line to normal mode.
EXTTST	The external test flag is used to force a jump to external memory located at hexadecimal address 800. This test is only made if SLFTST is also selected. The author of the self-test code must take care of either returning to the supplied standard program or handling all processing from that point on.
CURSOR SELECT	There are four types of cursors that can be selected on the Term-Mite: a solid underline, a blinking underline, a solid block, or a blinking block. The binary codes are as follows: 00 = solid underline cursor 01 = blinking underline cursor 10 = solid block cursor 11 = blinking block cursor
REVVID	A logic 0 will cause the display to switch to black characters on a white field. The default value of logic 1 specifies normal video, white on black.
EXTATT	A 0 in this position lets you use the external attribute memory, bits 8 through 15. A logic 1 sets the system up for internal attributes.
SLFTST	A logic 1 in this bit causes the self-test routine to be skipped on reset or power up. It must be set to logic 0 in order for the self-test to be performed. This does not affect the ESC, Q command to execute a self-test.

Table 4: *The functions of the switch positions in configuration switch 4.*

Switched Signal	Effect
AUTLF	This is the Auto-Linefeed flag. When set to a logic 0, a Linefeed character is sent whenever a Return character is detected. This operates for both the transmitted data and the displayed information. A logic 1 causes no Linefeed to be sent.
KEYENC	The encoded keyboard is selected by a logic 0. A logic 1 indicates that the scanned keyboard is to be handled.
LINLOC	This is the Online/Local default. When this flag is set to logic 0, Local mode is chosen, and no data is sent to the host machine. A logic 1 puts the terminal into Online mode and data is then transmitted.
FDXHDX	The full-duplex/half-duplex flag is used to set the communication protocol. When set to logic 0, the communication through the serial port is half duplex; data is both transmitted to the host and sent to the CRT display. A logic 1 will select full duplex. Data is then transmitted but not automatically sent to the display.
XMTRCV	The split-data-rate function uses this flag to decide whether the transmitter (logic 0) or the receiver (1) is to operate at the slower data rate.
SPLT BR	This field sets the split-data-rate divisor to one of four values. The divisor divides the affected rate by a factor of 1, 4, 16, or 32. More divisors are allowed, but they are not implemented in the standard program. 00 = divide by 32 10 = divide by 4 01 = divide by 16 11 = divide by 1 (default)
WRDLEN	The word-length flag specifies the number of bits in the serial character, either 7 or 8 bits excluding parity. A default of 1 selects 7 bits, and a logic 0 selects 8 bits.

Table 5: *The functions of the switch positions in configuration switch 5.*

When attribute bit 7 is set to a 0, the Term-Mite displays certain control characters as graphic symbols. These are shown in table 8 on page 105. The program draws the display shown in photo 4.

In Conclusion

Since building the Term-Mite ST, I've been finding all kinds of old and new uses for it. For instance, it makes a perfect status and command display for a Micromint Z8-BASIC Computer/Controller system (see reference 3). Also, I've been hesitating to spend $1000 for a regular video terminal that would be dedicated to a constant display of the time of day and my appointment schedule, but

Communication Parameter	Explanation
DATA RATE	Four bits select the data rate used by the serial-I/O section. The available rates are shown below:

position 8765		position 8765	
0000	110 bps	1000	3600 bps
0001	134.5 bps	1001	4800 bps
0010	150 bps	1010	7200 bps
0011	300 bps	1011	9600 bps
0100	600 bps	1100	19,200 bps
0101	1200 bps	1101	19,200 bps
0110	1800 bps	1110	19,200 bps
0111	2400 bps	1111	4800 bps

PARITY SELECT	There are four parity options:

position 43
11 = Parity is forced to a space level if enabled.
10 = Parity is forced to a mark level if enabled.
01 = Parity is even if enabled.
00 = Parity is off if enabled.

PARENE	The parity-enable flag is used to enable or disable the parity function. If the switch is a 0, parity is disabled. A value of 1 enables parity.
STPBIT	This flag specifies the number of stop bits to be sent with each character. A logic 1 will cause one stop bit to be sent; a 0 will send two stop bits.

Table 6: *The functions of the switch positions in configuration switch 6.*

Switch 4	Switch 5	Switch 6
11111100	10100111	00111010

Table 7: *For general use, typical settings of all three switches might be as shown here (most significant bits to the left).*

Listing 1: *A BASIC program to produce blinking characters on the Term-Mite.*

```
10 OPEN "COM:38N1D" FOR OUTPUT AS 1
20 PRINT #1,CHR$(12):REM CLEAR SCREEN
25 PRINT #1,CHR$(27);"M0+"
30 PRINT #1,"THIS IS HOW WE PRINT IN ";
40 PRINT #1,CHR$(27);"I";CHR$(251);"BLINKING ";
50 PRINT #1,CHR$(27);"I";CHR$(255);"CHARACTERS"
100 CLOSE #1
110 STOP
```

Listing 2: *A program to produce reverse-video characters on the Term-Mite.*

```
10 OPEN "COM:38N1D" FOR OUTPUT AS 1
20 PRINT #1,CHR$(12):REM CLEAR SCREEN
25 PRINT #1,CHR$(27);"M0+"
30 PRINT #1,"THIS IS HOW WE PRINT IN ";
40 PRINT #1,CHR$(27);"I";CHR$(254);"REVERSE VIDEO ";
50 PRINT #1,CHR$(27);"I";CHR$(255);"CHARACTERS"
100 CLOSE #1
110 STOP
```

Listing 3: *A BASIC program to demonstrate reverse-video, double-height, double-width characters on the Term-Mite.*

```
10  OPEN "COM:38N1D" FOR OUTPUT AS 1
20  PRINT #1,CHR$(12):REM CLEAR SCREEN
25  PRINT #1,CHR$(27);"M5&";
30  PRINT #1,"WE CAN PRINT THE REGULAR WAY"
40  PRINT #1,CHR$(27);"M *";CHR$(27);"I";CHR$(230);
50  PRINT #1,"OORR  TTHHIISS  WWAAYY  FFOORR  SSPPEECCIIAALL    EEMMPPHHAASSIISS"
60  PRINT #1,CHR$(27);"M +";CHR$(27);"I";CHR$(166);
70  PRINT #1,"OORR  TTHHIISS  WWAAYY  FFOORR  SSPPEECCIIAALL    EEMMPPHHAASSIISS"
80  PRINT #1,CHR$(27);"I";CHR$(255)
90  CLOSE #1
100 STOP
```

Listing 4: *A program to demonstrate all internal character attributes and the drawing of borders with line-graphics characters.*

```
10  OPEN "COM:38N1D" FOR OUTPUT AS 1
20  PRINT #1,CHR$(12)
30  PRINT #1,CHR$(27);"I";CHR$(127)
35  PRINT #1,CHR$(27);"M%#";
50  PRINT #1,STRING$(70,23)
60  PRINT #1,CHR$(27);"M$*";
70  PRINT #1,STRING$(71,23)
80  PRINT #1,CHR$(27);"M$-";
90  PRINT #1,STRING$(71,23)
100 PRINT #1,CHR$(27);"M$5";
110 PRINT #1,STRING$(71,23)
120 FOR I% = 3 TO 21
130 PRINT #1,CHR$(27);"M$";CHR$(32+I%);
140 PRINT #1,"J"
150 PRINT #1,CHR$(27);"Mk";CHR$(32+I%);
160 PRINT #1,"J"
170 NEXT I%
180 FOR I% = 10 TO 13
190 PRINT #1,CHR$(27);"M5";CHR$(32+I%);
200 PRINT #1,"J"
210 PRINT #1,CHR$(27);"MG";CHR$(32+I%);
220 PRINT #1,"J"
230 PRINT #1,CHR$(27);"MY";CHR$(32+I%);
240 PRINT #1,"J"
250 NEXT I%
251 PRINT #1,CHR$(27);"M$#X":CHR$(27);"Mk#L";
252 PRINT #1,CHR$(27);"M$5";CHR$(21);CHR$(27);"Mk5";CHR$(22);
253 PRINT #1,CHR$(27);"M$*";CHR$(28)
```

Listing 4 continued

Listing 4 continued:

```
254 PRINT #1,CHR$(27);"Mk*";CHR$(29)
255 PRINT #1,CHR$(27);"M$-";CHR$(28)
256 PRINT #1,CHR$(27);"Mk-";CHR$(29)
257 PRINT #1,CHR$(27);"M5*";CHR$(31);CHR$(27);"M5-";CHR$(30)
258 PRINT #1,CHR$(27);"MG*";CHR$(31);CHR$(27);"MG-";CHR$(30)
259 PRINT #1,CHR$(27);"MY*";CHR$(31);CHR$(27);"MY-";CHR$(30)
260 PRINT #1,CHR$(27);"I";CHR$(255)
270 PRINT #1,CHR$(27);"MF%";
280 PRINT #1,"THE"
285 PRINT #1,CHR$(27);"I";CHR$(231);
290 PRINT #1,CHR$(27);"M='";
295 PRINT #1,"TTEERRMM--MMIITTEE    SSTT"
300 PRINT #1,CHR$(27);"I";CHR$(167);
305 PRINT #1,CHR$(27);"M=(";
310 PRINT #1,"TTEERRMM--MMIITTEE    SSTT"
315 PRINT #1,CHR$(27);"I";CHR$(251);
320 PRINT #1,CHR$(27);"M*+";
325 PRINT #1,"BLINK"
330 PRINT #1,CHR$(27);"M(,";
335 PRINT #1,"ATTRIBUTE"
340 PRINT #1,CHR$(27);"I";CHR$(223);
345 PRINT #1,CHR$(27);"M:+";
350 PRINT #1,"UNDERLINE"
355 PRINT #1,CHR$(27);"M:,";
360 PRINT #1,"ATTRIBUTE"
365 PRINT #1,CHR$(27);"I";CHR$(253);
370 PRINT #1,CHR$(27);"MJ+";
375 PRINT #1,"HALF INTENSITY"
380 PRINT #1,CHR$(27);"ML,";
385 PRINT #1,"ATTRIBUTE"
390 PRINT #1,CHR$(27);"I";CHR$(254);
395 PRINT #1,CHR$(27);"M";CHR$(92);"+";
400 PRINT #1,"REVERSE VIDEO"
402 PRINT #1,CHR$(27);"M";CHR$(92);",";
404 PRINT #1,"  ATTRIBUTE  "
406 PRINT #1,CHR$(27);"I";CHR$(255);
408 PRINT #1,CHR$(27);"M)/";
415 PRINT #1,CHR$(27);"M00";
425 PRINT #1,CHR$(27);"I";CHR$(239);
430 PRINT #1,CHR$(27);"M'3";
435 PRINT #1,"DDOOUUBBLLEE    WWIIDDEE,,";
440 PRINT #1,CHR$(27);"I";CHR$(247);
445 PRINT #1,CHR$(27);"M@2";
450 PRINT #1,"DOUBLE HEIGHT,"
455 PRINT #1,CHR$(27);"I";CHR$(183);
460 PRINT #1,CHR$(27);"M@3";
465 PRINT #1,"DOUBLE HEIGHT,"
470 PRINT #1,CHR$(27);"I";CHR$(255);
475 PRINT #1,CHR$(27);"MO3";
480 PRINT #1,"and";
600 PRINT #1,CHR$(27);"I";CHR$(231);
605 PRINT #1,CHR$(27);"MS2";
610 PRINT #1,"DDOOUUBBLLEE    SSIIZZEE"
615 PRINT #1,CHR$(27);"I";CHR$(167);
620 PRINT #1,CHR$(27);"MS3";
625 PRINT #1,"DDOOUUBBLLEE    SSIIZZEE"
630 PRINT #1,CHR$(27);"I";CHR$(231);
650 PRINT #1,CHR$(27);"I";CHR$(191);
655 PRINT #1,CHR$(27);"MB";CHR$(34);"BLANK FAILURE";
660 PRINT #1,CHR$(27);"MG7";CHR$(27);"I";CHR$(255);
665 CLOSE 1
670 STOP
```

(3a)

(3b)

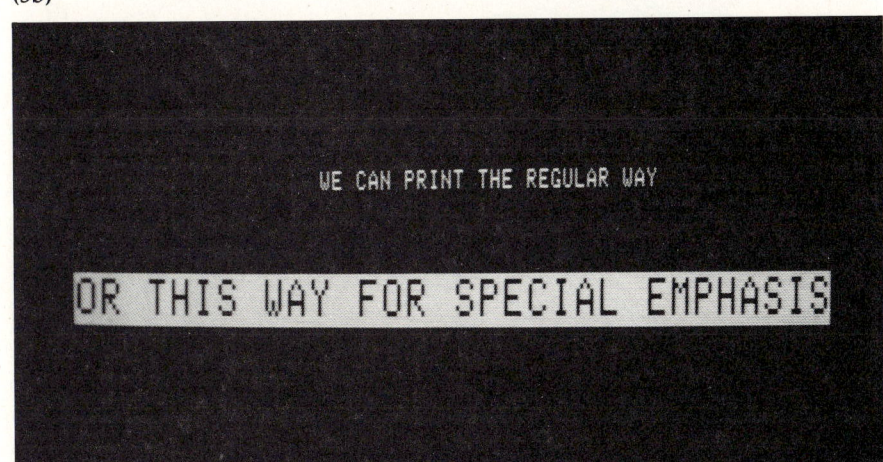

Photo 3: *The Term-Mite can display several character attributes singly or in combination, such as reverse video (3a) and reverse, double width, double height (3b).*

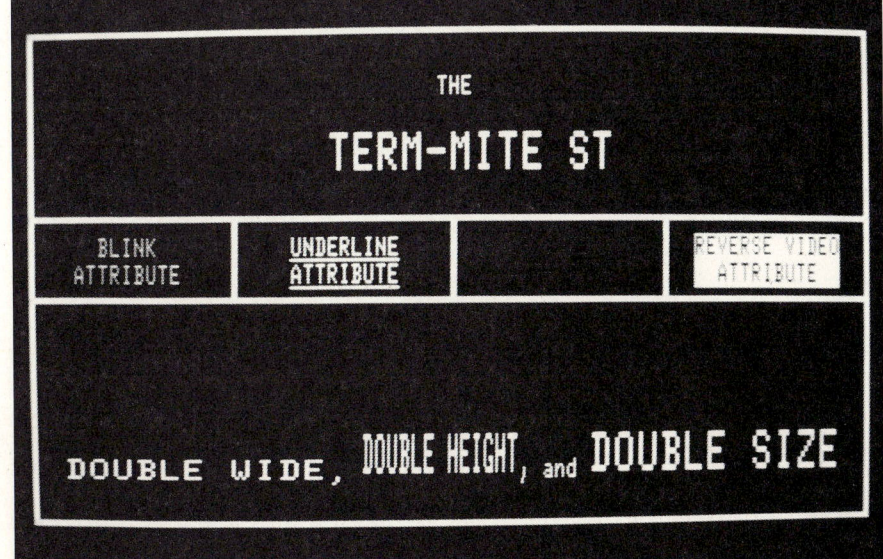

Photo 4: *The Term-Mite can display certain control characters such as line-graphics symbols for screen templates and other basic uses. This display was produced by the program of listing 4.*

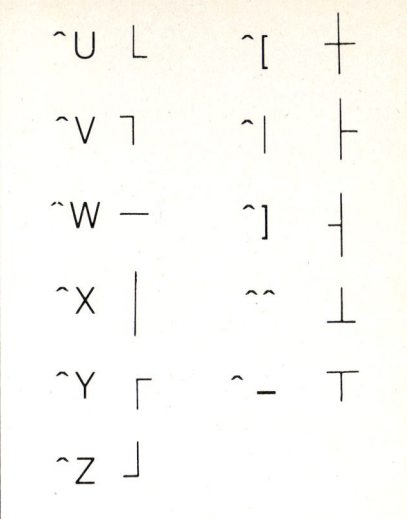

Table 8: *In the line-graphics mode, certain control characters produce visible displays of corners, lines, and crosses useful for forming borders on the screen display (^^ is Control-Caret).*

I can see using the Term-Mite's graphics and attributes for this and other applications. (I already have eight cathode-ray tubes staring at me in the Circuit Cellar. What's one more?)

Perhaps by building the Term-Mite you can put to good use that bargain keyboard and old monitor you've had sitting around for years. Be warned, however, that the unit's 25-line by 80-column display requires too great a bandwidth for satisfactory connection to a regular TV set. A 12-MHz monochrome monitor is the better choice, as shown in photo 5.

Since the Term-Mite's control software is stored in what is actually an EPROM (an erasable programmable ROM), it would be possible to add commands or modify the operation of its functions. National Semiconductor will eventually have complete documentation available for the NS455A TMP, including a listing of the standard supplied control program and the processor's instruction set. Within the 6K yet-unused bytes of program address space, some pretty fancy terminal software could be written, perhaps even to emulate the command protocols of various commercially sold terminals. The only

Photo 5: *The Term-Mite ST circuitry described in these articles is used with a separate keyboard and video monitor to form a complete functional terminal. Here a Jameco JE610 keyboard works with the Term-Mite, a NEC (Nippon Electric Company) green-phosphor monitor, and the Circuit Cellar MPX-16 computer.*

modification would be a simple EPROM change.

Special thanks to Bob Harbrecht of National Semiconductor Corporation for his help on this project.

References

1. *ANSI X3.64-1979: Additional Controls for Use with the American National Standard Code for Information Interchange.* New York: American National Standards Institute, 1979.
2. Ciarcia, Steve. "Build the Circuit Cellar Term-Mite ST Smart Terminal, Part 1: Hardware Description." BYTE, January 1984, page 37.
3. Ciarcia, Steve. "Build a Z8-Based Control Computer with BASIC." Part 1, BYTE, July 1981, page 38. Part 2, BYTE, August 1981, page 50.
4. Lancaster, Don. "TV Typewriter." *Radio Electronics*, September 1973, page 43.

The following items are available from:

The Micromint, Inc.
25 Terrace Drive
Vernon, CT 06266
For orders: (800) 635-3355
For information: (203) 871-6170

1. *Complete Term-Mite ST video-display terminal kit including NS455A, printed-circuit board, IC sockets, DB-25S connector, and all other components but without keyboard or CRT monitor. Board size is approximately 4½ inches by 6½ inches with a 0.156-inch 44-pin edge connector.*
   Price $239

2. *Assembled and tested Term-Mite ST video-display terminal, without keyboard or CRT monitor.*
   Price $279

Please add $5 for shipping in the continental United States, $25 elsewhere. New York residents please include 7 percent sales tax. Allow 4 to 6 weeks for delivery.

Build a Third-Generation Phonetic Speech Synthesizer

The idea for this month's project may seem familiar to many of you, and with good reason: I've done projects on building speech synthesizers five times before. Why? The integrated-circuit technology for doing it keeps getting better and better, and I like nothing more than experimenting with new chips.

The Shadow of the Past

The most successful of the past speech-synthesis projects were the Sweet Talker, presented in September 1981 (see reference 3), and the Microvox, presented in August and September 1982 (see reference 4). Many of you have built and used these as peripheral devices on your personal computers.

The original Sweet Talker was a simple, parallel-port-driven phoneme-based synthesizer. To make it talk, you just sent 6-bit phoneme codes to it. It's still useful; I've used it for broadcasting weather reports, among other things (see reference 1). The Microvox was an enhancement of the same basic design. Both these devices were built around the Votrax SC-01A chip, a member of the second generation of commercial speech-synthesis products. (The first generation consisted of hybrid modules containing many discrete components.)

Latest Technology

I suppose I would continue to tinker with improving the sound quality from this foundation if I had not been beaten to the punch by the appearance of an enhanced speech-synthesis chip. This month we'll look at the latest development in phonetic speech synthesis: the SSI263 integrated circuit from Silicon Systems Inc.

Speech-synthesis chips of the new third generation, such as the SSI263, produce much more intelligible

The latest development in phonetic speech synthesis is the Silicon Systems SSI263 chip

speech than did older devices, such as the SC-01A. The new chips achieve this through more flexible intonation, inflection, and filtration. With the SSI263, it is possible to vary these three effects on the fly as well as load new speech phonemes dynamically. When synthesizing in this way from sufficiently detailed data (at about 400 bits per second), the SSI263 generates the most human-sounding synthesized voice I've heard to date. (Systems that reproduce a digitized human voice can still sound better, however.) In its minimal operating mode, which requires data at about 50 to 70 bits per second, the quality of sound is comparable to that produced by the Votrax SC-01A.

Sweet Talker II

Seeing that the SSI263 could be easily interfaced to many different microprocessor-based systems, I decided to use an Apple II Plus as its host because the Apple's allocation of a separate address space for each expansion slot would eliminate address decoding on the speech card. The board, which contains only two chips, can be adapted for use with many other computers if you don't mind a little extra wiring. Out of sentiment for my earlier project, I decided to call the new package the "Sweet Talker II." It's shown in photo 1.

Programming the Sweet Talker II can be simple, if you'll settle for a monotonic, uninflected voice, or complicated, if you need the highest quality speech. Unlike its predecessor, the SC-01A, which used only a 6-bit phoneme input, the SSI263 contains five registers (totalling 40 bits) that influence the emitted sound. With the constant updating of all the registers (at a higher data rate and controlled by a more complex program), the SSI263 can even sing with vibrato.

Because the SSI263 is just out on the market, very little has been published on it. After a review of the basic techniques of computer speech, we'll go through some specific information about the device. Following that, we'll look at the simplicity of interfacing the SSI263 to a personal computer and discuss the software needed to support it.

Review of Computer Speech

There are three major techniques presently employed to allow computers to speak: formant synthesis, linear-predictive coding (LPC), and waveform digitization. The most noticeable difference between these three methods is the amount of data required to construct a word.

In *formant synthesis*, an electronic model of the human vocal tract is constructed. Driven with signals from frequency and noise generators, the model mimics the natural resonances of the vocal tract. The output spectrum contains bands of resonant frequencies called formants.

The most common variant of the formant technique is called *phoneme synthesis*, in which the spectral parameters are derived from basic word sounds—the phonemes. In such a circuit, each phoneme is assigned a digital code; the synthesizer circuit utters phoneme sounds corresponding to the codes it receives. Therefore, speech is produced simply by stringing the phoneme codes together.

The original electronic voices of this type were intelligible but had a slightly mechanical quality. The latest phoneme synthesizers, on the other hand, combine control of pitch, rate, amplitude, and filtration to achieve quite lifelike speech. Continuous speech using phoneme synthesis can generally be obtained with data at a rate of less than 100 bps (bits per second), using no extra control attributes. Even with all the embellishments, it never requires a data rate of more than 400 bps.

Linear-predictive coding (LPC) is similar to formant synthesis in that both techniques are based on the frequencies found in speech and use similar hardware to model the vocal tract. Instead of encoding phonemes, however, LPC uses stored filter coefficients, amplifier-gain settings, and excitation frequencies; the name of the method is derived from the programmed activities of the multistage lattice filters that produce the desired formants. Continuous speech can generally be achieved with data rates of 1200 to 2400 bps. LPC has been used in products from Texas Instruments (the Speak & Spell and the now-discontinued TI 99/4 Text-to-Speech Translator) and General Instrument (the VSM2032 Voice Synthesis Module).

The third technique of computerized speech is *waveform digitization*, which reproduces a voice waveform from its stored amplitude characteristics. The simplest form is uncompressed digital data recording by pulse-code modulation (PCM). A more complex method involving data compression is called adaptive differential pulse-code modulation (ADPCM).

In digital recording by PCM, the analog waveform from a real human voice is sampled at a frequency twice that of the highest frequency to be preserved from the voice; the samples are sent through an analog-to-digital (A/D) converter and stored. The digital signal is played back through a digital-to-analog (D/A) converter and a low-pass filter. Since it's essentially a recorded voice, the reproduced speech retains the original inflections and accents. Unfortunately, waveform digitization requires very high data rates, so the vocabulary is usually limited by the amount of data that can be stored.

For more detail on speech digitization, you can refer to some of my previous articles in BYTE. In June 1978 I published a simple project for digitizing and reproducing speech from uncompressed data; you might call it "brute-force digitization" (see reference 5). The second project,

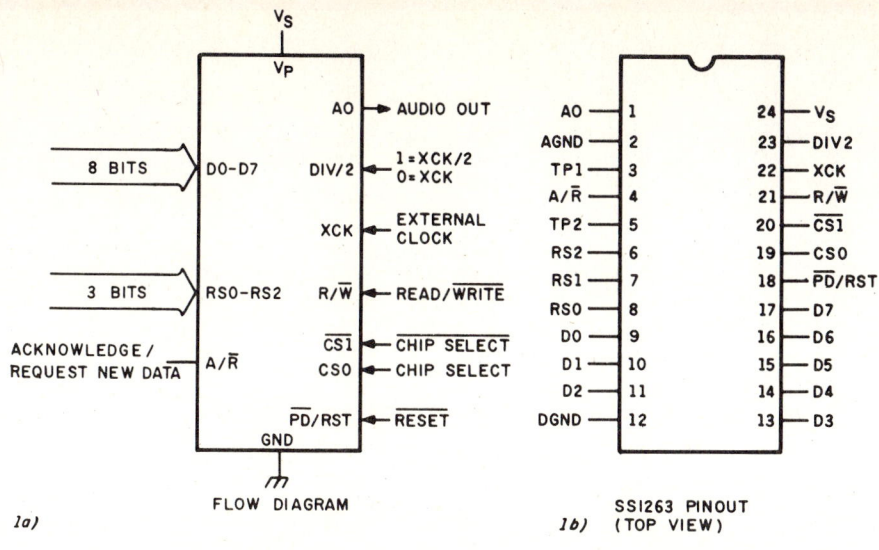

Figure 1: *Pinouts of the Silicon Systems SSI263 speech-synthesizer integrated circuit.*

1. appropriate control inputs (R/$\overline{\text{W}}$, CS0, and CS1) for address mapping with several buses
2. provision for resetting at power-down
3. five 8-bit internal registers
4. 256 phoneme equivalents (64 different phonemes each with four different duration settings)
5. four modes of handshaking
6. 4096 pitch variations (32 absolute levels with eight different speeds of inflection)
7. 16 speed settings
8. 16 amplitude levels
9. eight rates of articulation
10. 255 settings of the vocal-tract filter frequency response

Table 1: *Operational characteristics of the SSI263.*

published in June 1981, was for reproducing digitized speech from compressed data (reference 2), and the third article, appearing in June 1983, was a project that used ADPCM (see reference 6.)

SSI263 Integrated Circuit

The Silicon Systems SSI263 is a self-contained phoneme-based speech synthesizer. It consists of a single 24-pin CMOS (complementary metal-oxide semiconductor) integrated circuit that runs from a 5-V (volt) power source. It provides an analog output for music, sound effects, and continuous speech of unlimited vocabulary at low data rates. The SSI263 is easy to interface with any microprocessor; its principal characteristics are listed in table 1. Figure 1 contains a pinout diagram; a description of each pin's function is shown in table 2. The SSI263 can use a 3.59-MHz color-burst crystal divided by 2 or 4 as a timebase, or it can run off an external 1- or 2-MHz clock signal.

The SSI263 contains five 8-bit internal registers, which allow 256 phoneme-equivalents; 4096 pitch variations; and control of amplitude, rate of articulation, and vocal-tract filter frequency response (useful for sound effects). The individual registers are described in the text box on page 117.

A Simple Circuit

Connecting the SSI263 to any microcomputer system is not hard, but with the Apple II family of computers (II, II Plus, and IIe) the connection is simplicity itself (due largely to the address decoding provided on the Apple's motherboard). As you can see in figure 2 on page 111, the entire Sweet Talker II circuitry consists of only two integrated circuits and a few passive components.

In figure 2, IC1 is the SSI263, which can be directly connected to the Apple's data and address buses, operating on interrupts generated on the IRQ (normal interrupt request) line. The chip is selected when the A6 address line and the R/$\overline{\text{W}}$ and I/O-select status lines are active. Address lines A0, A1, and A2 select the proper register for data to be entered through the eight data-bus lines D0 through D7. The three low-order address lines are individually asserted to trigger one of three mutually exclusive register-select inputs on the voice-synthesis chip: RS0, RS1, or RS2.

When plugged into expansion slot 4 of any Apple II-series computer, the registers are addressed at hexadecimal C400 through C404 (decimal addresses 50240 through 50244).

The remaining components on the speech-synthesizer board constitute the amplifier and filter sections.

Capacitors C4 and C5 and resistors R2 and R3 form a simple low-pass filter. The audio signal is then amplified by an LM386 1-watt operational amplifier (IC2) to directly drive an 8-ohm speaker. Potentiometer R4 controls the volume on the external speaker (connected to the header provided). In addition to the +5-V supply needed by the SSI263, the board requires a +12-V supply to power the op amp.

Speaking in Phonemes

All the words in the English language can be written using only 26 alphabetic characters, but the language contains far more than 26 sounds—most letters of the alphabet (or combinations) can represent more than one sound. As a practical matter, English can be considered to contain 50 or so distinct sounds, called *phonemes*. These are listed in table 3 on page 112.

It's not as hard to use the phoneme list as it first appears. For example, the phonemes in the word "disk" are written as follows:

D I S K

Simple enough. From the table, the corresponding digital codes (in hexadecimal) are

25 07 30 29

These are the values fed into the speech synthesizer.

Few words are quite that easy, though. For instance, the five distinct phonemes in "hello" are

HF EH L O W

These are translated into the hexadecimal codes

2C 0A 20 11 23

It isn't necessary for you to become a linguist to use a phonetic speech synthesizer because many lists of words and their phoneme equivalents are available (see reference 3).

Programming for Phoneme Synthesis

The speech-synthesizer board speaks a word when it receives, in sequence, the codes for the constituent phonemes. The programming for this can be simple: POKE statements in BASIC suffice. If the synthesizer were plugged into slot 4 of an Apple II, the program of listing 1 would cause it to say "hello" in a monotonic voice. Observe that the program loads the stop phoneme (hexadecimal 0) after the end of the word; this makes the synthesizer stop sounding the last phoneme.

You'll also note that the first six executable statements in listing 1 load

Pin	Function
1	AO (Audio)—analog output, DC-biased at $V_{DD}/2$. Requires an external audio amplifier to drive speaker.
2	AG (Analog Ground)—must be connected to a good ground potential to eliminate noise in the output signal.
3	TP1 (Test Point 1)
4	A/$\overline{R}$ (Acknowledge/Request—Not)—digital open-collector output. When forced low it requests new data. It can respond on a frame or phoneme boundary condition or be deactivated (see DR0, DR1—page 38—and $\overline{PD}$/RST). This signal can also be read on D7 in an inverted state.
5	TP2 (Test Point 2)—normally not used.
6	RS2 (Register Select 2)—used in conjunction with RS1 and RS0, these three register select lines are used to select and write to one of five internal registers.
7	RS1 (Register Select 1)
8	RS0 (Register Select 0)
9	D0—first of data lines, listed in order of increasing significance
10	D1
11	D2
12	V_{SS} (ground)
13	D3
14	D4
15	D5
16	D6
17	D7—most significant bit of 8-bit data bus. D7 is bidirectional. When read, high is an active request for new data and low is an acknowledgment that data has been received (A/$\overline{R}$).
18	$\overline{PD}$/RST (Power Down—Not/Reset)—this input is active in the low state. When active, it powers down the SSI263 and silences the audio output and retains its DC bias without disturbing the internal registers. It also puts the SSI263 in a disabled-A/$\overline{R}$ mode.
19	CS0 (Chip Select 0)—control input that selects the SSI263 on a microprocessor control/mapping bus, active high.
20	$\overline{CS1}$ (Chip Select 1—Not)—active-low state.
21	R/$\overline{W}$ (Read/Write—Not)—control input, write is an active-low state (writes into D0 through D7), read is an active-high state (reads D7 only).
22	XCLK (External Clock Input)—input for externally supplied clock. Normal frequency input is in two ranges: 2.0 MHz/1.79 MHz or 1.0 MHz/895 kHz dependent upon level of DIV2 input.
23	DIV2 (External Clock Divide-by-2 Input)—when this input is high, XCLK = 2 or 1.79 MHz. When this input is low, XCLK = 1 or 0.895 MHz.
24	V_{DD} (Positive Voltage Supply)—V_{DD} operating range is +4.5 V DC to 6.6 V DC.

Table 2: *Pin functions of the SSI263.*

constant values into the registers in the speech-synthesis chip; the five attribute registers in the SSI263 must be properly initialized; once they have been, a program using this method can cause the chip to emit the sounds of essentially any word or series of words. But an interpreted BASIC program is not fast enough to operate in an interrupt mode or dynamically change the attribute registers with each phoneme.

So a program in BASIC can only scratch the surface of the SSI263's

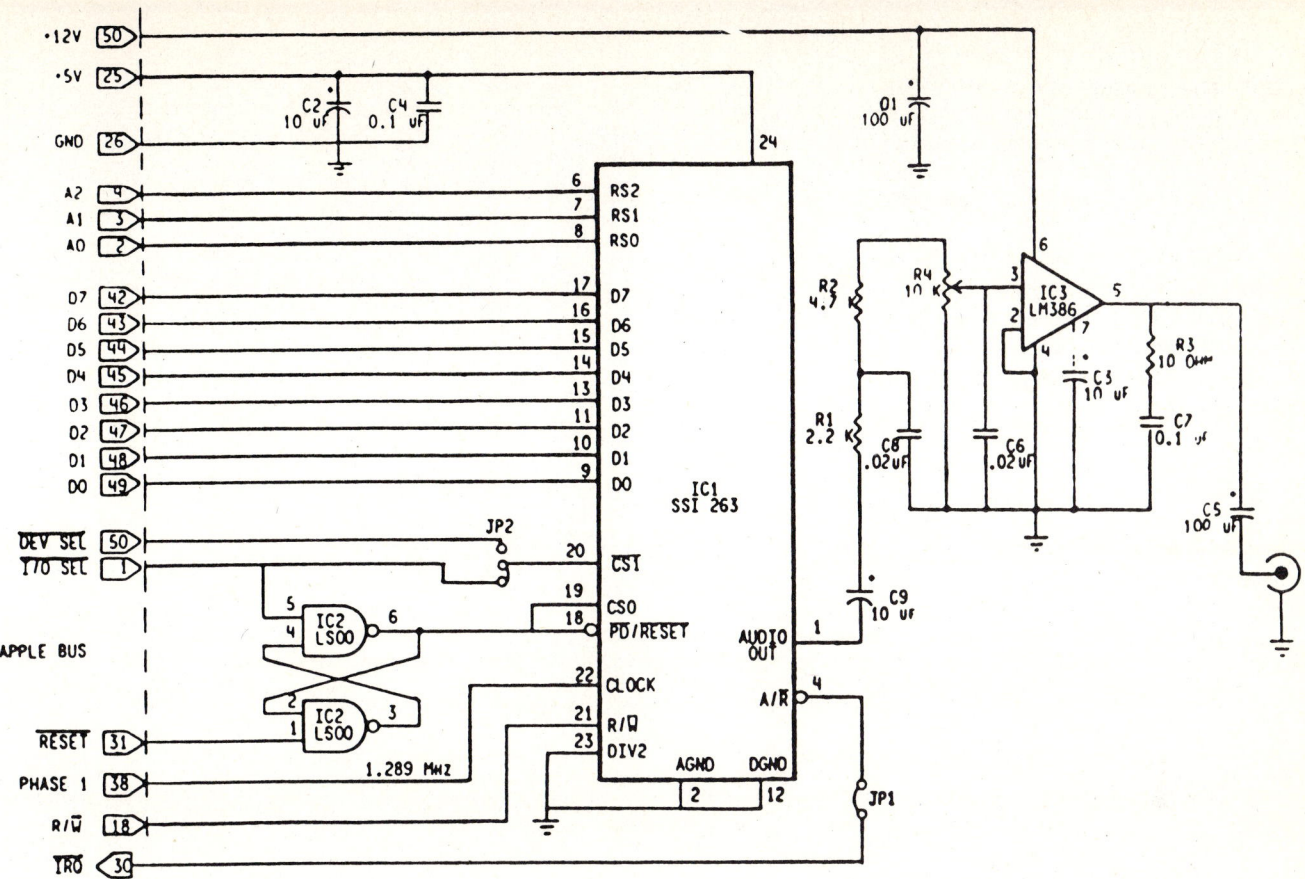

Figure 2: *Schematic diagram of the Sweet Talker II circuitry.*

Listing 1: *This BASIC program causes the Sweet Talker II to pronounce the word "hello" by sending it the minimal set of phoneme codes.*

```
10 REM SET UP SSI263 FOR TRANSITIONED INFLECTION
20 POKE 50243,255 :REM CONTROL BIT EQUALS 1
30 POKE 50240,192 :REM SET PHONEME DURATION
40 POKE 50243,116 :REM CONTROL BIT EQUALS 0 AND SET AMPLITUDE
50 POKE 50244,231 :REM SET FILTER FREQUENCY TO NORMAL
60 POKE 50242,168 :REM SET SPEECH RATE TO NORMAL
70 POKE 50241,127 :REM SET INFLECTION LEVEL
100 HOME
110 PRINT "HELLO"
120 DATA 44,10,32,17,35,0 :REM PHONEMES FOR WORD HELLO
130 FOR X=1 TO 6
140 READ A
150 POKE 50240,A :REM LOAD PHONEME INTO INPUT REGISTER
160 FOR T=0 TO 50 :NEXT T :REM DELAY TIL NEXT PHONEME TIME
170 NEXT X
180 END
```

capabilities. To really appreciate what it can do, you have to exercise the chip with an assembly-language program that adds inflection and intonation. The program in listing 2 on pages 113 and 114 also causes the system to say "hello" but with superior results. For speech applications requiring the most reliable intelligibility, an assembly-language word-list-to-phoneme output program is required.

But many applications require an essentially unlimited vocabulary, for instance, if you wanted your computer to read aloud the stories on the Associated Press newswire. (By the way, that's an acid test for intelligibility of synthesized speech.) In such cases, you'll need a program that can perform its own translation from normal spelling to phoneme codes using a text-to-speech algorithm.

Text-to-Speech Algorithm

A text-to-speech algorithm is a program that takes words spelled out in letters encoded in ASCII (American National Standard Code for Information Interchange) and analyzes them. It determines which characters are silent and which should produce sounds, and what kind of sounds, by following a set of general pronunciation rules. Most research in such synthesis by rule has been on English, but text-to-speech algorithms have been written for other languages as well.

The amount of program code

(3a)

Table 3: *Elementary speech sounds, or phonemes, that occur in English and a few other languages. Consonants are shown in 3a; vowels are listed in 3b. The four code columns for each phoneme represent spoken sounds of different length. You can choose the length that provides the best intelligibility or most pleasing sound. From left to right, the lengths decrease by about 25 percent in each column from the previous one. The code in column 1 can usually be used as the default value.*

Phoneme	Code 1	2	3	4	Examples
B	24	64	A4	E4	bat, tab
D	25	65	A5	E5	dub, bud
F	34	74	B4	F4	fat, ruff, photo, laugh
HV	2A	6A	AA	EA	eh
HVC	2B	6B	AB	EB	(post-B aspiration as in "tab")
HF	2C	6C	AC	EC	hat, home
HFC	2D	6D	AD	ED	(post-P aspiration as in "pad")
HN	2E	6E	AE	EE	ba-ba black sheep (voiceless glottal stop)
J	31	71	B1	F1	job, rage
K	29	69	A9	E9	kit, tick
KV	26	66	A6	E6	big, gag
L	20	60	A0	E0	lab, ball
L1	21	61	A1	E1	plan, club, slam
LB	3F	7F	BF	FF	il (French)
LF	22	62	A2	E2	bottle
M	37	77	B7	F7	mad, dam
N	38	78	B8	F8	not, ton
NG	39	79	B9	F9	ring, rang, drink, drank
P	27	67	A7	E7	pat, tap
R	1D	5D	9D	DD	rat
R1	1E	5E	9E	DE	(French)
R2	1F	5F	9F	DF	(German)
S	30	70	B0	F0	sat, lass
SCH	32	72	B2	F2	shop, push
T	28	68	A8	E8	tap, pat, baked
THV	35	75	B5	F5	bathe, the
TH	36	76	B6	F6	bath, theory
V	33	73	B3	F3	vow, pave
W	23	63	A3	E3	why, quake
Y	03	43	83	C3	(French)
Y	04	44	84	C4	you
Z	2F	64	A4	E4	zap, maze
(space)	00	40	80	C0	(pause)

(3b)

Phoneme	Codes 1	2	3	4	Examples	Comments
A	08	48	88	C8	day	
A1	09	49	89	C9	care	
AE	0C	4C	8C	CC	dad, plaid	
AE1	0D	4D	8D	CD	ask	
AH	0E	4E	8E	CE	top, father	
AH1	0F	4F	8F	CF	about	
AW	10	50	90	D0	saw, caught	
E	01	41	81	C1	beet, be	
E1	02	42	82	C2	advent	
EH	0A	4A	8A	CA	leg, said	
EH1	0B	4B	8B	CB	silent	
ER	1C	5C	9C	DC	third, urn, heard	
I	07	47	87	C7	sit, bid	
O	11	51	91	D1	boat	
OO	13	53	93	D3	put, pull, look	
OU	12	52	92	D2	orb	
U	16	56	96	D6	boot, you	
U1	17	57	97	D7	poor	
UH	18	58	98	D8	cup	
UH1	19	59	99	D9	nation, circus	
UH2	1A	5A	9A	DA	nation, circus	
UH3	1B	5B	9B	DB	nation, circus	
Foreign Sounds						
AY	05	45	95	C5	français	French
A	3A	7A	BA	FA	être	French, or umlauted a (ä) in German
E2	3E	7E	BE	FE	schön	German
IE	06	46	86	C6	il	French
IU	14	54	94	D4	peut	French
IU1	15	55	95	D5	Goethe	German
OH	3B	7B	BB	FB	menu, tu	French
U	3C	7C	BC	FC	Fühlen	German
UH	3D	7D	BD	FD	menu, tu	French

needed to implement a text-to-speech algorithm varies, with longer programs usually performing better. Typical microprocessor routines are in the 4K- to 8K-byte range, but some of the more sophisticated programs take up to 80K bytes. The main difference between the most common algorithms is the number of pronunciation exceptions and the length of word tables. An 80K-byte routine, for example, is often 90 percent look-up tables of words that are pronounced in unpredictable ways.

The Sweet Talker II speech synthesizer outlined in this article can be made to speak by direct input of individually selected phonemes, as demonstrated above, or through use of a text-to-speech algorithm. While two of my previous Circuit Cellar speech-synthesis projects, the Sweet Talker I and the Microvox, had some form of text-to-speech capability, they were built around the Votrax SC-01A; the software is not compatible with the SSI263 in the Sweet Talker II.

At first, I thought I would have to

Listing 2: *An assembly-language program for the 6502 microprocessor that causes the synthesizer to say "hello" with better inflection and intonation by sending more data than the BASIC program.*

```
                        1          * SSI-263 COMPOSITE DATA DRIVER
                        2          *
                        3          *
                        4                   ORG     $8000
                        5          *
                        6          *
                        7          OUTPTR   EQU     $FB             ;POINTER TO START OF DATA
                        8          ENDPTR   EQU     $FD             ;POINTER TO END OF DATA
                        9          BUSY     EQU     $FF             ;BUSY FLAG
                       10          IRQL     EQU     $03FE           ;IRQ VECTOR, LOW BYTE
                       11          IRQH     EQU     $03FF           ;IRQ VECTOR, HIGH BYTE
                       12          BASE     EQU     $C440           ;REGISTER 0 OF SSI-263
                       13          DURPHON  EQU     BASE
                       14          INFLECT  EQU     BASE+$01        ;REGISTER 1 OF SSI-263
                       15          RATEINF  EQU     BASE+$02        ;REGISTER 2 OF SSI-263
                       16          CTTRAMP  EQU     BASE+$03        ;REGISTER 3 OF SSI-263
                       17          FILFREQ  EQU     BASE+$04        ;REGISTER 4 OF SSI-263
                       18          *
                       19          *
                       20          * SET-UP ROUTINE *
                       21          *
                       22          *
8000: 78               23                   SEI                     ;DISABLE INTERRUPTS
8001: A9 2E            24                   LDA     #<INTERR        ;INT SERVICE ROUTINE, LOW ADDRESS
8003: 8D FE 03         25                   STA     IRQL
8006: A9 80            26                   LDA     #>INTERR        ;INT SERVICE ROUTINE, HIGH ADDRESS
8008: 8D FF 03         27                   STA     IRQH
                       28          *
800B: A9 80            29                   LDA     #>TABLE         ;DATA TABLE, HIGH ADDRESS
800D: 85 FC            30                   STA     OUTPTR+1
800F: 85 FE            31                   STA     ENDPTR+1
8011: A9 69            32                   LDA     #<TABLE         ;DATA TABLE, LOW ADDRESS
8013: 85 FB            33                   STA     OUTPTR
8015: A9 9B            34                   LDA     #<TABLE+$32
8017: 85 FD            35                   STA     ENDPTR
                       36          *
                       37          *
8019: A9 FF            38                   LDA     #$FF            ;SET BUSY FLAG
801B: 85 FF            39                   STA     BUSY
801D: A9 80            40                   LDA     #$80            ;SET SSI-263 TO
801F: 8D 43 C4         41                   STA     CTTRAMP         ;TRANSITIONED INFLECTION MODE
8022: A9 C0            42                   LDA     #$C0
8024: 8D 40 C4         43                   STA     DURPHON
8027: A9 70            44                   LDA     #$70
8029: 8D 43 C4         45                   STA     CTTRAMP
802C: 58               46                   CLI                     ;CLEAR INTERRUPT MASK
802D: 60               47                   RTS                     ;RETURN TO CALLER
                       48          *
                       49          *
                       50          * INTERRUPT SERVICE ROUTINE *
                       51          *
                       52          *
802E: 8A               53          INTERR   TXA                     ;SAVE X REGISTER
802F: 48               54                   PHA
8030: 98               55                   TYA                     ;SAVE Y REGISTER
8031: 48               56                   PHA
8032: A0 00            57                   LDY     #$00            ;INIT Y REGISTER
8034: A2 04            58                   LDX     #$04            ;INIT X REGISTER
8036: A5 FB            59                   LDA     OUTPTR          ;CHECK FOR END OF DATA
8038: C5 FD            60                   CMP     ENDPTR
803A: D0 1B            61                   BNE     CONT            ;NO, SO CONTINUE
803C: A5 FC            62                   LDA     OUTPTR+1
```

Listing 2 continued

Listing 2 continued:

```
803E: C5 FE      63           CMP   ENDPTR+1
8040: D0 15      64           BNE   CONT          ;NO, SO CONTINUE
8042: A9 00      65           LDA   #$00          ;END, SO TURN OFF SSI-263
8044: 8D 40 C4   66           STA   DURPHON
8047: A9 70      67           LDA   #$70
8049: 8D 43 C4   68           STA   CTTRAMP
804C: A9 00      69           LDA   #$00          ;RESTORE BUSY FLAG
804E: 85 FF      70           STA   BUSY
8050: 68         71    RET    PLA                 ;RESTORE Y REGISTER
8051: A8         72           TAY
8052: 68         73           PLA                 ;RESTORE X REGISTER
8053: AA         74           TAX
8054: A5 45      75           LDA   $45           ;RESTORE ACCUMULATOR
8056: 40         76           RTI                 ;RETURN FROM INTERRUPT
                 77    *
                 78    *
8057: B1 FB      79    CONT   LDA   (OUTPTR),Y    ;GET DATA
8059: 9D 40 C4   80           STA   BASE,X        ;AND PASS IT TO SSI-263
805C: E6 FB      81           INC   OUTPTR        ;INCREMENT POINTER
805E: D0 02      82           BNE   CONT1
8060: E6 FC      83           INC   OUTPTR+1
                 84    *
8062: CA         85    CONT1  DEX                 ;NEXT REGISTER
8063: E0 FF      86           CPX   #$FF          ;LAST REGISTER?
8065: D0 F0      87           BNE   CONT          ;NO, SO CONTINUE
8067: F0 E7      88           BEQ   RET           ;EXIT INTERRUPT
                 89    *
                 90    *
                 91    *
                 92    *
8069: E7 3B A8
806C: 7A AC      93    TABLE  HEX   E73BA87AAC
806E: E8 4D A8
8071: 7B CA      94           HEX   E84DA87BCA
8073: E8 5D A8
8076: 74 CA      95           HEX   E85DA874CA
8078: E7 6C A8
807B: 64 E0      96           HEX   E76CA864E0
807D: E7 7B A8
8080: 53 A0      97           HEX   E77BA853A0
8082: E7 7A A8
8085: 5A 11      98           HEX   E77AA85A11
8087: E7 79 A8
808A: 61 63      99           HEX   E779A86163
808C: E7 70 A8
808F: 60 00      100          HEX   E770A86000
8091: E7 70 A8
8094: 58 00      101          HEX   E770A85800
8096: E7 70 A8
8099: 50 00      102          HEX   E770A85000
                 103   *
                 104   *
```

find a new text-to-speech program. Fortunately, Silicon Systems had foreseen the demand for a text-to-speech routine for the SSI263 and arranged for the necessary software to be available under license.

Customizing Software

The Sweet Talker II's text-to-speech routine was produced by Sweet Micro Systems of Cranston, Rhode Island. It was derived from an algorithm originally developed at the Naval Research Laboratory and uses a similar rule-definition format. But it's much more versatile, and it nicely uses the talents of the SSI263. (At this writing, the software has been implemented only for the Apple II computer family.)

One problem of text-to-speech translation is of personal interest to me. There isn't an algorithm written that can digest and properly pronounce "Ciarcia" (it should sound like "see-ARE-see-uh") unless there is a specific rule for that character string that outputs a predetermined set of phonemes when the string is detected.

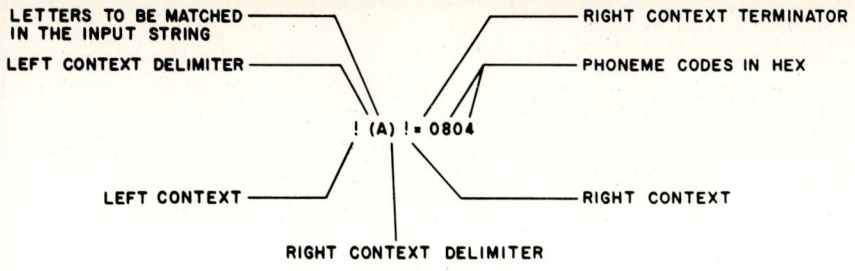

Figure 3: *Format of the rules in the text-to-speech algorithm.*

Symbol	Function
!	represents any nonalphabetic character in the input string
#	represents one or more vowels
:	represents zero or more consonants
+	represents a front vowel (E, I, Y)
$	represents one consonant
.	represents a voiced consonant (B, D, G, J, L, M, N, R, V, W, Z)

Table 4: *Context symbols used in the rules of the text-to-speech algorithm.*

Most text-to-speech routines don't let you modify the rule tables or expand the number of exceptions. The Sweet Micro Systems program employs a rule-based method that encompasses whole-word, morpheme, and letter rules in character-specific subtables. Furthermore, this new program also comes with a utility routine for changing, editing, or redefining those rules. The utility can even adjust the size of the main routine depending on whether rules or characters in the rules have been inserted or deleted. This flexibility in design allows you to totally redefine the rule table for foreign languages and dialects, to prepare for certain proper names, or simply to improve pronunciation of a specific frequent word.

The original Naval Research Laboratory algorithm did not include any facility for varying intonation and stress, but the software for the SSI263 uses a simple yet effective system of stress markers to allow intonation and stress to be specified. (In addition, a more extensive utility program is available that provides total control over parametric changes on an individual phoneme basis. This utility routine allows you to create high-quality words or phrases that can be easily used in any application, without the need for the text-to-speech algorithm and rule table to be resident in memory at the time of execution.)

How the Algorithm Works

Before I describe the rule-based text-to-speech algorithm, an explanation of the rule format is in order. Figure 3 shows that each rule consists of several different components. The letter or letters contained within the parenthetic delimiters are the characters to be matched in the input string. The symbols to the left of the delimiters define the left context (what comes before the characters to be matched), and the the symbols to the right define the right context. The context symbols' meanings are listed in table 4. The equals symbol (" = ") denotes the end of the rule definition, and the numbers to the right are the phoneme codes in hexadecimal. Most rules have a left or right context for the matched characters, but some situations do not require either.

The routine begins by converting the entire input string to uppercase characters to maintain uniformity in conversion. During this period, all characters in the input string are categorized and marked according to the symbol groupings described in table 4, and corresponding rule symbols are stored for later use. Also, stress markers are found and marked for use in the determination of intonation and stress characteristics.

The first input character is then compared to the first applicable rule in the subtable for that character. Since all rules in a single subtable begin with the same character, the algorithm attempts to match the left context first, then if the match succeeds, it attempts to match the second and following characters, if any, remaining in the parenthetic string. If that match succeeds, the algorithm attempts to match the right context. Should any match attempt fail, the algorithm proceeds to the next rule in the subtable and attempts to match that rule.

The last rule in every rule subtable for a given letter contains a parenthetic string of just that letter with no left or right context, thus guaranteeing a match for any letter in the input string. Once a match has been achieved, the algorithm then places the phoneme codes designated in the rule (to the right of the equals sign) into a phoneme buffer.

The rule table also allows the definition of punctuation marks and numbers. This allows a description of those characters to be pronounced. Some rules, for silent letters, are defined with no phoneme codes. The algorithm simply places nothing in the phoneme buffer and proceeds to the next character in the input string.

When the routine attempts a match on either the left or right context, it may encounter special symbols in the rule, those shown in table 4. These symbols cause the algorithm to compare each context symbol in the rule to the categorical symbols assigned earlier to the characters in the input string. Again, should the algorithm encounter a failure in a match attempt, it will proceed to the next rule.

When the phoneme conversion is complete, the routine collects values for intonation and stress characteristics in other buffers. The parameters involved are those for pitch inflection, output amplitude, speech rate, filter frequency, rate of inflection transition (slope), and rate of articulation transition (slope). These last two parameters are not stored in separate buffers but as one nybble of parameters in the speech-rate and amplitude parameter buffers, respectively. A detailed explanation of these parameters appears in table 5.

Interpreting the Rules

Let's look at a few sample cases of the phoneme rules. For instance:

!(A)! = 0804

This rule states that a letter A preceded and followed by a nonalphabetic character (the "!" context) is to be pronounced by the phonemes A and Y1.

The exclamation points indicate that on both sides of the A are one or more nonalphabetic characters, which can be a space, punctuation mark, number, or any symbol other than an alphabetic letter. For example, the A in the string "1A " would be pronounced according to the above rule since it meets the conditions indicated. The strings " A ", "A%", and "$A?" would all match the above pattern.

When putting another rule into a subtable, you must carefully consider the exact position of that rule because the algorithm takes the rules in sequence. If the conditional parameters of two rules were similar, it would be possible for sequential preference to cause the wrong rule to be selected for a given input string.

For example, it is possible that "YOUR" would be pronounced like "you-are" if the rule for "YOU" preceded a specific rule for "YOUR", because the rule for "YOU" would be accepted before the rule for "YOUR" was even considered, so the rule for "YOUR" would never be evaluated. (After accepting "YOU", the algorithm would next search the "R" subtable for a rule to satisfy "R" in the last position.)

To avoid such errors, clearly the rule for "YOUR" must precede the rule for "YOU"; "YOU" would not be mispronounced by the first rule since the trailing "R" would cause a no-match situation. The algorithm would continue to search the subtable until it reached the rule for "YOU".

The example given above for the letter A is tightly defined, and the pronunciation appears to be an exception to a rule that is more generally applicable, which comes next in the table:

Duration/Phoneme
This parameter contains the actual phoneme code. The upper 2 bits of this value designate the duration of the phoneme. The real-time duration varies depending on the value of the speech-rate parameter.

Inflection
Of the several modes of inflection that could be used, only the transitioned inflection mode is used by the text-to-speech algorithm. In this mode, there are 32 target values for the pitch of the spoken output. The output signal moves smoothly from one target value to the next.

Speech-Rate Inflection Slope
This controls the rate at which speech is produced. The lower 3 bits control the rate of inflection transition.

Control Articulation Amplitude
The upper bit controls the mode of operation of the chip. The next 3 bits control the rate of articulation transition. The lower 4 bits control amplitude.

Filter Frequency
Controls the vocal-tract filters. Different values will instantaneously shift the frequency of all formant filters.

Table 5: *A detailed explanation of the parameter bytes used by the text-to-speech routine for pitch inflection, output amplitude, speech rate, filter frequency, rate of inflection transition, and rate of articulation transition.*

1!(A)! = 0804
2(A)! = 1A

The second rule states that "A" in the input string followed by a nonalphabetic character is to be pronounced as the phoneme UH (hexadecimal 1A represents UH). The second rule does not specify anything in particular to the left of "A". Therefore, as long as "A" is merely followed by a nonalphabetic character, this pronunciation will be used. Note that if the rules were reversed, the rule !(A)! would never be reached because the (A)! rule would match its patterns.

Although rule 1 above is rather narrow in scope, the class of nonalphabetic characters comprises a number of different symbols, and even more specific rules might sometimes be needed. For example, if you found that the pronunciation for the input string "A?" was too short, a rule could be created to deal with this, as follows:

!(A?) = 080404

This rule would have to be placed before !(A)! in the table because having a question mark following the "A" is an exceptional condition. Any other nonalphabetic symbol would not match the question mark, and the algorithm would proceed to the next rule.

In addition to adding new rules, you may delete any rule within the range of the subtable, with the exception of the last rule. (Each subtable must have at least one rule, even if its input character is to produce no sound.) All rules following the deleted rule will move up one position, reducing the rule count and number of bytes accordingly.

The text-to-speech software provides a convenient test mode as part of the rule editor to let you evaluate any changes made to the rule tables.

The Rule Editor

The rule-table editor provided with the text-to-speech algorithm allows complete control over the rules that govern the conversion of input text into phonemes.

The rule table contains all alphabetic characters, all numbers, all punctuation marks, and all printable special-purpose symbols. (Control characters are not included.) Although the number of rules in a table is unlimited (insofar as the disk storage and memory capacity will allow), extremely large tables will cause the conversion process and the output of speech to slow down.

When the rule editor is in use, the

(6a)	Key	Function	Key	Function
	D	delete an entry	S	select new subtable
	E	edit an entry	T	test mode
	H	help menu	U	update main rule table
	I	insert a new rule	Control-P	print current subtable
	Q	quit or exit program	Control-S	save rule table to disk
	(space)	advance to next page of current subtable (if any)		

(6b)	Keys	Command	Function
	Control-A	(amplitude)	set amplitude level (0–11)
	Control-F	(filter frequency)	set filter frequency (0–253)
	Control-I	(inflection)	set inflection level (0–25)
	Control-R	(speech rate)	set speech rate (0–13)
	Control-X	(help)	display brief introductions
	Control-Z	(return to editor)	exit test mode

Table 6: *The rule editor provides a set of editing functions (6a) and a test mode with several commands (6b) consisting of control characters.*

screen shows up to 10 of the rules in the current subtable (see photo 2). If the subtable contains more than 10 rules, the program lets you scroll through the rules 10 at a time, wrapping around to the first set when the end of the rule list is reached. A list of rule-table editing functions is shown in table 6a.

The test mode allows you to enter any word or phrase up to 239 characters long, including punctuation marks. The algorithm then converts the input phrase into phoneme codes, based on the rules currently in the rule table, and instructs the Sweet Talker II to speak the codes. Commands available in the test mode are control characters as shown in table 6b. To make the rule-table changes permanent, the rule table must be saved to disk after an editing session.

Intonation

Perhaps the best way to see how the text-to-speech routine deals with indications of varying intonation and stress is to provide examples of usage. When entering text to be spoken, you can indicate what words or syllables are to differ in stress by setting them off with slash marks ("/").

For example, the string of characters

HELLO

would be pronounced with monotone inflection. But the string

/HE/LLO

would produce stress on the first syllable, while the string

SSI263 Registers

The registers discussed below are listed with their bits in order of decreasing significance.

Duration/Phoneme Register 0
DR1 DR0 P5 P4 P3 P2 P1 P0

The duration/phoneme register, D/P, is an 8-bit register where the 6 low-order bits (D5 through D0) designate one of 64 phonemes (P5 through P0). Table 3 on page 33 lists the 64 phonemes produced by the SSI263. The 2 high-order bits of the D/P register (DR1 and DR0) control the duration (timing) of the phoneme called out by the 6 low-order bits (P5 through P0).

Inflection Pitch Register 1
I10 I9 I8 I7 I6 I5 I4 I3

The inflection register is an 8-bit register (D7 through D0) where all 8 bits (I10 through I3) set or determine the rate of movement of inflection pitch. There are two modes of implementing inflection: transitioned *inflection* and immediate or *instantaneous inflection*. Immediate inflection causes the current output to instantly take on the value corresponding to the 8-bit code in the inflection register and the 4-bit code in the R/I register. The total of 12 bits in these two inflection registers gives a range of seven octaves on an even-tempered scale. The immediate mode is useful for singing, musical sound effects, and for fine-tuning a duplication of human inflection.

Rate/Inflection Register 2
R3 R2 R1 R0 I11 I2 I1 I0

The rate/inflection register, R/I, is an 8-bit register where the 4 high-order bits (D7 through D4) designate one of 16 (R3 through R0) overall settings of speech/sound-effects rates and A/R timing response. The 4 low-order bits (D3 through D0) of the R/I register (I11, I2, I1 and I0) are 4 bits of inflection that are always in the immediate mode. In combination with the 8 bits of inflection from the I inflection register, I2, I1 and I0 are the 3 low-order bits of the 12-bit counter-inflection chain, and I11 is the MSB (most significant bit).

Control/Articulation/Amplitude Register 3
CTL TR2 TR1 TR0 A3 A2 A1 A0

This 8-bit register, C/A/A, has three functions. The 4 low-order bits (D3 through D0) designate one of 16 (A3 through A0) amplitude levels of the analog audio output (AO, pin 1). The MSB (D7) of this register (C/A/A) is the control (CTL) bit. When this bit is found to be high, the SSI263 is powered down. The remaining 3 bits (D6, D5, and D4) of the C/A/A register (TR2, TR1, and TR0) determine the rate of movement of the formant position for articulation.

Filter-Frequency Register 4
FF7 FF6 FF5 FF4 FF3 FF2 FF1 FF0

The file-frequency register is an 8-bit register (D7 through D0) in which all 8 bits (FF7 through FF0) instantaneously set or shift the frequency of all the vocal-tract filters (formants), which can produce an effect similar to slowing down or speeding up a record player, but with a greater range and without affecting inflection (pitch) or other SSI263 timing. With a setting other than the normal one for speech, the filter-frequency register can raise or lower the vocal-tract filter frequencies to create sound effects.

HELL/O/

would put stress on the final "O". The program is smart enough to consider punctuation in setting inflection. The text string

HELLO?

would be pronounced with a rise in pitch at the end, and the string

HELLO.

would be sounded with a drop in pitch at the end.

The algorithm looks for pairs of stress markers; it can ordinarily be overridden only by proper terminating punctuation, such as a period or question mark. The type of terminator determines the slope of pitch inflection at the end of a phrase or sentence. Multiple pairs of stress markers can be used to set the starting level of inflection and to specially emphasize any subsequent stressed portions of the remaining phrase or sentence.

Figure 4 depicts the inflection pattern for a single pair of stress markers. The inflection level begins at the default of level 2; it continues at this level until the program encounters the first stress marker, where it raises the pitch to level 3. The inflection level remains at level 3 until the program finds the second marker. After the second marker is found, the program looks for the terminator at the end of the clause. The inflection level will be raised or lowered at the end depending on the type of terminator. If no terminator exists, then the algorithm assumes that it should glide the pitch down (as in a declarative statement), but the glide will begin at a later point than otherwise. The maximum value of the up or down glide is predetermined; if the limit is reached prior to the end of the phrase or sentence, it will continue at that level.

Let's take a look at some real examples:

/HOW/ ARE YOU?

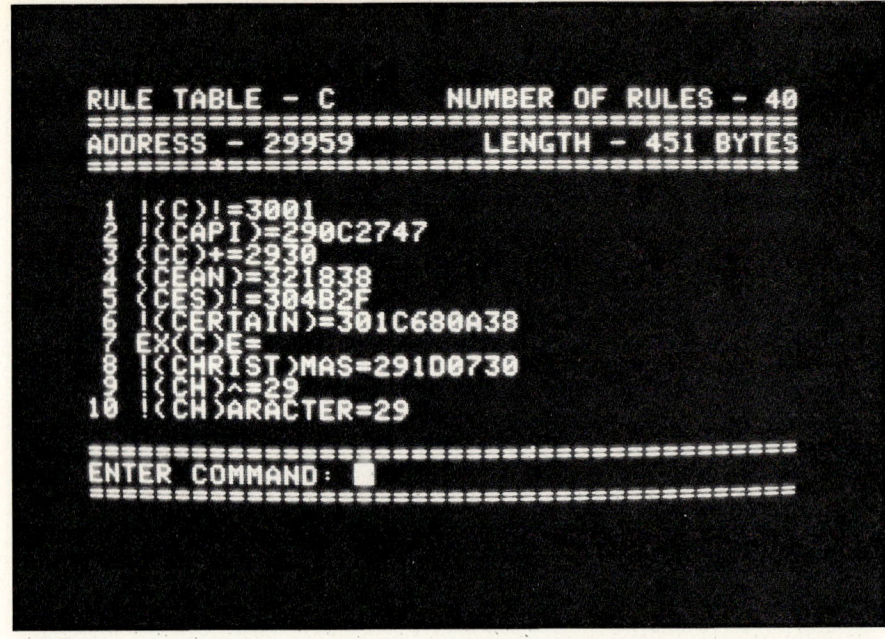

Photo 2: *The rule editor for the text-to-speech software displays the phoneme-translation rules for each letter in groups of 10.*

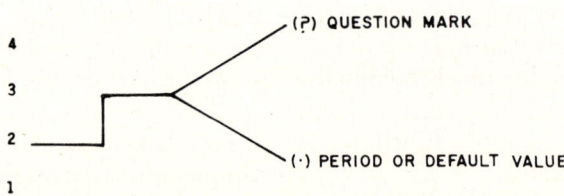

Figure 4: *The inflection pattern for a single pair of stress markers, as used by the text-to-speech routine to identify where to change pitch.*

"HOW" is stressed; its pitch glides up to the fourth level until the limit is reached or the terminator is encountered.

HOW ARE /YOU?

A simple system of stress markers allows specification of intonation and stress.

The word "YOU" is stressed in this case; the pitch then continues to glide up until the end of the sentence. Since the algorithm finds the interrogative terminator before anything else happens, the glide also stops at that point. In this case, the sound produced for the word "YOU" is very short. It is so short that the algorithm may not have time to react to the stress marker and initiate a glide before it encounters the terminator. The result may be an abrupt rise in pitch. Greater and more significant differences may be seen in the following example:

DID YOU /SEE/ THAT?

The effect of the stress markers and question mark works extremely well in this particular case.

The question-mark terminator has two possible effects on a sentence. In general, the question mark causes the pitch inflection to glide up at the end of a sentence, with one exception. The algorithm looks for questions whose first word begins with

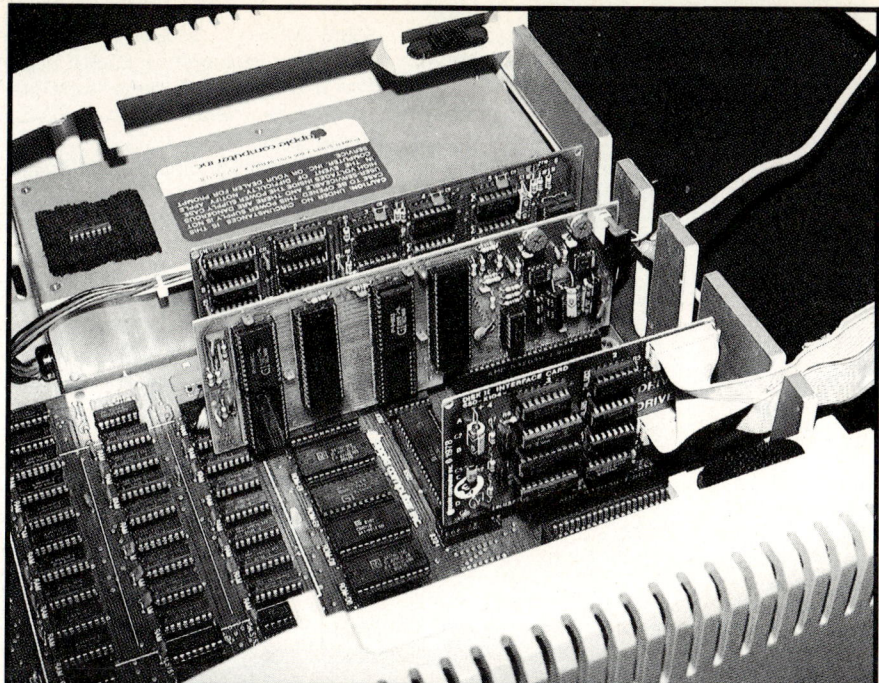

Photo 3: *The latest version of the Mockingboard music-synthesis system has two sockets into which SSI263s can be plugged. The speech synthesizers can then sing along in stereo. The Mockingboard is a product of Sweet Micro Systems, 50 Freeway Dr., Cranston, RI 02920, (401) 461-0530.*

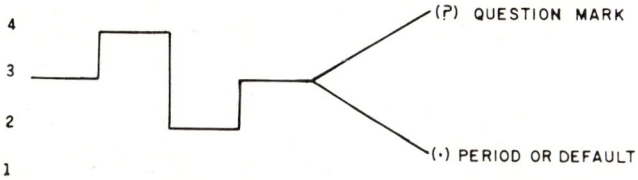

Figure 5: *The pitch inflection follows the pattern shown here when the algorithm encounters multiple pairs of stress markers.*

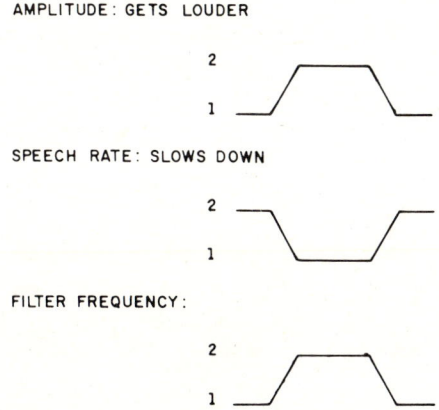

Figure 6: *Not only is the pitch raised for stressed syllables, but the other speech parameters are altered as well.*

"WH". In such cases, the glide at the end of the sentence should be down. For example:

WHAT /WAS/ THAT?

When the algorithm encounters multiple pairs of stress markers, the inflection follows the pattern in figure 5. In figure 5, we can see the inflection pattern beginning at level 3 and continuing until the text-to-speech routine finds the first stress marker. The level then goes to level 4 until the routine finds the second marker. At that point, the inflection level drops to 2, where it stays until the third marker is reached. The inflection level then goes up to level 3 again and remains until another marker or the terminator comes along.

As a general rule, the first syllable in a clause is stressed to a greater degree than subsequent syllables. This inflection pattern does not take effect if a compound sentence contains multiple pairs of stress markers or clauses separated by commas. Commas are also considered terminators, so the clause preceding the comma will be treated separately.

The other speech parameters besides pitch are also altered, in a rather consistent and straightforward manner, when a syllable is being stressed. Their changes are shown in figure 6.

In Conclusion

I haven't mentioned many uses for speech synthesis, but I'm sure you have a few ideas for what you could do with a speech synthesizer. I hope you now appreciate the amount of vocal power currently available at a relatively low price. The SSI263's highly intelligible speech promises new applications for computers, and the Sweet Talker II will likely enjoy a lot of indirect software support because of its compatibility with other SSI263-based speech synthesizers.

The invention of SSI263 does not mean the demise of other approaches to computer speech. But it introduces low-cost speech output into fields that could never previously have justified the expense.

References

1. Ciarcia, Steve. "Build a Computerized Weather Station." BYTE, February 1982, page 38.
2. Ciarcia, Steve. "Build a Low-Cost Speech-Synthesizer Interface." BYTE, June 1981, page 46.
3. Ciarcia, Steve. "Build an Unlimited-Vocabulary Speech Synthesizer." BYTE, September 1981, page 38.
4. Ciarcia, Steve. "Build the Microvox Text-to-Speech Synthesizer." Part 1, BYTE, September 1982, page 64. Part 2, BYTE, October 1982, page 40.
5. Ciarcia, Steve. "Talk to Me: Add a Voice to Your Computer for $35." BYTE, June 1978, page 142.
6. Ciarcia, Steve. "Use ADPCM for Highly Intelligible Speech Synthesis." BYTE, June 1983, page 35.
7. Ciarcia, Steve. "Use Voiceprints to Analyze Speech." BYTE, March 1982, page 50.
8. Elovitz, Honey Sue, Rodney W. Johnson, Astrid McHugh, and John E. Shore. "Automatic Translation of English Text to Phonetics by Means of Letter to Sound Rules." *United States Naval Research Laboratory Report 7948*, 1976.
9. Kuecken, John A. *Talking Computers and Telecommunications*. New York: Van Nostrand Reinhold, 1983.

Special thanks to Rod Nakamoto for his contributions to this project. Technical data on the SSI263 speech synthesizer reprinted by permission of Silicon Systems Inc., 14351 Myford Rd., Tustin, CA 92680, (714) 731-7110.

The following item is available from:

The Micromint, Inc.
25 Terrace Drive
Vernon, CT 06266
For orders: (800) 635-3355
For information: (203) 871-6170

Sweet Talker II: Apple-II-compatible synthesizer board. Assembled and tested with SSI263 speech-synthesis chip, demonstration software, user's manual, and text-to-speech algorithm on DOS 3.3 floppy disk $100

SSI263 and Algorithm $79
SSI263 Chip $42

Please include $4 for shipping and handling in the continental United States, $12 elsewhere. Residents of New York state please include 7 percent sales tax.

Photo 1: *The two printed-circuit boards needed to form one character segment in the display. One board contains the 35 LEDs and the current-limiting resistors; the other holds the shift-register sections and buffer/drivers.*

Build a Scrolling Alphanumeric LED Display

Individual character arrays can be linked together to show longer messages

The helicopter banked around, and someone gasped audibly as we beheld the spectacle of the Grand Canyon. The computer-show exhibitor who had sponsored this special excursion for our group was getting his money's worth. But my thoughts, dulled by visits to seven hotel hospitality suites, strayed from the scenery to another computer show....

In Atlantic City in the sleepy precasino year of 1977, I had wandered the aisles of Personal Computing '77, the second such event organized by John Dilks. Dozens of garage-based entrepreneurs were showing off their answers to the MITS Altair 8800 in display booths that mostly contained prototypes propped up on card tables. I was looking for new ideas that might help me in my work as a full-time engineering consultant or might inspire me to write a second article for Carl Helmers, who was editing a new magazine called BYTE. It would still be many months before he would suggest that I write regularly.

That early convention contained none of the hype and high-tech glossiness of today's extravaganzas. The idea of setting up a lavish hospitality suite or of renting Disneyland for a night hadn't yet occurred to any microcomputer companies, none of which had hired image-conscious advertising executives.

But technical ideas came thick and furious in those days, I mused. Although a few clever inventors had used their inspirations to vault into prosperity, some ideas I had seen weren't the kind on which fortunes are built. Like the guy I had talked to that year who was convinced that his hand-soldered scrolling LED (light-emitting diode) display would make him a millionaire....

Wait a minute! Maybe it did. Just yesterday I had seen scads of scrolling LED display signs in the exhibit hall. These signs, which are generally designed as a single message line 10 to 20 characters long, consist of a multitude of LEDs that are wired in such a way that the text scrolls in sequence from right to left.

I had stopped at one company's booth to admire its LED display. The LED unit consisted of twenty 5- by 7-element dot-matrix characters and a hand-held controller that allowed you to enter messages that would be displayed on the sign. It came housed in an attractive wooden enclosure with a power supply. The price was $2000, and the exhibitor was doing land-office business—if I wanted one, I'd have to wait three months. Two aisles away the real pièce de résistance resided: at the Hewlett-Packard booth was a 4- by 5-foot LED graphics display (not for sale). The LEDs in the character segments were spaced so that the characters could be expanded both horizontally and vertically. I had inspected HP's display and estimated that there were approximately 10,700 LEDs in it.

The helicopter pitched, and my thoughts were jerked back to the present. But my next project was decided. I wanted a scrolling LED sign for the Circuit Cellar.

Design Alternatives

You don't have to have an engineering degree to design a scrolling LED display. But when a project contains so many components, it's best to

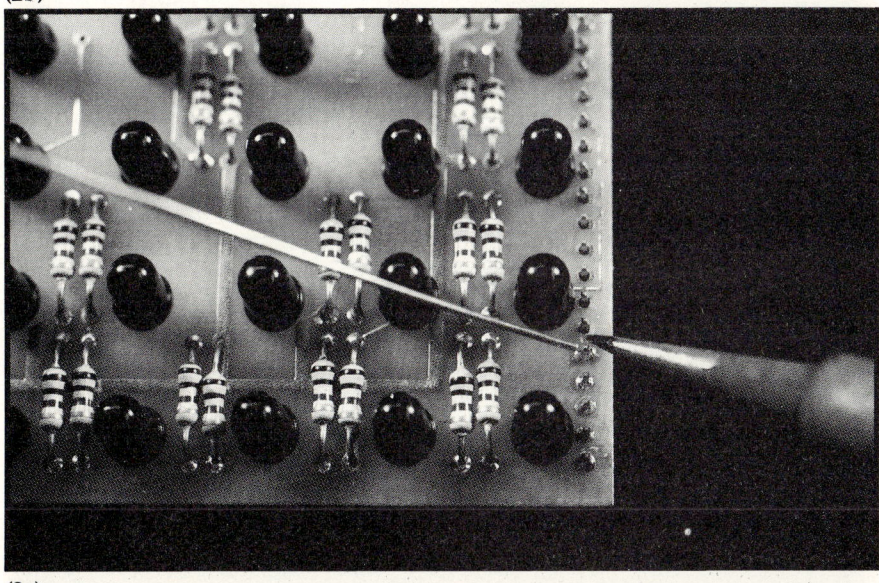
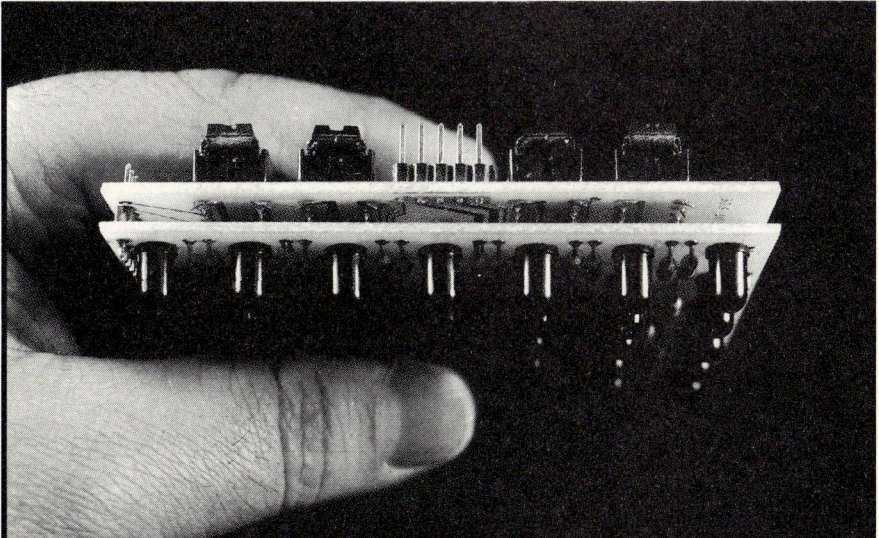

Photo 2: *The two boards are mounted back to back (2a). Two 18-pin Berg-type connectors are soldered together (2b) to electrically and mechanically join the two halves into a sandwich (2c). The 35 LEDs in the character segment are placed on half-inch centers in both dimensions for possible use in graphics.*

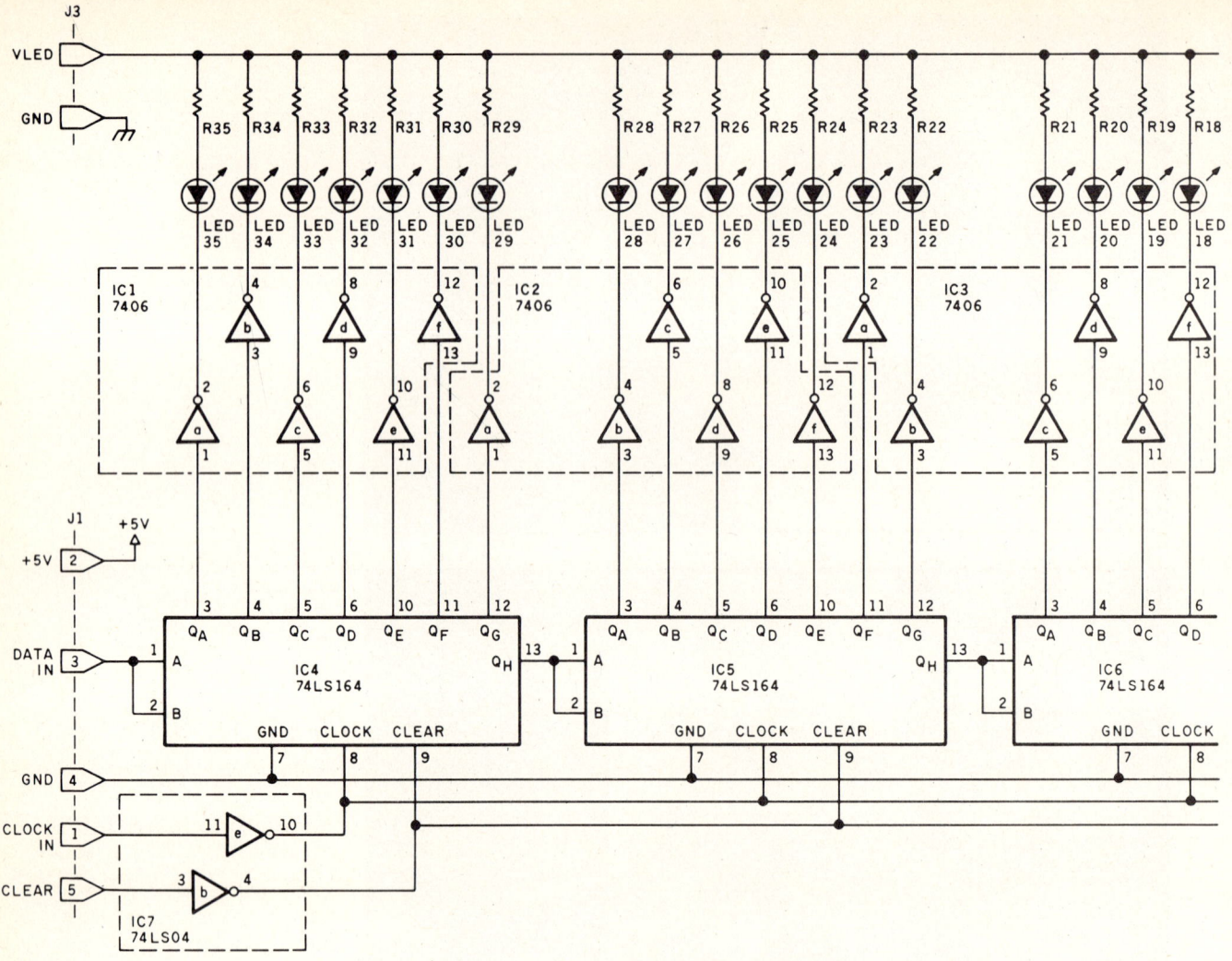

Figure 1: *Schematic diagram of one of the character-segment subassemblies for the scrolling LED display.*

spend some time early in the design process considering how to keep the final product easy to make and use.

When I looked at that first LED sign seven years ago, I thought, "What a simple product! How hard is it to drive a few LEDs?" But now that I was close to building a sign, I noticed the roadblock that had kept other people from doing the same: there are a *lot* of LEDs in it.

The most direct approach to controlling all those LEDs is *direct drive*, in which a wire for each individual LED is connected to one line of a parallel output port. When the appropriate bit in the output port is turned on, the LED connected to it (through an LED-driver component) will light up. A typical direct-drive setup requires little sophistication,
but it takes a lot of hardware. Controlling 100 LEDs requires 100 separate output lines and drivers.

The best alternative to direct drive is the *multiplexed* display. In this arrangement, the LEDs are wired at the intersections of a series of rows and columns. Each LED is illuminated when voltages appear on both its row connection and its column connection and current flows through it. Component count is thus reduced because only a single set of row and column drivers is required. Unfortunately, figuring out the proper scanning speed, the peak power dissipation, and the average intensity makes designing the control circuit relatively complicated. A few years ago I worked through a multiplexed display design and wrote an article en-
titled "Self-Refreshing LED Graphics Display" (see reference 2), which you can read if you want to know more on that topic.

The LED display I built for this article had to be simple enough to be wired by hand (by someone with time to spare) and yet also be sufficiently sophisticated to duplicate the intelligent features of commercial units. I decided to use a version of the direct-drive technique.

Design Specifics

Each character segment must consist of at least 35 LEDs (for a 5 by 7 matrix) to display a complete set of alphanumeric characters, even for a single-line display. A 6 by 9 matrix would need 54 LEDs. The support circuitry—drivers, current-limiting

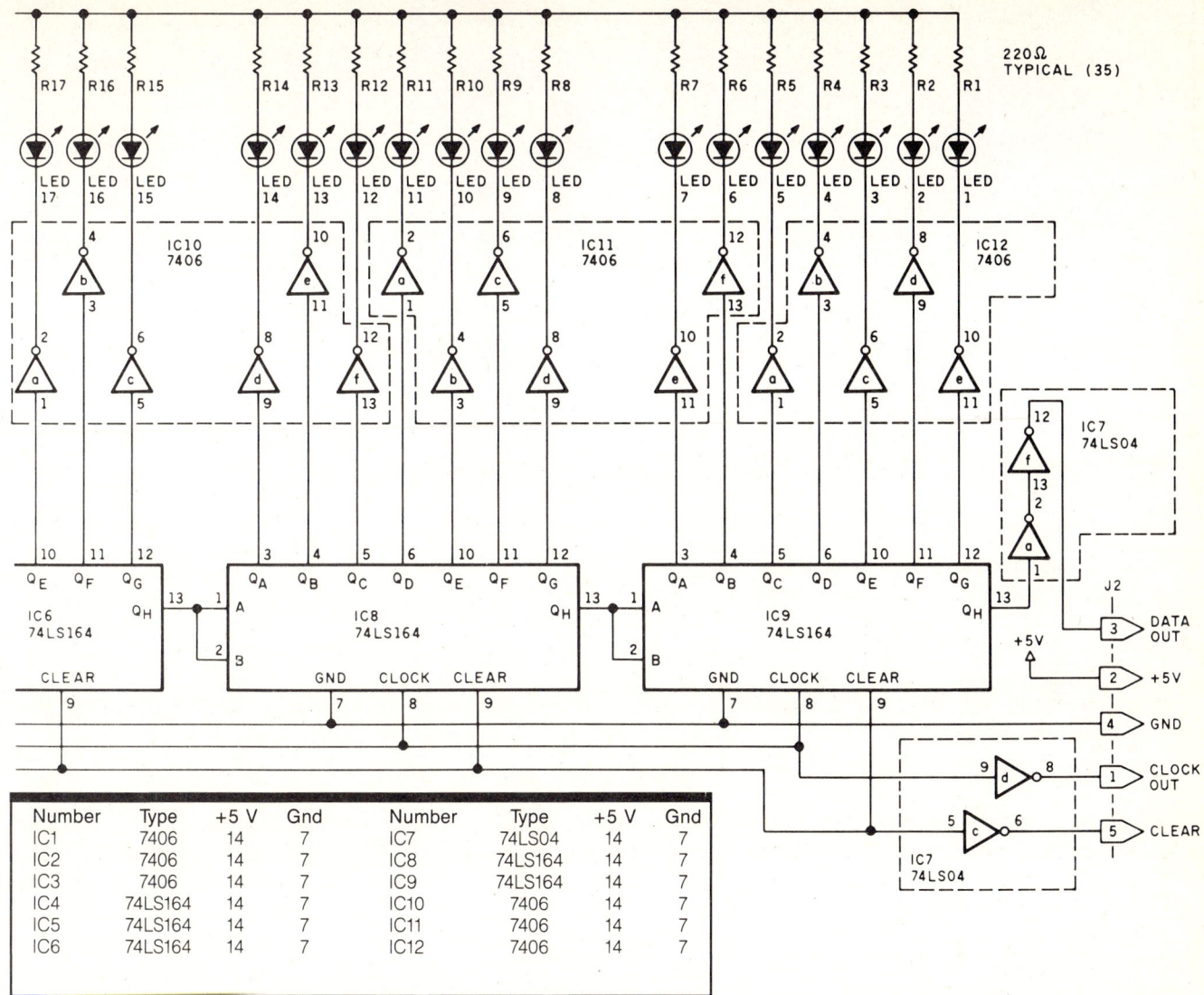

resistors, etc.—necessitates hundreds of connections for each segment. While I considered wiring one segment by hand, I didn't have the time to construct two, let alone 20. Even though an LED sign is relatively uncomplicated, the sheer number of connections makes it hard to put together. I definitely needed to display between 10 and 20 characters for most applications, so I decided to lay out a printed-circuit board even before building my first prototype. (I don't recommend this shortcut to anyone who doesn't possess years of experience with digital electronics.)

The configuration I came up with for a character segment, shown in photos 1 and 2, is an amalgam of two printed-circuit boards mounted back to back. The front board contains 35 LEDs and their current-limiting resistors, while the rear board contains 12 integrated circuits (the shift registers and drivers); all these parts form a 5- by 7-element character matrix. Segments are linked together in daisy-chain fashion to build linear arrays of many characters, as shown in photo 3. And while a single-line textual display is the most probable application, the LEDs are spaced on half-inch centers both horizontally and vertically so that graphics images can be displayed.

Instead of using conventional parallel direct-drive activation, I put in a serial shift register, effectively 35 bits long, to control the 35 LEDs. As a bit of data is shifted through the register, the LED to which it corresponds is illuminated or extinguished depending upon the bit's logic level, high or low.

Figure 1 is a schematic diagram of one of the character-segment subassemblies; figure 2 shows how the LEDs are placed to form the matrix. A serial shift register, physically 40 bits in length, is formed by IC4, IC5, IC6, IC8, and IC9: five type-74LS164 8-bit registers. The first seven positions in each of these chips (0 through 6) are connected to LEDs through the open-collector driver sections of IC1, IC2, and IC3 (type-7406 inverting buffer/drivers); the eighth position (bit 7) is not connected to an LED but is reserved for connecting adjacent register stages.

Data is entered into the shift register one bit at a time by setting the appropriate logic level for the bit

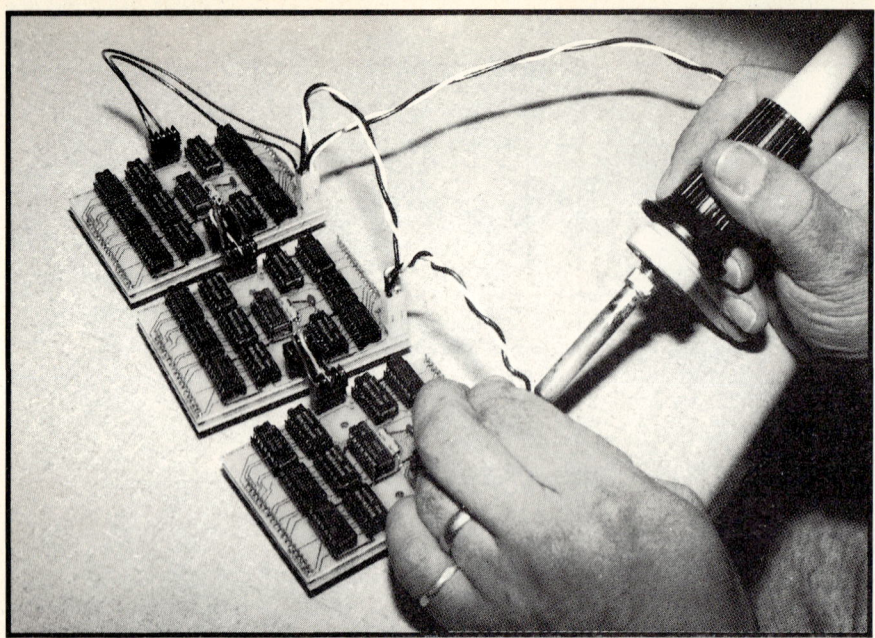

Photo 3: *Multiple character segments are connected together in a daisy chain to form a linear display capable of holding several words at once. The red connectors at the center of the board edges carry the CLOCK and DATA signals, while the black and white wires carry the power to drive the LEDs.*

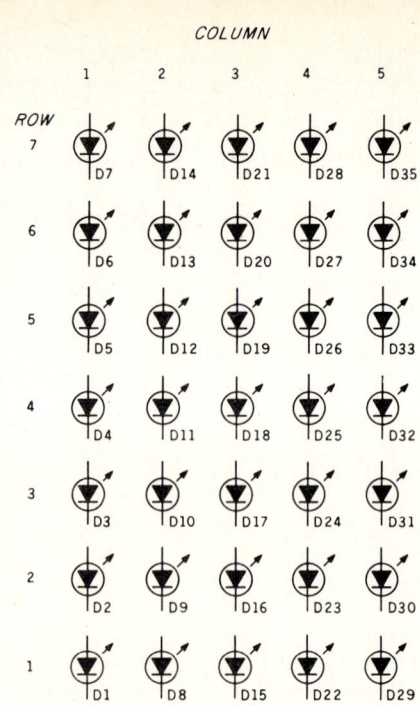

Figure 2: *The LEDs are placed to form a 5-by 7-element matrix.*

on the data input line and then pulsing the clock input line to a logic 0. Data is synchronously shifted through the register on the occurrence of the negative-going edge of the clock pulse, with the result that the light pattern on the LED matrix also shifts. To clear the display, a high level is set on the CLEAR line, which normally operates at a logic 0 level.

As figure 3 shows, the bits enter the rightmost column of the display (column 5), filling it from top to bottom. Once column 5 is filled, additional data entering the register causes the contents of column 5 to be shifted into column 4, again starting from the top. In essence, data enters the segment matrix from element D35, progressing through to D1.

Although 35 LEDs are in the display, it takes 40 clock pulses to fill the five 8-bit shift-register stages. Each 7-bit word of column data is preceded by a filler bit (usually a 0) that does not show up in the display. After eight clock pulses have occurred, the

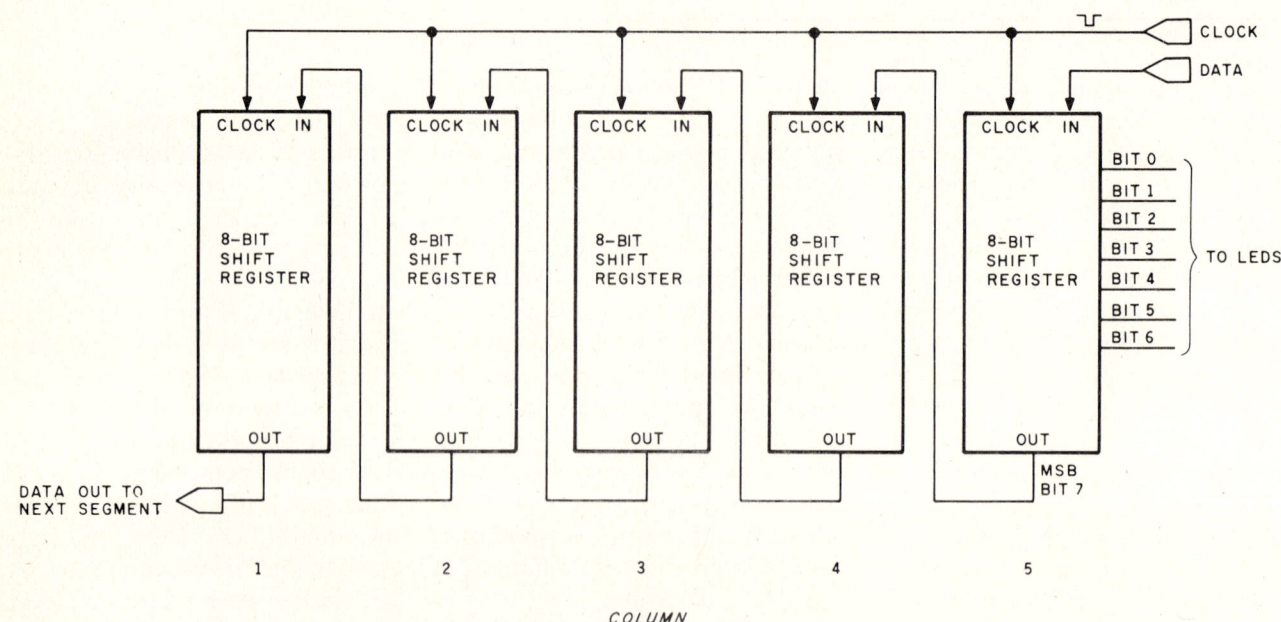

Figure 3: *The shift-register sections are wired so that bits being shifted out of the rightmost column (at the most significant bit) enter the next column, and so on to the end of the character segment. Each low-order register position is connected to an LED driver. (Bits can also be shifted from one segment to another.)*

7 data bits will be displayed in LEDs D29 through D35. After 8 more bits have been shifted in, the contents of column 5 have been completely moved to column 4, and so on. (To eliminate confusion, it is best to think of this as a column display where data is always entered 8 bits at a time.)

The shifting takes place too fast to see, so if the display contents are to be viewed by human eyes, the shifting must be halted occasionally for a viewing interval. If the 8 data bits are shifted in quickly relative to the amount of time they are left stationary for viewing, all the display contents will appear to scroll smoothly from right to left, one column at a time. How smoothly the display scrolls depends mostly on the speed of the program controlling it.

How to Use the Display

The Circuit Cellar LED display is simple to use. Because of its serial input, it requires use of only a single output bit from a computer. It's convenient to use one bit of a conventional parallel printer port as a source for this signal; figure 4 shows how the printer port on an IBM Personal Computer's Monochrome Display Adapter can be wired to the LED display. Another benefit arises because of this arrangement: when a bit is output through the printer port, the port's 5-microsecond (μs) STROBE signal is automatically triggered. These STROBE signals can be used as clock pulses to shift the bits through the registers. (I used only the least significant bit, bit 0 or the LSB, of the 8 available, but a second data line could have been connected to the display's CLEAR line to control that as well.) If multiple character segments are used, the data input of the first segment is connected to the computer; the other segment inputs are chained from the leftmost column of the preceding segments.

Once you're set up to send bits into the display, you need to know what bits to send. For textual displays, the bits should be set according to matrix patterns that form alphanumeric characters, such as the pattern of figure 5 that forms the numeral "2"

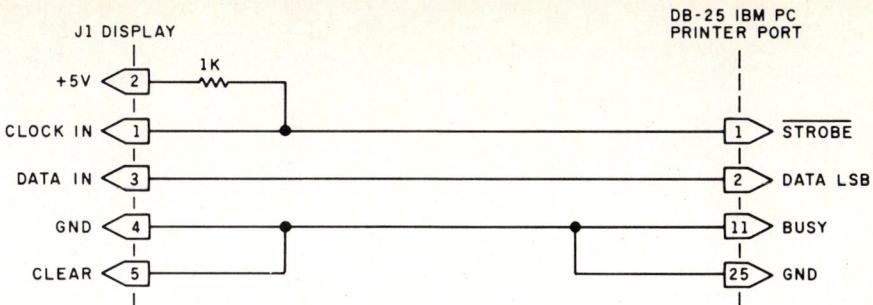

Figure 4: *The parallel printer port on an IBM PC's Monochrome Display Adapter can be wired to drive the LED display using only the least significant data bit.*

Listing 1: *BASIC program that causes a "4" to appear on the scrolling LED display, sending data through a parallel printer port on the IBM PC. If you turn the page sideways, you can see the image pattern in the DATA values.*

```
100 DIM X(50)
110 DATA 0,0,1,1,0,0,0,0
120 DATA 0,0,1,0,1,0,0,0
130 DATA 0,0,1,0,0,1,0,0
140 DATA 1,1,1,1,1,1,1,1
150 DATA 0,0,1,0,0,0,0,0
200 FOR C=1 TO 40
220 READ X(C)
240 NEXT C
250 FOR C=1 TO 40
260 LPRINT CHR$(36+X(C));
270 NEXT C
999 END
```

Figure 5: *A bit pattern that forms the numeral "2" in a 5- by 7-element representation.*

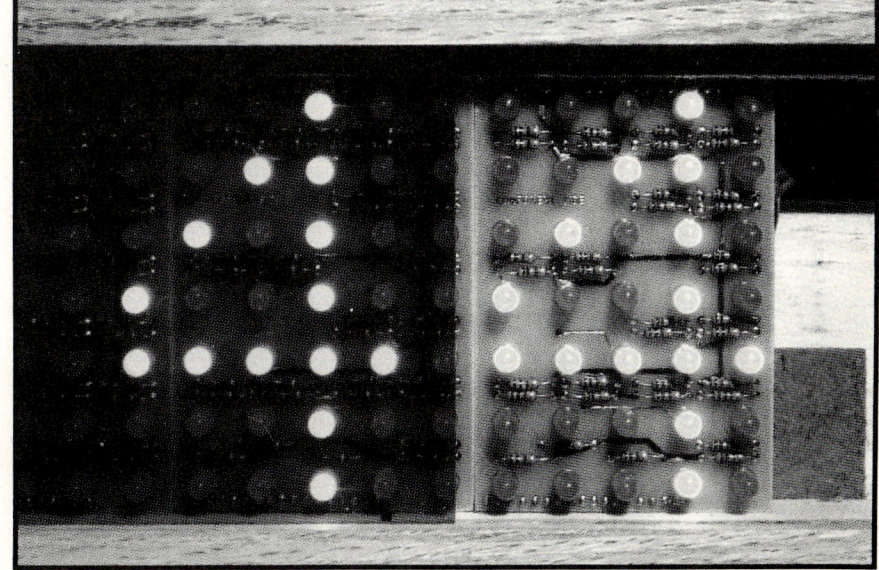

Photo 4: *These examples of the numeral "4" were produced by the same bit pattern used in listing 1, but under the control of the Z8-BASIC System Controller. The red plastic filter over the left segment greatly improves the appearance of the display.*

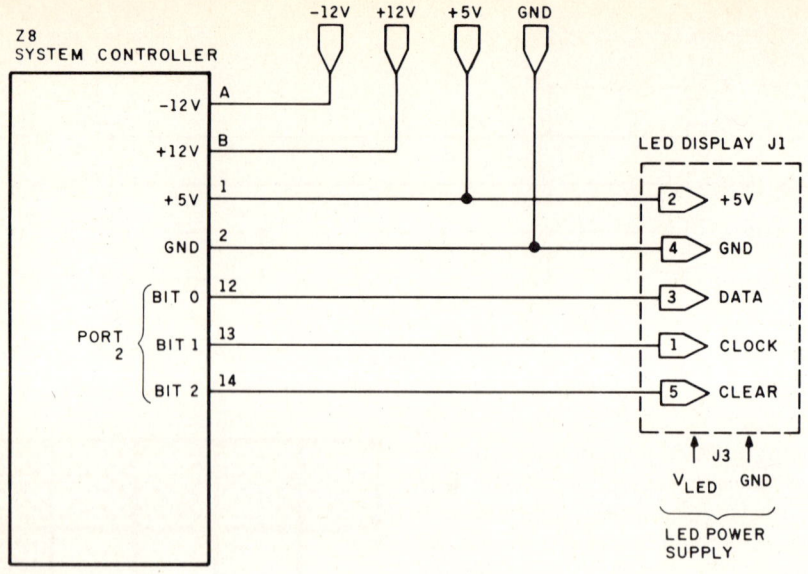

Figure 6: *Diagram for connecting the scrolling LED display to the Z8-BASIC System Controller.*

in a 5 by 7 representation. In the most simple case, data can be sent to the display using LPRINT statements from a BASIC interpreter.

A short BASIC program that displays "4" on the LEDs is shown in listing 1. The program begins with DATA statements that contain the bit patterns for each of the five columns. These values of 0 or 1 are read into the array X in lines 200, 220, and 240. The second FOR...NEXT loop sends the data to the display one bit at a time with an LPRINT statement. The expression CHR$(36+X(C)) includes an offset value of 36, which does not affect the LSB but avoids the unpredictability of sending values in the control-code range. If the value of the current element of the X array is 0, the output value of decimal 36 ("$") is emitted at the output port. If the X value is 1, a 37 ("%") is emitted. If a 36 is sent, the LSB is 0, a value that causes its corresponding LED to be darkened as the bit is shifted through the display. If a 37 is sent, the LSB is a logic 1, and the LED will be lit. Two examples of a displayed "4" appear in photo 4.

This is admittedly a rudimentary way to drive the display. But the method is quite effective if you don't mind taking time to code the program. It allows use of the scrolling LED display in a very flexible and economical manner, using the power of your existing computer system.

Control by a Dedicated Processor

While you can tie the scrolling LED display to your main computer, as we have seen, using the display is much less trouble if a small, dedicated control processor running custom software is used. I decided to drive my display with a Z8-BASIC System Controller, the successor to the Z8-BASIC Control Computer I developed almost three years ago (see reference 1). Based on the Zilog Z8 self-contained microprocessor, this small computer is well suited for jobs like this. It has an on-board tiny-BASIC interpreter, space for 2K bytes of RAM (random-access read/write memory) and 4K bytes of EPROM (erasable programmable read-only memory), an RS-232C serial port, two parallel output ports, and one parallel input port.

The LED display is attached to one of the parallel output ports on the Z8 board, as shown in figure 6 and photo 5. The message-storage space can hold 256 characters using the unexpanded Z8 board. With the addition of one or more memory-expansion boards, messages up to 50,000 characters long can be stored.

The Z8 board can be programmed to provide several intelligent display features using a combination of tiny-BASIC statements, shown in listing 2, and assembly-language subroutines, shown in listing 3. (A flowchart of the controller software is shown in figure 7.) The software, which is modular in design, uses a table for storage of the character-display bit patterns. Shown in figure 8, these patterns can be easily modified to allow special characters to be displayed. Finally, listing 4 is a sample run of the controller software with the external-terminal mode selected.

You'll need to have a video-display

Photo 5: *In my first attempt, several display segments were attached to the Z8 board; the message content of the display can be changed interactively from the video terminal.*

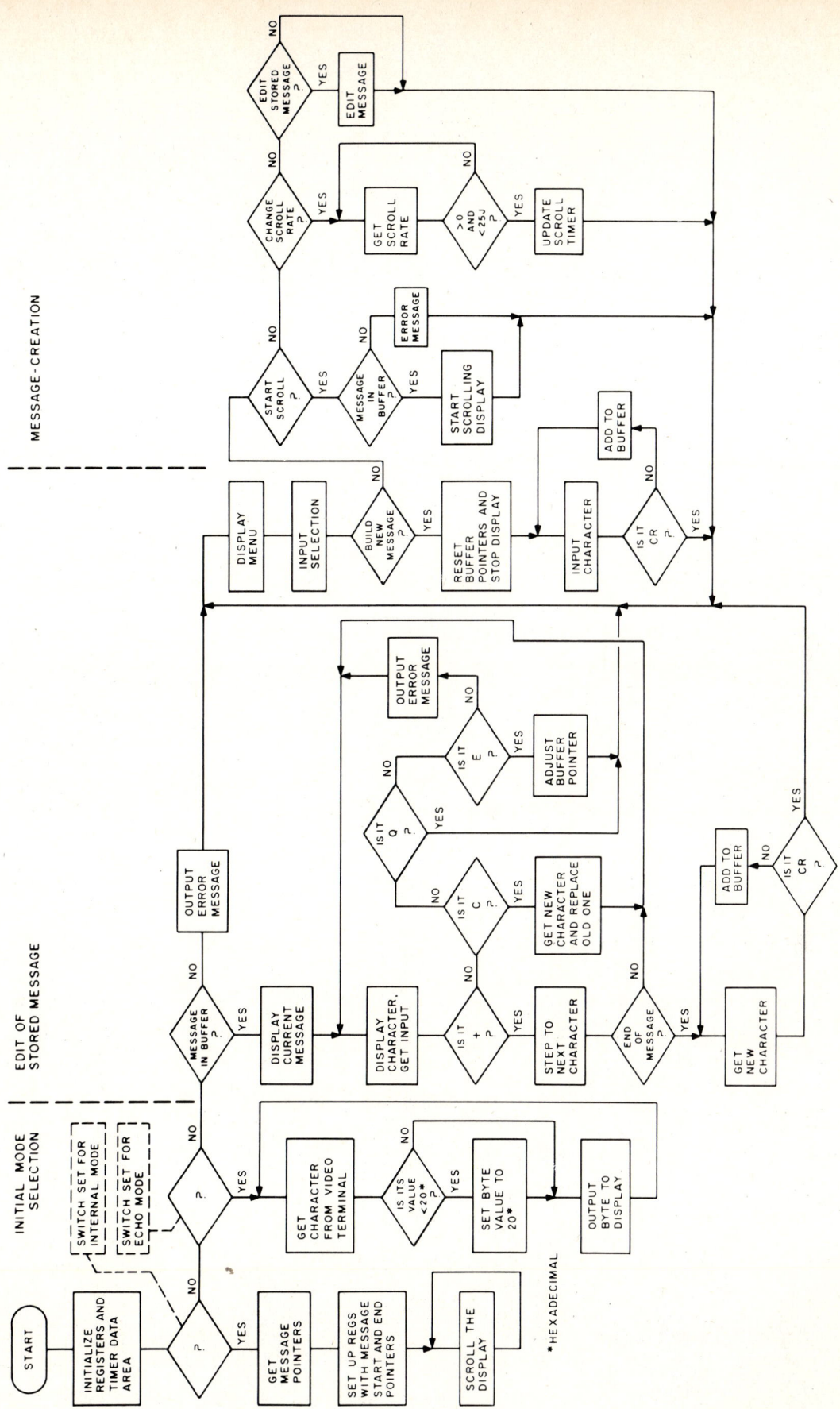

Figure 7: *Flowchart of the display-controller software for the Z8-BASIC System Controller. Some modes of operation are selected by DIP-switch settings on the Z8 board.*

Listing 2: *BASIC program written for the Z8-BASIC System Controller to control the scrolling LED display. Some actions are controlled by machine-language subroutines called by BASIC.*

```
10  GO @%1900
15  B=%0810:E=%080F
20  REM IF I GOT HERE, EXTERNAL MODE IS SELECTED
30  @%F4=%40
40  IF AND(@%FFD,2)=0 THEN GOTO 10000
50  REM BUILD MESSAGE MENU
60  PRINT
70  PRINT "CREATE A DISPLAY MENU"
80  PRINT:PRINT "1 - BUILD A NEW MESSAGE"
90  PRINT "2 - START MESSAGE SCROLLING"
100 PRINT "3 - CHANGE SCROLL RATE OF DISPLAY"
110 PRINT "4 - EDIT THE STORED MESSAGE"
140 PRINT:INPUT A
150 IF A<1 THEN 50
160 IF A>4 THEN 50
170 GOSUB A*1000
180 GOTO 50
1010 REM ENTER NEW DATA FOR DISPLAY
1015 GO @%1900
1020 PRINT "ENTER DISPLAY CHARACTERS. END WITH RETURN OR ENTER KEY"
1040 GO @%1900
1050 PRINT
1060 GOTO 1030
1070 RETURN
2000 REM START MESSAGE SCROLLING ON DISPLAY
2010 IF E<>%080F THEN 2040
2020 PRINT "NO MESSAGE IN BUFFER - DISPLAY NOT STARTED"
2030 RETURN
2040 ^%38=%010
2050 @%0801=%08:@%0802=%10
2060 ^%3A=E
2070 GO @%198E
2080 RETURN
3010 REM EDIT THE STORED MESSAGE
3010 PRINT "ENTER NEW SCROLL RATE"
3020 INPUT A
3030 IF A<1 THEN 3010
3040 IF A>255 THEN 3010
3050 X=USR(%1A76,A)
3060 RETURN
3010 REM EDIT THE STORED MESSAGE
4010 GO @%1900
4020 IF E<>%080F THEN 4050
4030 RETURN
4040 PRINT "NO MESSAGE IN BUFFER - EDIT NOT POSSIBLE"
4040 RETURN
4050 PRINT "LISTING OF CURRENT MESSAGE"
4060 X=B
4070 C=@%X:GO @%61,C
4080 X=X+1
4090 IF X>=E THEN 4110
4100 GOTO 4070
4110 X=B
4120 PRINT:PRINT
4130 PRINT "AS EACH CHARACTER IS DISPLAYED, ENTER + TO SKIP,C TO CHANGE"
4135 PRINT "TO QUIT LEAVING REST OF MESSAGE,"
4137 PRINT "TYPE E TO QUIT ENDING MESSAGE AT THIS CHARACTER"
4140 GO @%61,@%X
4150 PRINT "      ? ";
4160 C=USR(%54)
4165 PRINT
4170 IF C=%2B THEN X=X+1:GOTO 4220
4180 IF C=%51 THEN RETURN
4190 IF C=%45 THEN E=X:RETURN
4200 IF C<>%43 THEN PRINT "PLEASE TYPE +,C,Q, OR E":GOTO 4140
4205 PRINT "NEW CHARACTER ? ";
4207 C=USR(%54):@%X=C:PRINT:GOTO 4140
4210 PRINT "NEW CHARACTER ? ";:C=USR(%54):@%X=C:GOTO 4140
4220 IF X<=E THEN 4140
4230 PRINT "END OF ORIGINAL MESSAGE"
4240 PRINT "TYPE ADDITIONAL CHARACTERS AND END WITH RETURN OR ENTER"
4250 C=USR(%54)
4260 IF C=13 THEN RETURN
4270 E=E+1:@E=C
4280 GOTO 4250
10010 REM ECHO MODE
10010 @%38=%08
10020 @%3A=%10
10030 X=USR(%54)
10035 IF X<%20 THEN X=%20
10040 @%0810=X
10050 GO @%1A12
10060 GOTO 10030
```

Listing 3: *Assembly-language listing of the machine-language subroutines needed to control the LED display.*

```
1900                   0010 ;  Z-8 LED DISPLAY DRIVER PROGRAM
1900                   0020 ;  FOR THE Z-8 SYSTEM CONTROLLER ONLY
1900                   0030 ;  VERSION 1.1 12/31/83
1900                   0040 ;
1900                   0045 ;  INITIALIZATION ROUTINE
1900                   0060 ;
1900                   0070         ORG     1900H
1900          8F       0080 DINIT   EQU     $
1900          8F       0090         DI
1901 70   FD           0110         PUSH    RP
1903 31   30           0110         SRP     #MYREG
1905 CC   10           0120         LD      R12,#10H
1907 DC   0F           0130         LD      R13,#0FH
1909 FC   8D           0140         LD      R15,#8DH
190B 92   FC           0150         LDE     @RR12,R15
190D A0   EC           0160         INCW    R12
190F FC   19           0170         LD      R15,#SCROLLH
1911 92   FC           0180         LDE     @RR12,R15
1913 A0   EC           0190         INCW    R12
1915 FC   C1           0200         LD      R15,#SCROLLL
1917 92   FC           0210         LDE     @RR12,R15
1919                   0220 ;
1919                   0230 ;  REGISTER LAYOUT -
1919                   0240 ;  RR0 - POINTER TO COLUMN TO OUTPUT
1919                   0250 ;  R2  - COUNT OF COLUMNS LEFT TO DISPLAY
1919                   0260 ;  R3  - RIPPLE TIMER CONSTANT FOR SCROLLING
1919                   0270 ;  R4  - FLAGS BYTE :
1919                   0280 ;          BIT 0 - RUN MODE
1919                   0290 ;          BIT 1 - UNDEFINED
1919                   0300 ;          BIT 2 - BLANK COLUMN SUPPRESS IF SET
1919                   0310 ;          BITS 3-7 - UNDEFINED
1919                   0320 ;  RR6 - POINTER TO SWITCHES
```

Listing 3 continued:

```
                                                                    ADD    R13,R15           ; ADD IN MESSAGE # OFFSET
1919                                                        1060
1919                                                        1070  ; NOTE: THE ABOVE CODE WILL ONLY WORK IF THE
1919  E6 F6 00                    LD     0F6H,#0           ; SET PORT 2 AS ALL OUTPUT
                                                            1080  ;        LOW ORDER BYTE DOES NOT OVERFLOW WHEN
191C  E6 02 04                    LD     2,#4              ; DATA=0, CLOCK=0, CLEAR=1
                                                            1090  ;        0FH IS ADDED TO IT.
191F  E6 02 00                    LD     2,#0              ; DATA,CLOCK,CLEAR=0
                                                            1100
1922  4C 00              ; MODE BYTE TO *RUN, BLANKS ENABLED
                                                            1110
1924                     ; SET BLANK FLAG FROM SWITCH                             1120       LDE    R8,@RR12          ; GET MESSAGE BEGIN AND
1924  82 F6 00                    LDE    R15,@RR6          ; PICK UP SWITCHES     1130       INCW   RR12
1926  76 EF 40                    TM     R15,#40H          ; IS BLANK SUPPRESS DESIRED ?  1140  LDE    R9,@RR12          ; END POINTERS
1929  6B 03                       JR     Z,TIMER           ; NO, PROCEED          1150       INCW   RR12
192B  46 E4 04                    OR     R4,#4H            ; YES, SET BLANK SUPPRESS FLAG  1160  LDE    R10,@RR12         ; INTO THEIR
192E                     ; SET UP TIMER, BUT CLEAR INTERRUPT MASK FOR IT          1170       INCW   RR12              ; RESPECTIVE REGISTERS
192E  TIMER     EQU    $                                                          1180       LD     R11,@RR12
192E  6C FF                       LD     R6,#SWITCH!H      ; HIGH ORDER BYTE OF SWITCHES   1190       LD     R12,#BGNBUF!H     ; GET ADDRESS OF BEGIN
1930  7C FD                       LD     R7,#SWITCH!L      ; LOW ORDER BYTE OF SWITCHES    1200       LD     R13,#BGNBUF!L     ; MESSAGE SAVE AREA
1932  0C 30                       LDE    R3,#30H           ; GET THE BYTE         1210       LDE    @RR12,R8          ; AND SAVE THE
1934  56 E3 30                    AND    R3,#30H           ; ISOLATE RIPPLE RATE BITS      1220       INCW   RR12              ; POINTER THERE
1937  06 E3 10                    ADD    R3,#10H           ; ADD 16 TO BYTE       1230       LDE    @RR12,R9
193A  CC 00                       LD     R12,#TIMEIH       ; GET ADDRESS OF       1240       CALL   SCRST             ; START THE SCROLL PROCESS
193C  DC 00                       LD     R13,#TIMEIL       ; TIMER DATA AREA      1250  LOOP EQU    $
193E  92 3C                       LDE    @RR12,R3          ; STORE RATE THERE     1260       JR     LOOP              ; DO AN INFINITE LOOP
1940  06 E3 03                    LDE    PRE1,#03H         ; SET PRESCALE1 TO 64  1270
1943  E6 F2 20                    LD     T1,#20H           ; START TIMER1 COUNT TO 32      1280
1946  F1 0C                       OR     TMR,#0CH          ; START T1 A'COUNTIN'  1290
1949  E6 FB 00                    LD     IMR,#0            ; RESET ALL INTERRUPTS.        1300
194C                     ; RESET BUFFER BEGIN DATA AREA TO 0                      1310  ; SCRST - SCROLL START ROUTINE
194C  CC 08                       LD     R12,#BGNBUF!H                                   1320
194E  DC 01                       LD     R13,#BGNBUF!L                                   1330  ; ASSUMES :
1950  FC 00                       LD     R15,#0                                          1340  ;
1952  92 FC                       LDE    @RR12,R15                                       1350  ;           ALL REGS SET EXCEPT COLUMN POINTER
1954  A0 EC                       INCW   R12                                             1360  ;           AND COLUMN COUNTER
1956  92 FC                       LDE    @RR12,R15                                       1370  ;
1958                     ; TEST FOR INTERNAL/EXTERNAL MODE                        1380
1958                                                                              1390  SCRST     EQU    $
1958  82 F6                       LDE    R15,@RR6          ; GET SWITCHES         1400       PUSH   RP                ; SAVE OLD REG POINTER
195A  76 EF 01                    TM     R15,#1            ; TEST INT/EXT SWITCH  1410       SRP    #MYREG            ; POINT TO MY WORK REGS
195D  6B 04                       JR     Z,INTERNAL        ; IF 0 THEN INTERNAL MODE      1420       LDE    R15,@RR8          ; GET 1ST DATA BYTE FROM BUFFER
195F                     ; EXTERNAL MODE MEANS JUST RETURN TO BASIC ?             1430  COLSE     EQU    $
195F                                                                              1440       AND    R15,#07FH         ; STRIP OFF H.O. BIT
195F  50 FD                       POP    RP                ; RESTORE REG POINTER  1450       LD     R0,#CHRTABIH      ; PICK UP CHRTAB
1961  9F                          EI                       ; ENABLE INTERRUPTS BEFORE EXIT 1460  LD     R1,#CHRTABIL      ; BASE ADDRESS
1962  AF                          RET                      ; RETURN TO CALLER     1470       LD     R15,#0            ; IS DATA BYTE = 0 ?
1963                                                                              1480       CP     EQ,COLCNT         ; YES, CHRTAB ADDRESS OK
1963  D6 19 66           INTER    EQU    $                                        1490  SKIP1     EQU    $
1963                              CALL   INTMSG                                   1500       INCW   RR0               ; STEP OUT
1966                                                                              1510       INCW   RR0               ; TO THIS BYTE'S
1966                     ; END OF INIT ROUTINE                                    1520       INCW   RR0               ; TRANSLATE
1966                                                                              1530       INCW   RR0               ; AREA
1966                                                                              1540       INCW   RR0               ; IN THE
1966                     ; INTMSG - SET UP AND SCROLL ONE OF                      1550       INCW   RR0               ; CHRTAB TABLE.
1966                     ;         THE INTERNALLY STORED MESSAGES                 1560       DJNZ   R15,SKIP1         ; DO UNTIL 0
1966                                                                              1570  COLCN     EQU    $
1966                     ; ASSUMES :                                              1580       LD     R2,#6             ; ASSUME 6 COLUMNS FOR NOW
1966                     ;    INIT HAS BEEN EXECUTED, OR REGS SET                 1590       TM     H4,#4H            ; ARE WE BLANK SUPPRESSING ?
1966                     ;        BY SOME OTHER PROGRAM                           1600       JR     Z,START           ; NO, ALL IS WELL. CRANK IT UP
1966                                                                              1610       DEC    R2                ; ONLY 5 COLS PER CELL PLEASE
1966                     INTMS    EQU    $                                        1620       INCW   RR0               ; SKIP BLANK COLUMN IN TABLE
1966  82 F6                       LDE    R15,@RR6          ; PICK UP SWITCHES     1630
1966  56 EF 0C                    AND    R15,#0CH          ; ISOLATE THE MESSAGE SELECT BITS  1640  ; START THE INTERRUPTS FOR SCROLLING
196B  CC 1A                       LD     R12,#INTTBL!H     ; GET MESSAGE TABLE    1650
196D  DC 85                       LD     R13,#INTTBL!L     ; BASE ADDRESS         1660  START     EQU    $
                                                                                  1670       DI                       ; SHUT DOWN INTERRUPTS (SHOULD BE OFF
                                                                                  1680  ANYWAY)
                                                                                  19BA  46 FB 20             OR     IMR,#20H          ; ALLOW INTERRUPT IRQ5 (T1)
                                                                                  19BD  50 FD                POP    RP                ; RESTORE REG POINTER
                                                                                  19BF  9F                   EI                       ; TURN ON INTERRUPTS
                                                                                  19C0  AF                   RET                      ; RETURN TO CALLER
                                                                                  19C1
                                                                                  19C1
                                                                                  19C1
```

Listing 3 continued

131

Listing 3 continued:

```
                    1750  ; SCROLL - INTERRUPT DRIVEN SCROLL ROUTINE
                    1760  ;
                    1770  ; ASSUMES :
                    1780  ;
                    1790  ;       ALL POINTER REGS CORRECTLY INITIALIZED
                    1800  ;       REG 3 HAS TIME INCREMENTS REMAINING TO NEXT OUTPUT
                    1810  ;       TIME CONSTANT IS STORED AT "TIME" DATA AREA
                    1820  ;
19C1                1830  ;
19C1                1840  SCROL EQU     $
19C1 70 FD          1850        PUSH    RP                  ; SAVE OLD REG POINTER
19C1 31 30          1860        SRP     #MYREG              ; POINT TO MY REGS
19C1                1870        DJNZ    R3,RETRN            ; IF INTERVAL HAS NOT EXPIRED, GO AWAY
19C5 3A 40          1880        LD      R12,#TIME!H         ; POINT TO TIMER
19C7 CC 0B          1890        LD      R13,#TIME!L         ; CONSTANT DATA AREA
19C9 DC 00          1900        LDE     R3,@RR12            ; RESTORE FULL COUNT
19CD 82 5C          1910        CALL    COLWRT              ; WRITE A COLUMN OF DATA
19D0 A0 E0          1920        INCW    RR0                 ; STEP TO NEXT COLUMN
19D2 2A 3B          1930        DJNZ    R2,RETRN            ; IF NOT LAST COLUMN, GO AWAY
19D4                1940  ;
19D4                1950  ; STEP TO NEXT CHARACTER IN BUFFER
19D4                1960  ;
19D4 A0 E6          1970        INCW    R6                  ; ADVANCE CURBUF BY 1
19D6 A2 6A          1980        CP      R6,R10              ; COMPARE CURBUF AND ENDBUF HIGH BYTE
19D8 7B 0E          1990        JR      ULT,SETCH           ; IF NOT GE THEN NO WRAP
19DA A2 9B          2000        CP      R9,R11              ; COMPARE LOW BYTES
19DC 3B 0A          2010        JR      ULE,SETCH           ; IF NOT GT THEN NO WRAP
19DE                2020  ;
19DE                2030  ; WRAP TO BEGINNING OF MESSAGE
19DE                2040  ;
19DE CC 08          2050        LD      R12,#BGNBUF!H       ; PICK UP POINTER TO
19E0 DC 01          2060        LD      R13,#BGNBUF!L       ; BEGIN BUFFER REMEMBERANCE
19E2 C2 8C          2070        LDE     R8,@RR12            ; GET HIGH BYTE
19E4 A0 EC          2080        INCW    RR12
19E6 C2 9C          2090        LDE     R9,@RR12            ; AND LOW BYTE
19E8                2100  ;
19E8                2110  ; PICK UP CHARACTOR AND PROCESS -
19E8                2120  ; SET UP POINTER TO TRANSLATE
19E8                2130  ; TABLE ADDRESS AS WELL AS # OF COLUMNS TO PROCESS
19E8                2140  ;
19E8                2150  SETCH EQU     $
19E8 82 F0          2160        LDE     R15,@RR8            ; GET DATA BYTE FROM BUFFER
19EA 0C 1A          2170        LD      R0,#CHRTAB!H        ; SET UP POINTER TO
19EC 1C 95          2180        LD      R1,#CHRTAB!L        ; DATA TABLE BASE
19EE 56 EF 7F       2190        AND     R15,#7FH            ; MASK OFF H.O. BIT
19F1 A6 EF 00       2200        CP      R15,#0              ; IS DATA BYTE 0?
19F4 6B 0E          2210        JR      EQ,SETCOL           ; YES, POINTER IN TABLE IS OK.
19F6                2220  ;
19F6                2230  ; ADJUST CHRTAB POINTER
19F6                2240  ;
19F6                2250  ;
19F6                2260  SKIP2 EQU     $
19F6 A0 E0          2270        INCW    RR0
19F8 A0 E0          2280        INCW    RR0
19FA A0 E0          2290        INCW    RR0
19FC A0 E0          2300        INCW    RR0
19FE A0 E0          2310        INCW    RR0
1A00 A0 E0          2320        INCW    RR0
1A02 FA F2          2330        DJNZ    R15,SKIP2
1A04                2340  ;
1A04                2350  ; DETERMINE # OF COLUMNS TO WRITE
1A04                2360  ; ADJUST CHRTAB POINTER IF NECESSARY
1A04                2370  ;
1A04                2380  SETCO EQU     $
1A04 2C 06          2390        LD      R2,#6               ; ASSUME 6 COLS TO START
1A06 76 E4 04       2400        TM      R4,#04H             ; TEST COLUMN BLANK FLAG
1A09 6B 04          2410        JR      Z,RETRN             ; IF 0, THEN BLANK. NO ADJUST
1A0B 00 E2          2420        DEC     R2                  ; IF 1, NO BLANK. ADJUST COL. COUNT
1A0D A0 E0          2430        INCW    RR0                 ; AND CHRTAB POINTER
1A0F                2440  RETRN EQU     $
1A0F D0 FD          2450        POP     RP                  ; RESTORE REG POINTER
1A11 BF             2460        IRET                        ; RETURN FROM INTERRUPT
1A12                2470  ;
1A12                2480  ; CHARACTER OUTPUT ROUTINE - NON-INTERRUPT DRIVEN
1A12                2490  ;
1A12                2500  ; ASSUMES :
1A12                2510  ;
1A12                2520  ;       BUFFER POINTER IS ON XTER TO BE OUTPUT.
1A12                2530  ;       CHARACTOR IS INDEX TO LED TABLE
1A12                2540  ;       SWITCH POINTER REGS ARE INITIALIZED
1A12                2550  ;
1A12                2560  CHOUT EQU     $
1A12 70 FD          2570        PUSH    RP                  ; SAVE OLD REG POINTER
1A14 31 30          2580        SRP     #MYREG              ; POINT TO MY WORK REGS
1A16 82 F8          2590        LDE     R15,@RR8            ; GET CHARACTOR TO BE DISPLAYED
1A18 56 EF 7F       2600        AND     R15,#07FH           ; STRIP OFF H.O. BIT
1A1B                2610  ;
1A1B                2620  ; BUILD INDEX INTO TABLE FOR COLUMNS OF DATA
1A1B                2630  ;
1A1B 0C 1A          2640        LD      R0,#CHRTAB!H        ; PICK UP TABLE
1A1D 1C 95          2650        LD      R1,#CHRTAB!L        ; BASE ADDRESS
1A1F A6 EF 00       2660        CP      R15,#0              ; IS THIS CELL 0 ?
1A22 6B 08          2670        JR      EQ,SUPRS            ; YES, GO TO BLANK SUPPRESS
1A24 EC 06          2680        LD      R14,#6              ; SET UP 6 COLUMNS PER CELL
1A26                2690  INDEX EQU     $
1A26 A0 E0          2700        INCW    RR0                 ; ADD 1 TO COLUMNS POINTER
1A28 EA FC          2710        DJNZ    R14,IDXLP           ; DO FOR 6 COLUMNS
1A2A FA F0          2720        DJNZ    R15,INDEX           ; DO FOR R15 TIMES
1A2C                2730  IDXLP EQU     $
1A2C                2740  ;
1A2C                2750  ; NOW CHECK TO SEE IF FIRST COLUMN BLANK IS WANTED
1A2C                2760  ;
1A2C                2770  SUPRS EQU     $
1A2C 2C 06          2780        LD      R2,#6               ; ASSUME 6 COLUMNS PER CELL NOW
1A2E 76 E4 04       2790        TM      R4,#04H             ; TEST BLANK BIT
1A31 6B 04          2800        JR      Z,BLANK             ; SKIP BLANK COLUMN
1A33 00 E2          2810        DEC     R2                  ; DROP 1 COLUMN COUNT
1A35 A0 E0          2820        INCW    RR0                 ; SKIP BLANK COLUMN
1A37                2830  BLANK EQU     $
1A37                2840  ;
1A37                2850  ; THIS IS THE OUTPUT ROUTINE FOR THE CELL
1A37                2860  ;
1A37 D0 1A 41       2870        CALL    COLWRT              ; WRITE THIS COLUMN
1A3A A0 E0          2880        INCW    RR0                 ; STEP TO NEXT ONE
1A3C 2A F9          2890        DJNZ    R2,BLANK            ; IF NOT DONE, DO IT AGAIN
1A3E 50 FD          2900        POP     RP                  ; RESTORE OLD REG POINTER
1A40 AF             2910        RET                         ; RETURN TO CALLER
1A41                2920  ;
1A41                2930  ; ROUTINE TO WRITE A COLUMN OF LED INFORMATION
1A41                2940  ;
1A41                2950  ; ASSUMES :
1A41                2960  ;
1A41                2970  ;       REGS R0-R1 POINT TO THE BYTE TO OUTPUT
1A41                2980  ;       REG4 BIT 2 INDICATES INVERSE VIDEO IF SET
1A41                2990  ;       THE CLOCK BIT IS 0 ON ENTRY
1A41                3000  ;       REG 2 IS CORRECTLY SET FOR OUTPUT
1A41                3010  ;
1A41                3020  COLWR EQU     $
1A41 70 FD          3030        PUSH    RP                  ; SAVE OLD REG POINTER
1A43 31 30          3040        SRP     #MYREG              ; POINT TO MY WORK REGS
1A45 82 F0          3050        LDE     R15,@RR0            ; PICK UP DATA BYTE
1A47 90 EF          3060        RL      R15                 ; ADJUST COLUMN CORRECTLY
1A49 EC 08          3070        LD      R14,#8              ; DO THIS 8 TIMES
```

Listing 3 continued:

```
1A4B                              DW      #MSG1BGN
1A4B 1C D6                        DW      #MSG1END
1A4D 1C E6                        DW      #MSG2BGN
1A4F 1C E7                        DW      #MSG2END
1A51 1D 17                        DW      #MSG3BGN
1A4D 56 ED 01          3130 BITLO EQU     $
1A50 D9 02             3140       LD      R13,R15       ; SAVE DATA BIT. RESET ALL OTHERS
1A52 40 ED 02          3150       AND     R13,#1        ; AND PUT DATA BIT ON PORT 2
1A55 FF                3160       LD      2,R13         ; TURN ON CLOCK BIT
1A56 FF                3170       OR      R13,#2        ; WAIT A LITTLE BIT
1A57 FF                3180       NOP
1A58 FF                3190       NOP
                       3200       NOP
                       3210       NOP
1A59 D9 02             3220       LD      2,R13         ; AND WRITE TO PORT 2
1A5B 56 ED 01          3230       AND     R13,#1        ; TURN OFF CLOCK BIT
1A5E FF                3240       NOP                   ; WAIT A LITTLE BIT
1A5F FF                3250       NOP
1A60 FF                3260       NOP
1A61 FF                3270       NOP
1A62 D9 02             3280       LD      2,R13         ; AND WRITE TO PORT 2
1A64 E0 EF             3290       RR      R15           ; SET REG 15 TO NEXT BIT
1A66 FF                3300       NOP                   ; WAIT A LITTLE WHILE
1A67 FF                3310       NOP
1A68 FF                3320       NOP
1A69 FF                3330       NOP
1A6A EA DF             3340       DJNZ    R14,BITLOOP   ; DO UNTIL ALL BITS FINISHED
1A6C 50 FD             3350       POP     RP            ; RESTORE OLD REG POINTER
1A6E AF                3360       RET                   ; GO BACK TO CALLER
                       3370 ;
                       3380 ;
                       3390 ;
                       3400 ; CLEAR THE DISPLAY ROUTINE
                       3410 ;
                       3420 ; ASSUMES :
                       3430 ;
                       3440 ;          PORT 2 IS CORRECTLY CONFIGURED FOR OUTPUT
                       3450 ;
1A6F                   3460 ;
1A6F                   3470 CLEAR EQU    $
1A6F E0 02 04          3480       LD     2,#04H        ; CLEAR ON, CLOCK AND DATA LOW
1A72 E0 02 00          3490       LD     2,#0H         ; TURN OFF CLEAR
1A75 AF                3500       RET                  ; RETURN TO CALLER
                       3510 ;
                       3520 ;
                       3530 ;
                       3540 ; RIPPLE - SET RIPPLE CLOCK RATE FOR
                       3550 ; SCROLL MODE
                       3560 ;
                       3570 ; ASSUMES :
                       3580 ;
                       3590 ;          NEW RIPPLE RATE IN REG 15 (ABSOLUTE)
                       3610 ;
1A76                   3620 RIPPL EQU    $
1A76                   3630       PUSH   RP
1A76 70 FD             3640       SRP    #MYREG        ; SAVE OLD REG POINTER
1A78 31 30             3650       LD     R12,#TIME1H   ; SET REG POINTER TO MY REGS
1A7A CC 00             3660       LD     R13,#TIME1L   ; POINT TO TIMER
1A7C DC 00             3670       LD     R15,1         ;   DATA AREA
1A7E F0 13             3680       LDE    @RR12,R15     ; STORE NEW RIPPLE RATE
1A82 92 FC             3690       POP    RP            ; RESTORE OLD REG POINTER
1A84 50 FD             3700       RET                  ; RETURN TO CALLER
1A86 AF                3710 ;
                       3720 ;
                       3730 ; DATA AREAS AND CONSTANTS
                       3740 ;
1A85                   3750 MYREG EQU    30H
1A85                   3760 SWITC EQU    0FFFDH
1A85                   3770 TIME  EQU    0800H
1A85                   3780 BGNBU EQU    TIME+1
1A85                   3820 INTTB EQU    $
1A85 1C D6                        DW     #MSG1BGN
1A87 1C E6                        DW     #MSG1END
1A89 1C E7                        DW     #MSG2BGN
1A8B 1D 17                        DW     #MSG2END
1A8D 1D 18                        DW     #MSG3BGN
1A8F 1D 52                        DW     #MSG3END
1A91 1D 63                        DW     #MSG4BGN
1A93 1D A6                        DW     #MSG4END
1A95 FF                3910 CHRTA EQU    $
1A95                   3920       DS     192           ; (20H * 6)
1B55                   3930 SPECI EQU    $
1B55 00                3940       DB     0,0,0,0,0
1B5B 00 00 00 00 00    3950       DB     0,0,0,7DH,0,0
1B61 00 00 00 70 00    3960       DB     0,0,70H,0,70H,0
1B67 00 14 7F 14 7F    3970       DB     0,14H,7FH,14H,7FH,14H
1B6D 00 12 2A 7F 2A    3980       DB     0,12H,2AH,7FH,2AH,24H
          24
1B73 00 62 64 06 13    3990       DB     0,62H,64H,06H,13H,23H
          23
1B79 00 36 49 35 02    4000       DB     0,36H,49H,35H,02H,05H
1B7F 05                4010       DB     0,00,00,70H,00,00
1B85 00
1B8B 00 1C 22 41 00    4020       DB     0,1CH,22H,41H,0,0
1B91 00                4030       DB     0,0,0,41H,22H,1CH
          1C
1B97 00 22 14 7F 14    4040       DB     0,22H,14H,7FH,14H,22H
          22
1B9D 00 08 08 3E 08    4050       DB     0,08H,08H,3EH,08H,08H
1BA3 00                4060       DB     0,0,1,6,0,0
          06
1BA9 00 08 08 08 08    4070       DB     0,8,8,8,8,8
1BAF 00 02 04 06 10    4080       DB     0,0,0,1,0,0
          20
1BB5                   4090       DB     0,2,4,8,10H,20H
                       4100 ; NUMBERS
1BB5 00 3E 45 49 51    4110       DB     0,3EH,45H,49H,51H,3EH
          3E
1BBB 00 00 21 7F 01    4120       DB     0,0,21H,7FH,01,0
          00
1BC1 00 23 45 49 49    4130       DB     0,23H,45H,49H,49H,31H
          31
1BC7 00 42 41 49 59    4140       DB     0,42H,41H,49H,59H,66H
          66
1BCD 00 0C 14 24 7F    4150       DB     0,0CH,14H,24H,7FH,04H
          04
1BD3 00 72 51 51 51    4160       DB     0,72H,51H,51H,51H,4EH
          4E
1BD9 00 1E 29 49 49    4170       DB     0,1EH,29H,49H,49H,46H
          46
1BDF 00 40 47 48 50    4180       DB     0,40H,47H,48H,50H,60H
          60
1BE5 00 36 49 49 49    4190       DB     0,36H,49H,49H,49H,36H
          36
1BEB 00 31 49 49 4A    4200       DB     0,31H,49H,49H,4AH,3CH
          3C
                       4210 ; MORE SPECIAL CHARACTORS
1BF1 00 00 00 14 00    4220       DB     0,0,0,14H,0,0
1BF7 00 00 01 1B 00    4230       DB     0,0,1,16H,0,0
1BFD 00 06 14 22 41    4240       DB     0,8,14H,22H,41H,0
```

Listing 3 continued

Listing 3 continued:

```
1C03  00                            4250           DB      0,14H,14H,14H,14H,14H
1C09  00 14 14 14 14 14
1C0F  00 00 41 22 14                4260           DB      0,0,41H,22H,14H,08H
      08
1CCF  00 20 40 4D 50                4270           DB      0,20H,40H,40H,50H,20H
      20
                                    4280 ; AT SIGN AND LETTERS (UPPER CASE)
1C15  00 3E 41 5D 4D                4290           DB      0,3EH,41H,5DH,4DH,39H
      39
1C1b  1F                            4300           DB      0,1FH,24H,44H,24H,1FH
1C21  00 1F 24 44 24
      1F
1C27  00 7F 49 49 49                4310           DB      0,7FH,49H,49H,49H,36H
      36
1C2D  00 3E 41 41 41                4320           DB      0,3EH,41H,41H,41H,22H
      22
1C33  00 7F 41 41 41                4330           DB      0,7FH,41H,41H,41H,3EH
      3E
1C39  00 7F 49 49 49                4340           DB      0,7FH,49H,49H,49H,41H
      41
1C3F  00 7F 48 48 48                4350           DB      0,7FH,48H,48H,48H,40H
      40
1C45  00 3E 41 41 45                4360           DB      0,3EH,41H,41H,45H,47H
      47
1C4B  00 7F 08 08 08                4370           DB      0,7FH,08H,08H,08H,7FH
      7F
1C51  00 00 41 7F 41                4380           DB      0,00H,41H,7FH,41H,00H
      00
1C57  00 02 01 01 01                4390           DB      0,02H,01H,01H,01H,7EH
      7E
1C5D  00 7F 08 14 22                4400           DB      0,7FH,08H,14H,22H,41H
      41
1C63  00 7F 01 01 01                4410           DB      0,7FH,01H,01H,01H,01H
      01
1C69  00 7F 20 10 20                4420           DB      0,7FH,20H,18H,20H,7FH
      7F
1C6F  00 7F 10 00 04                4430           DB      0,7FH,10H,08H,04H,7FH
      7F
1C75  00 3E 41 41 41                4440           DB      0,3EH,41H,41H,41H,3EH
      3E
1C7B  00 7F 46 48 48                4450           DB      0,7FH,48H,48H,48H,30H
      30
1C81  00 3E 41 45 42                4460           DB      0,3EH,41H,45H,42H,3DH
      3D
1C87  00 7F 48 4C 4A                4470           DB      0,7FH,48H,4CH,4AH,31H
      31
1C8D  00 32 49 49 49                4480           DB      0,32H,49H,49H,49H,26H
      26
1C93  00 40 40 7F 40                4490           DB      0,40H,40H,7FH,40H,40H
      40
1C99  00 7E 01 01 01                4500           DB      0,7EH,01H,01H,01H,7EH
      7E
1C9F  00 7C 02 01 02                4510           DB      0,7CH,02H,01H,02H,7CH
      7C
1CA5  00 7F 02 0C 02                4520           DB      0,7FH,02H,0CH,02H,7FH
      7F
1CAB  00 63 14 08 14                4530           DB      0,63H,14H,08H,14H,63H
      63
1CAB  00 60 10 0F 10                4540           DB      0,60H,10H,0FH,10H,60H
      60
1CB1  00 43 45 49 51                4550           DB      0,43H,45H,49H,51H,61H
      61
1CB7  00 7F 7F 41 41                4560           DB      0,7FH,7FH,41H,41H,41H
      41
1CBD  00 20 10 08 04                4570           DB      0,20H,10H,08H,04H,02H
      02
1CC3  00 41 41 41 7F                4580           DB      0,41H,41H,41H,7FH,7FH
      7F
1CC9  00 04 08 10 08                4590           DB      0,04H,08H,10H,08H,04H
      04
1CCF  00 01 01 01 01                4600           DB      0,01H,01H,01H,01H,01H
      01
                                    4610 ;
                                    4620 ; END OF CHRTAB
                                    4630 ;
                                    4640 ;
                                    4650 ;
                                    4660 ;
                                    4670 ; TEST MESSAGES FOR INTERNAL USE
                                    4680 ;
1CD5                                4690 MSG1B   EQU     $
1CD5  54 45 53 54 20                4700           DB      'TEST MESSAGE 1'
      4D 45 53 53 41
      47 45 20 31
1CE5                                4705 MSG1E   EQU     $-1
1CE7                                4710 MSG2B   EQU     $
1CE7  2E 2E 2E 2E                   4720 MSG2B   DB      '....'
1CEB  54 48 45 20 51                4730           DB      'THE QUICK BROWN FOX JUMPED OVER THE LAZY DOG '
      55 49 43 4B 20
      42 52 4F 57 4E
      20 46 4F 58 20
      4A 55 4D 50 45
      44 20 4F 56 45
      52 20 54 48 45
      20 4C 41 5A 59
      20 44 4F 47 20
1D14  2E 2E 2E 2E                   4735           DB      '....'
1D18                                4740 MSG2E   EQU     $-1
1D1B                                4750 MSG3B   EQU     $
1D1B  30 20 31 20 32                4760           DB      '0 1 2 3 4 5 6 7 8 9 '
      20 33 20 34 20
      35 20 36 20 37
      20 38 20 39 20
1D2C  21 20 22 20 23                4770           DB      '! " # $ & ( ) * : = - @ + ; , . / ? '
      20 24 20 26 20
      28 20 29 20 2A
      20 3A 20 3D 20
      2D 20 40 20 2B
      20 3B 20 2C 2F
      20 2F 20 3F 20
1D52  27 20 5B 20 5D                4780           DB      027H,02OH,05BH,02OH,05DH,02OH
1D58  3C 20 3D 20 3E                4790           DB      03CH,02OH,03DH,02OH,03EH,02OH
      20
1D5F                                4800 MSG3E   EQU     $-1
1D63                                4810 MSG4B   EQU     $
1D63  20 43 49 41 52                4820           DB      ' CIARCIA',027H,'S CIRCUIT CELLAR IS '
      43 49 41 27 53
      20 43 49 52 43
      55 49 54 20 43
      45 4C 4C 41 52
      20 49 53 20
1D80  53 55 50 50 4F                4830           DB      'SUPPORTED BY CURLEW',027H,'S SOFTWARE CELLAR ....'
      52 54 45 44 20
      42 59 20 43 55
      52 4C 45 57 27
      53 20 53 4F 46
      54 57 41 52 45
      20 43 45 4C 4C
      41 52 20 2E 2E
      2E
1DA9                                4840 MSG4E   EQU     $-1
1DA9                                4850 ;
1DA9                                4860 ; END OF TEST MESSAGES
1DA9                                4870 ;

                                         SYMBOL TABLE
```

Listing 3 continued:

```
BGNBG 0001   BITLO 1A4B   BLANK 1A57   CHFIX 1A18   CHOUT 1A12
CHRTA 1A95   CLEAR 1A9F   COLCH 19AE   COLSE 1994   COLWK 1A41
DINIT 1900   IDALP 1A20   INDEX 1A24   INTER 1963   INTMS 1966
INTTB 1A65   LOOP  19DC   MSG1B 1CD5   MSG1E 1CE6   MSG2B 1CE7
MSG2E 1D17   MSG3B 1D16   MSG4B 1D62   MSG4E 1DA0
MYREG 0030   RETRN 1A0F   RIPPL 1A76   SCROL 19C1   SCRST 19BE
SETCH 19C0   SETCO 1A04   SKIP1 19A0   SKIP2 19F6   SPECI 1B55
START 1989   SUPRS 1A2C   SWITC FFFD   TIME  0800   TIMER 192E
```

Listing 4: *The display-control program operating in the echo mode produces output on the video-display terminal like that shown here.*

```
CREATE A DISPLAY MENU

1 - BUILD A NEW MESSAGE
2 - START MESSAGE SCROLLING
3 - CHANGE SCROLL RATE OF DISPLAY
4 - EDIT THE STORED MESSAGE

? 1

ENTER DISPLAY CHARACTERS.  END WITH RETURN OR ENTER KEY
TUST

CREATE A DISPLAY MENU

1 - BUILD A NEW MESSAGE
2 - START MESSAGE SCROLLING
3 - CHANGE SCROLL RATE OF DISPLAY
4 - EDIT THE STORED MESSAGE

? 4

LISTING OF CURRENT MESSAGE
TUST

AS EACH CHARACTER IS DISPLAYED, ENTER + TO SKIP,C TO CHANGE
TYPE Q TO QUIT LEAVING REST OF MESSAGE,
TYPE E TO QUIT ENDING MESSAGE AT THIS CHARACTER
T ? +
U ? C
NEW CHARACTER ? E
E ? +
S ? +
T ? +
END OF ORIGINAL MESSAGE
TYPE ADDITIONAL CHARACTERS AND END WITH RETURN OR ENTER
 MESSAGE FOR DISPLAY ....

CREATE A DISPLAY MENU

1 - BUILD A NEW MESSAGE
2 - START MESSAGE SCROLLING
3 - CHANGE SCROLL RATE OF DISPLAY
4 - EDIT THE STORED MESSAGE

? 4

LISTING OF CURRENT MESSAGE
TEST MESSAGE FOR DISPLAY ....

AS EACH CHARACTER IS DISPLAYED, ENTER + TO SKIP,C TO CHANGE
TYPE Q TO QUIT LEAVING REST OF MESSAGE,
TYPE E TO QUIT ENDING MESSAGE AT THIS CHARACTER
T ? Q

CREATE A DISPLAY MENU

1 - BUILD A NEW MESSAGE
2 - START MESSAGE SCROLLING
3 - CHANGE SCROLL RATE OF DISPLAY
4 - EDIT THE STORED MESSAGE

? 2

CREATE A DISPLAY MENU

1 - BUILD A NEW MESSAGE
2 - START MESSAGE SCROLLING
3 - CHANGE SCROLL RATE OF DISPLAY
4 - EDIT THE STORED MESSAGE

? 3

ENTER NEW SCROLL RATE
? 2

CREATE A DISPLAY MENU

1 - BUILD A NEW MESSAGE
2 - START MESSAGE SCROLLING
3 - CHANGE SCROLL RATE OF DISPLAY
4 - EDIT THE STORED MESSAGE

? 3

ENTER NEW SCROLL RATE
? 20

CREATE A DISPLAY MENU

1 - BUILD A NEW MESSAGE
2 - START MESSAGE SCROLLING
```

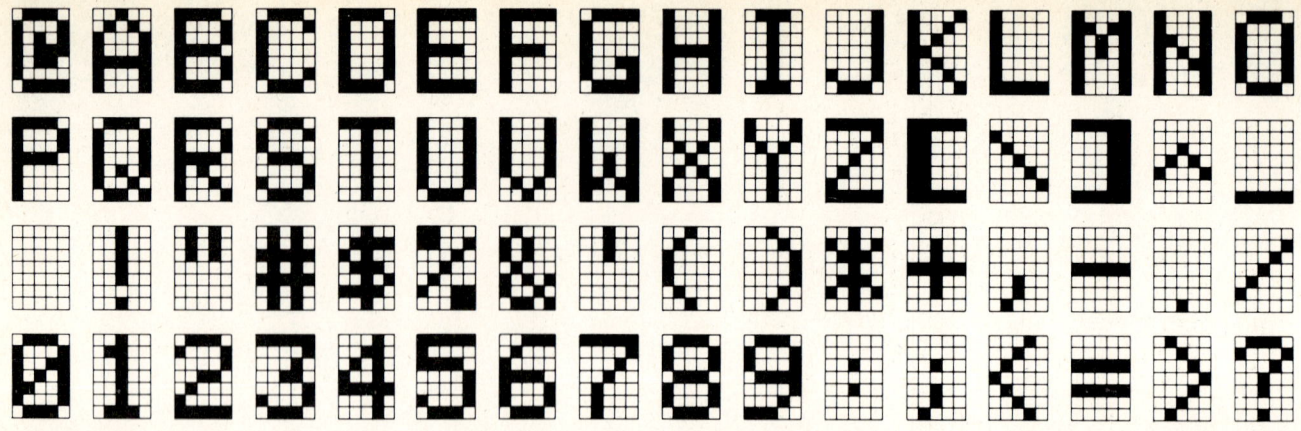

Figure 8: *An entire set of alphanumeric characters can be represented by these bit patterns in a 5- by 7-element matrix, that formed by the LED arrangement in figure 2.*

Photo 6: *In the finished project, multiple character segments are linked and housed together in an attractive wooden enclosure. Either canned or live messages can be written or continuously scrolled across the array under control of the Z8 software.*

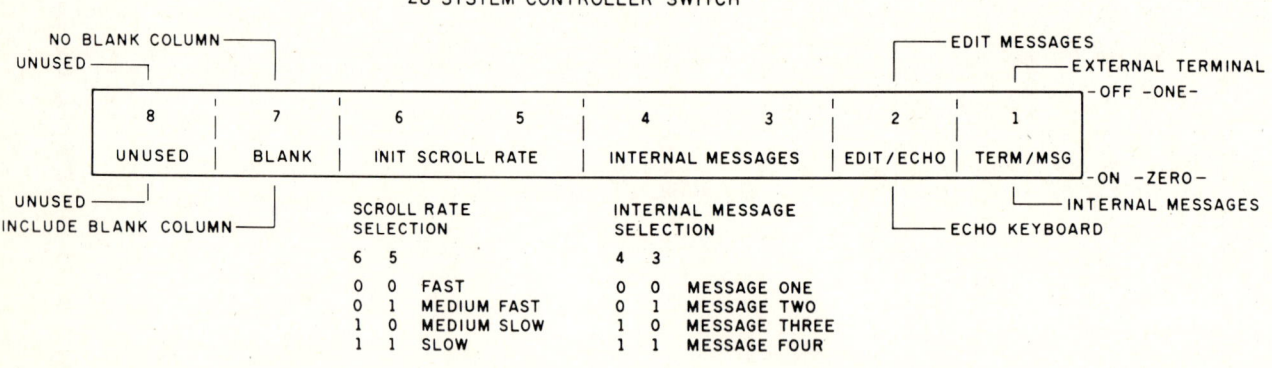

Figure 9: *Options that can be selected by DIP-switch settings on the Z8 board. When used without a terminal, the controller can be thereby set up to start automatically and display one of four prestored messages at one of four scrolling rates.*

terminal attached to the Z8 board to set up or change the scrolled message, but not for mere continuous operation with a single message. When a terminal is used and the echo mode is selected, input data or other keyboard entries appear in real time on the display. In the nonecho mode, the terminal is used to enter complete messages for subsequent display, to set the scroll rate, or to edit existing messages.

Several options can be selected by DIP (dual-inline pin)-switch settings on the Z8 board, as shown in figure 9. When used without a terminal, the controller can be switch-selected to start automatically and display one of four prestored messages at one of four scrolling rates. Automatic operation would be appropriate if the LED display were used for advertising, housed in an attractive wooden enclosure just like a commercial unit,

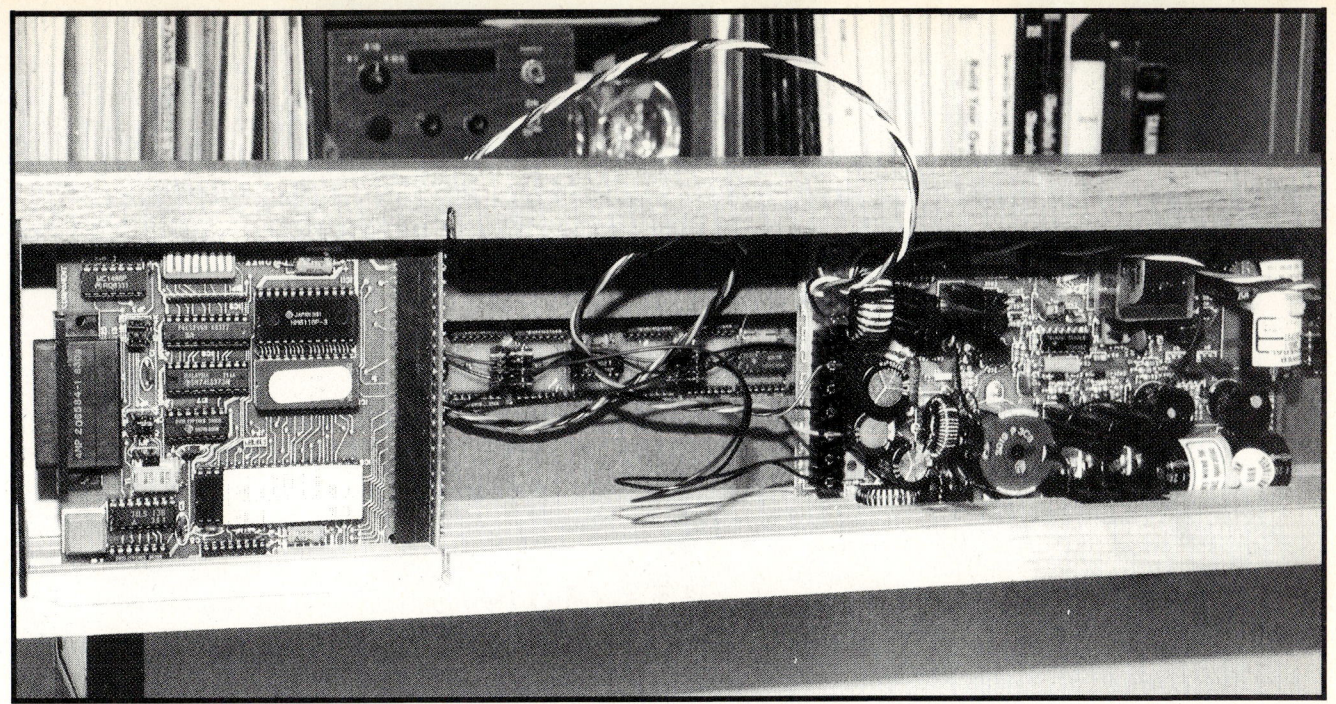

Photo 7: *The Z8-BASIC System Controller and a hefty switching-type power supply reside in the rear of the wooden box.*

as shown in photo 6. (In the rear of the box are the Z8 board and a high-current switching power supply, as photo 7 reveals.)

In Conclusion

The assembly-language routines written for the Z8 could be converted to other processors without too much trouble. If you get ambitious and perform this conversion or devise some clever ways to use the LED display, let me know. I'd like to see your ideas and make them available to others.

I've only begun to find uses for my scrolling LED display. In another month or so, I should be able to add 20 or 30 more character segments. I wonder if I'll ever get to build a 24-line by 80-character display screen?

References

1. Ciarcia, Steve. "Build a Z8-Based Control Computer with BASIC." Part 1, BYTE, July 1981, page 38. Part 2, BYTE, August 1981, page 50.
2. Ciarcia, Steve. "Self-Refreshing LED Graphics Display." BYTE, October 1979, page 58.

Special thanks to Bill Curlew and Ray Long for their contributions to this project.

The following items are available from

The Micromint, Inc.
25 Terrace Drive
Vernon, CT 06266
For orders: (800) 635-3355
For information: (203) 871-6170

1. *Kit for a single-character-segment LED display, 5 by 7 matrix. Includes LEDs, integrated circuits, printed-circuit boards, and all components necessary to display one 5 by 7 character.*

   single-segment kit SSK1, each..$45
   10 or more segment kits, each...$42

2. *Pair of blank printed-circuit boards for a single character segment, includes two Berg-type connectors.*

   single-segment board pair, PSS1...$17

3. *Z8-BASIC System Controller (scrolling LED software on EPROM optionally available).*

   BCC11, assembled and tested...$149

Please add $4 for shipping and insurance in the continental United States, $16 elsewhere. New York residents please include 7 percent sales tax.

Trump Card
Part 1: Hardware
Speed up your IBM PC with 16-bit coprocessing power

When asked what computer language I prefer, I generally reply, "Solder." This response is not an effort to be cute but rather to express a preference for dealing in the terms I know best. I don't avoid software. I just try to minimize my involvement.

When it is necessary to write simulation and test programs, I bite the bullet. Unless the function is time-critical, I most often choose BASIC because it comes closest to being a universal programming language. Virtually all personal and business computers support it, and if I confine my command choices to the more common instructions, the demonstration programs that I compose on an IBM PC should also run on your Cromemco Z2.

With few exceptions, you can compute your accounts receivable or type in and play a game equally well with an Apple or IBM PC using BASIC. The fact that one has a 6502 microprocessor and the other uses an 8088 is irrelevant. The output will be the same.

The value of high-level languages is that they isolate the user from microprocessor peculiarities and facilitate transportable software. Unfortunately, the average ROM (read-only memory)-resident BASIC interpreter was never written with perfor-

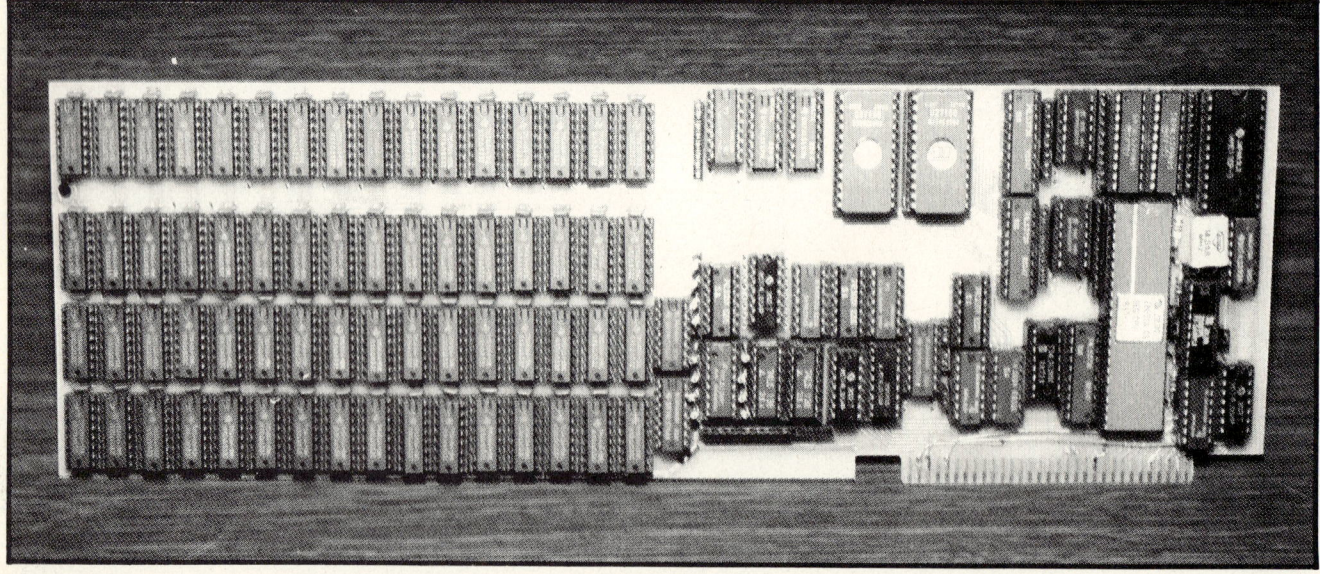

Photo 1: *The wire-wrapped prototype of the Trump Card, shown from the front. The left side of the board contains 512K bytes of type-4164 dynamic RAM; the right side contains the Zilog Z8001 and an interface to the IBM PC I/O-expansion bus.*

mance in mind. Usually taking 5 to 10 milliseconds (ms) to execute an individual instruction, it can seem like forever when running long programs.

As a writer, I have grown to appreciate the universality of BASIC, even with its shortcomings. By treating the computer as a black box with I/O (input/output) ports and BASIC, I have been able to provide projects that can be implemented on most systems directly. As an engineer/designer, however, I am aggravated by its slowness and feel no animosity toward critics who have converted to languages such as Pascal or C to gain processing speed.

Rather than make further excuses, I decided to solve the problem in classic Circuit Cellar tradition—simply build a black box that improves system throughput and runs BASIC programs faster.

Photo 2: *The rear of the Trump Card prototype. To save time, the memory section was laid out as a printed-circuit board, with wire-wrapping saved for the processor side. As shown here, the Trump Card is installed for testing in an MPX-16 computer, which has I/O slots compatible with the IBM PC.*

Processors and Performance

Generally speaking, most people confuse microprocessor benchmarks with system throughput. The comparison of microprocessor-instruction execution speeds is not really indicative of a computer's capabilities. Performance is more often governed by the operating system and magnitude of the application program. It is a false assumption that all software written for a 16-bit microprocessor will necessarily run faster than on an 8-bit microprocessor. Machine-language fast Fourier transforms (FFT) run quickly on a 6502, but an accounting package that has to constantly interleave a program into and out of disk may be encumbered by 64K bytes of operational memory in the Apple. In all likelihood, large spreadsheets and accounting programs will run more efficiently in the larger memory space provided on an 8088 system such as the IBM PC.

Raising the performance of a high-level language such as BASIC takes more than raising a microprocessor's clock rate. Instead, it involves a combination of decisions that can ultimately affect the entire system throughput. We can expand the memory available to application programs in an effort to limit repeated disk accesses and configure a portion of memory as a RAM (random-access read/write memory) disk drive to expedite disk operations when they are required. We can optimize the effi-

(3a)
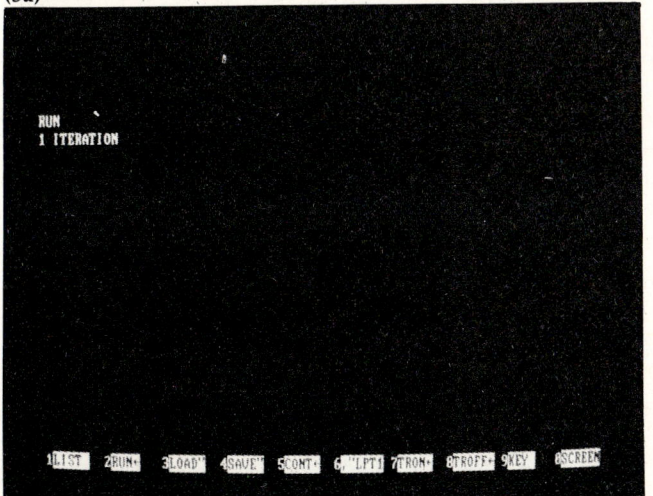

(3b)
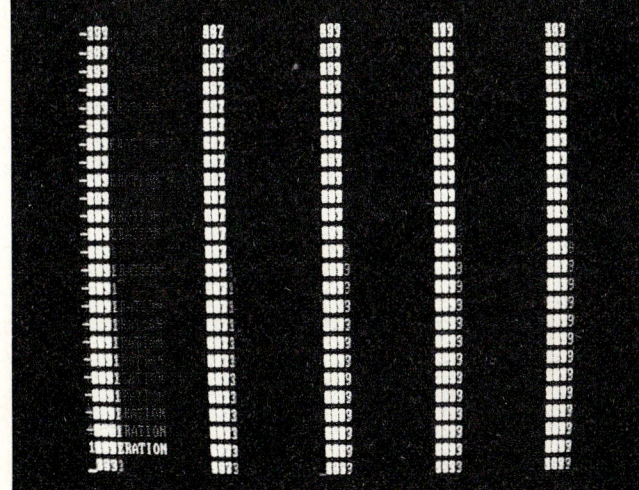

Photo 3: *Execution-time visual comparison. (3a) Without Trump Card—a two-second exposure of the display while running the BASICA program in listing 1. The program has executed the PRINT statement and still is dimensioning the arrays. (3b) With Trump Card—the same two-second exposure of the program execution (with PRINT statement added) shows the arrays have been dimensioned and the prime numbers are being overprinted so fast that they blur.*

Apple II	Apple III	TRS-80 Model II	IBM PC	IBM PC (with Trump Card)
224	222	189	190	2.4

Table 1: *A comparison of execution times (in seconds) of the benchmark program in listing 1.*

Listing 1: *Sieve of Eratosthenes prime-number-generator program.*

```
  5  DEFINT A-Z
 10  SIZE = 8190
 20  DIM FLAGS(8191)
 30  PRINT "Only 1 iteration"
 50  COUNT = 0
 60  FOR I = 0 TO SIZE
 70  FLAGS(I) = 1
 80  NEXT I
 90  FOR I = 0 TO SIZE
100  IF FLAGS(I) = 0 THEN 180
110  PRIME = I+I +3
120  K = I + PRIME
130  IF K> SIZE THEN 170
140  FLAGS(K) = 0
150  K = K + PRIME
160  GOTO 130
170  COUNT = COUNT + 1
180  NEXT I
190  PRINT COUNT, " PRIMES"
```

ciency of the high-level language by operating it in a compiled mode rather than as a repeatedly interpreted task. Finally, if the functional throughput of a particular application becomes dependent upon direct microprocessor intervention, for those tasks, substitute a faster microprocessor or help it with a coprocessor.

A Black Box Called Trump Card

This article is not about building a classic speed-up board for the IBM PC. The word "speed-up" implies replacing the 8088 with an 8086 or 80186. Instead, visualize your PC as a black box with an input, output, and crank. Rather than simply turning the crank faster, think of adding another black box, in the same path between input and output, that performs selective tasks more efficiently and faster than the 8088 alone. To increase the relative throughput of the system, I have designated an alternate path for specific program functions.

I've named this separate box Trump Card. It is a functionally independent 10-MHz Zilog Z8001-based computer with its own 512K bytes of memory. Designed specifically as a compiled high-level-language computer, Trump Card is addressed as an I/O device that communicates through the expansion bus (see photos 1 and 2).

Among the specific functions that Trump Card supports are BASIC, C, CP/M-80, text editing, Z8000 assembly-language programming, and a RAM disk. It does not directly execute programs written in 8088 assembly code, such as Lotus 1-2-3. It instead executes programs written in high-level languages such as BASIC or C (a Pascal compiler and 8088-to-Z8000 translator are in the works). Alternatively, it can enhance the function of programs such as 1-2-3 by expanding available memory and speeding disk functions. The ultimate purpose of Trump Card is to improve system throughput.

This month, I will outline the basic functions of Trump Card and describe its hardware in detail. This is, of course, a Circuit Cellar construction project, and you are encouraged to build your own Trump Card. More on that later. Next month, I'll describe some of the software in detail and do a little benchmarking.

First, a little about Trump Card and the Z8001.

Trump Card

Trump Card is a peripheral board that plugs into any expansion slot on an IBM PC or PC-compatible computer. It contains a 10-MHz Z8001 and up to 512K bytes of memory. To use it, you simply load a BASIC, CP/M-80, or C program from PC-DOS and type "RUN." Its memory can also be used as a RAM disk.

Trump Card comes with software that translates existing BASIC and other high-level-language programs to run with reduced overhead. To speed the execution of BASICA, Trump Card compiles the code with a special version of BASIC called TBASIC. Unlike other compilers, this has no separate compiled-code disk files (unless you specifically want them) and no long delays. TBASIC instantly compiles the program in a few tenths of a second when you load the file into Trump Card. In appearance, it looks like any old, slow interpreted BASIC, but it runs with the speed of a compiler.

TBASIC is PC BASICA-compatible. You can use either the Trump Card screen editor or BASICA's editor to write your programs. Then run the same program using either Trump Card or BASICA. Depending upon the instructions you use, Trump Card provides a tenfold to hundredfold increase in program performance (see photo 3). Table 1 shows typical results of what Trump Card can do with the prime-number Sieve of Eratosthenes program (September 1981 BYTE, page 180) frequently used to benchmark computer systems (see listing 1).

Though I conceived of Trump Card initially as a BASICA enhancement, it didn't take me long to realize that a Z8001 with 512K bytes of memory has some real computing power and deserves proper support. For that reason, the software supplied with this project is much more extensive than usual. With the utilities and languages included, you should have little trouble using the vast software base of Z80 and Z8000 programs.

Trump Card includes the following software:

BASIC Compiler—TBASIC is PC BASICA-compatible. The differences between the BASICA interpreter and the TBASIC compiler are minimal. Most instructions are implemented without modification.

CP/M-80 Emulator—Trump Card can run your CP/M-80 Z80 assembly-language programs directly without special disk headers or translation programs. Simply download your Z80 programs and run them.

C-Compiler—Trump Card includes the industry standard version of C that is described in *The C Programming Language* by Kernighan and Ritchie.

Debugger—Intended to aid in program development. With it, you can examine and replace memory and register contents, set breakpoints, or single-step through programs.

Screen Editor—Incorporating many of the features included in word processors, the editor enables you to write or examine ASCII text files for either the PC or Trump Card's use.

Multilevel Language Compiler—This is a structured assembler that allows Pascal-like control and data types, arithmetic expressions with automatic or specified allocations of registers, and procedure calls with parameter passing.

RAM Disk—Trump Card can allocate 128K to 387K bytes of its on-board memory to function as an intelligent RAM disk (DOS 2.0 only). This memory is separate from and in addition to any already existing on the PC bus. Trump Card's other functions can run concurrently.

The Z8000 Microprocessor

A block diagram of the Z8000's internal structure appears here as figure 1. As the programmer sees it, the Z8000 contains sixteen 16-bit general-purpose registers (for addresses or data) that may also be used in groups to form as many as eight 32-bit registers or four 64-bit registers. The low-order halves of the registers may be used for byte operations, thus the Z8000 is able to manipulate data in 8-, 16-, 32-, and 64-bit pieces.

The eight addressing modes are register, indirect-register, direct-address, indexed, immediate, base-address, base-indexed, and relative-address. The instruction set utilizes data types ranging from single bits to a 32-bit-long word. The processor executes 110 distinct instruction types that, when permuted by all the addressing modes and data types, create a set of more than 400 instructions.

The Z8000 has two different modes of operation: system and normal. Which mode of operation is in effect is controlled by a bit in the flag-and-control word (FCW). The main difference between the operating modes is that some of the control/interrupt and I/O instructions work only in the system mode. To simplify the design of the Trump Card, I chose to use only the system mode.

The Z8001 (see photo 4) is the memory-segmented version of Zilog's chip; it comes in a 48-pin DIP (dual-

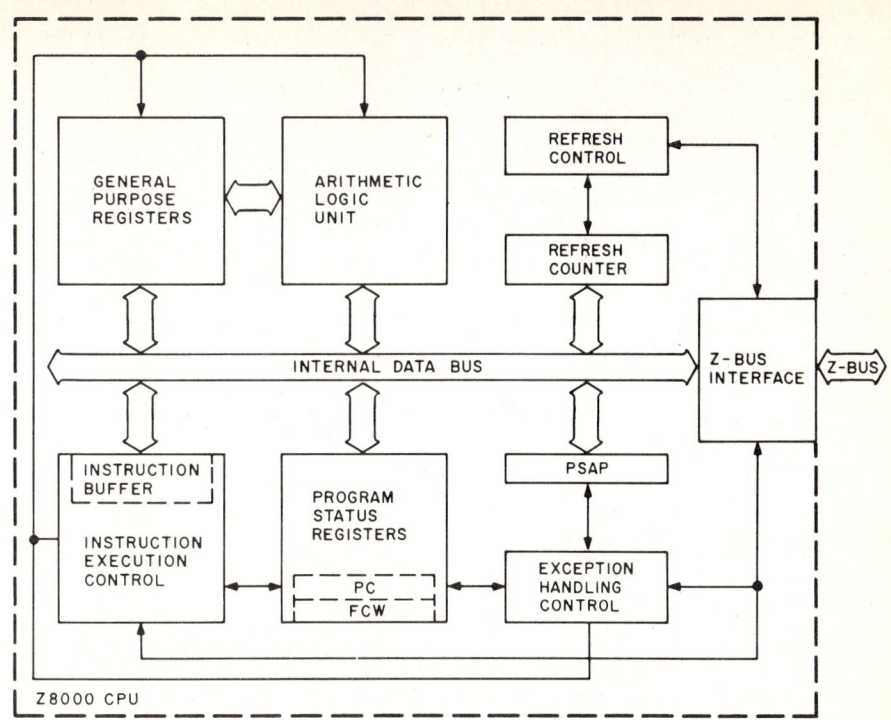

Figure 1: *Block diagram of the internal structure of the Zilog Z8000 family of microprocessors.*

Photo 4: *The 48-pin dual-inline package that houses the Zilog Z8001 microprocessor, the heart of the Trump Card.*

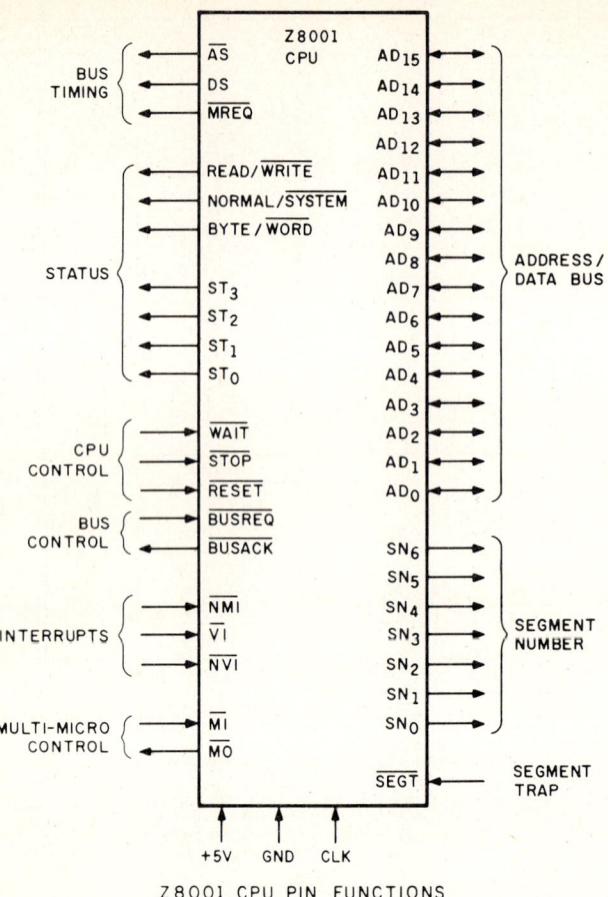 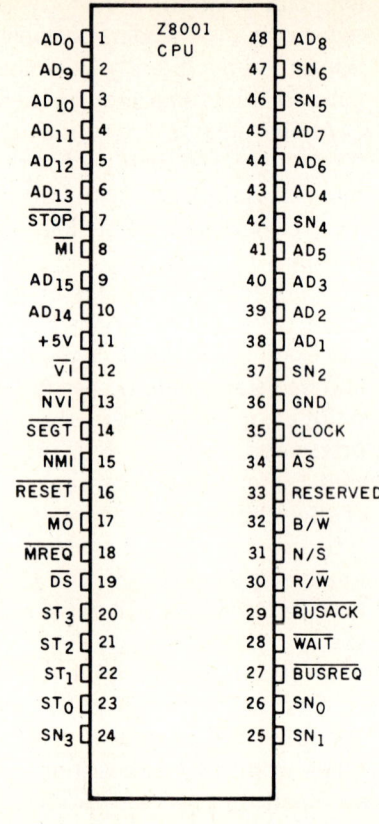

Figure 2: *Pinout arrangement of the Z8001 memory-segmented version.*

inline package), the pinouts of which are shown in figure 2. (The nonsegmented 40-pin version is called the Z8002.) By memory segmentation, the directly addressable 8-megabyte memory space is divided into as many as one hundred twenty-eight 64K-byte regions. Seven segment-selection lines coming out of the Z8001 control the high-order memory addressing. When the Z8001 is reset, the segment addressing automatically reverts to segment 0, the lowest 64K-byte block of memory. Transfer of control between segments is done by jumps, calls, and returns.

Inside the Trump Card

The schematic diagram of figure 3 shows the Trump Card's circuitry. It can be plugged into any expansion slot of an IBM PC or into any other computer with compatible I/O slots and operating system.

Five of the Z8001's seven segment-selection lines, SN0 through SN4, are used in the Trump Card to decode addresses for up to 1 megabyte of RAM (512K bytes fit on the board) and 4K bytes of ROM (read-only memory). Segment line 4 selects between the ROM, mapped into segments 0 through 15, and the RAM, residing in segments 16 through 31. The states of the segment lines are latched by IC3; segment line 4 is named RAM/$\overline{\text{PROM}}$.

Address/Data Bus

The address/data lines coming from the Z8001 (AD0 through AD15) are a time-multiplexed address and data bus, which can address a range of 65,536 (64K) bytes of memory or a like number of I/O addresses. Since the Z8001 can form addresses at either word or byte boundaries, the least significant bit AD0 is used in byte operations to determine if the upper or lower byte is to be operated upon. The address on the AD lines becomes valid when the Z8001 asserts the $\overline{\text{AS}}$ (address strobe—active low) line; it remains that way for a short hold time after $\overline{\text{AS}}$ returns to its idle high state. The address from the Z8001 is latched by two type-74LS373 transparent latches, IC5 and IC6, that are always enabled. The use of transparent latches allows for maximum address-setup time to the memories.

The latched addresses (LA0 through LA15) come out of the 74LS373s with LA0 combined with the signal B/$\overline{\text{W}}$ (byte or word address) to form the $\overline{\text{EVEN}}$ or $\overline{\text{ODD}}$ byte-bank-select signal for memory. When B/$\overline{\text{W}}$ is low, it signifies that a 16-bit memory word is being referenced; this causes the outputs of the two AND gates at IC8 pin 3 and IC8 pin 6 to be active irrespective of the state of LA0. By doing byte operations in this manner, it is possible for the Z8001 to do single-byte memory writes without first reading an entire word location.

The Trump Card contains a pair of type-2716 EPROMs (erasable programmable read-only memories),

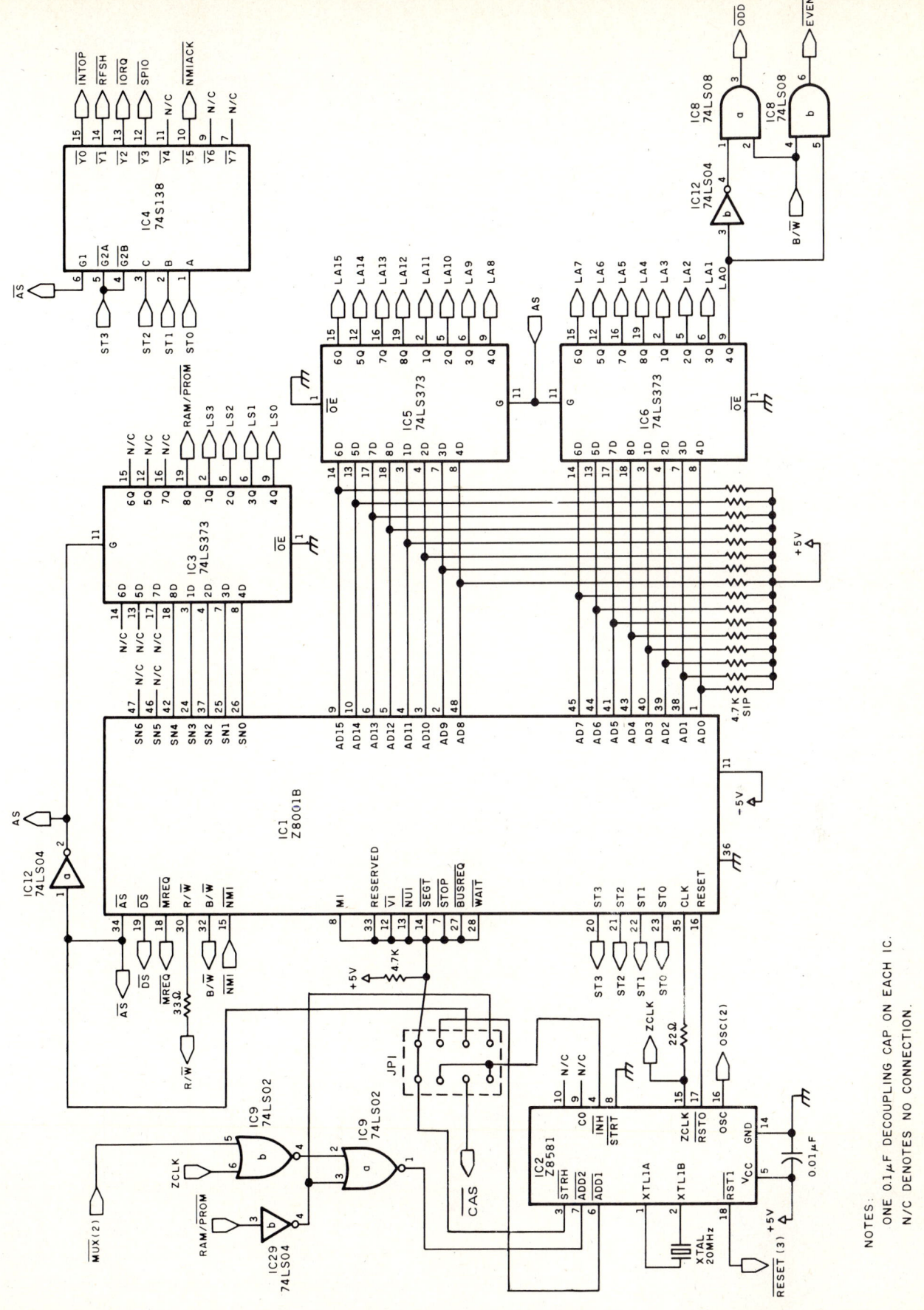

Figure 3: *Schematic diagram of the Trump Card. Support is provided for 512K bytes of dynamic RAM in the form of type-4164 chips. If a 6-MHz Z8001A is used, the crystal frequency must be reduced to 12 MHz.*

Figure 3 continued

Figure 3 continued:

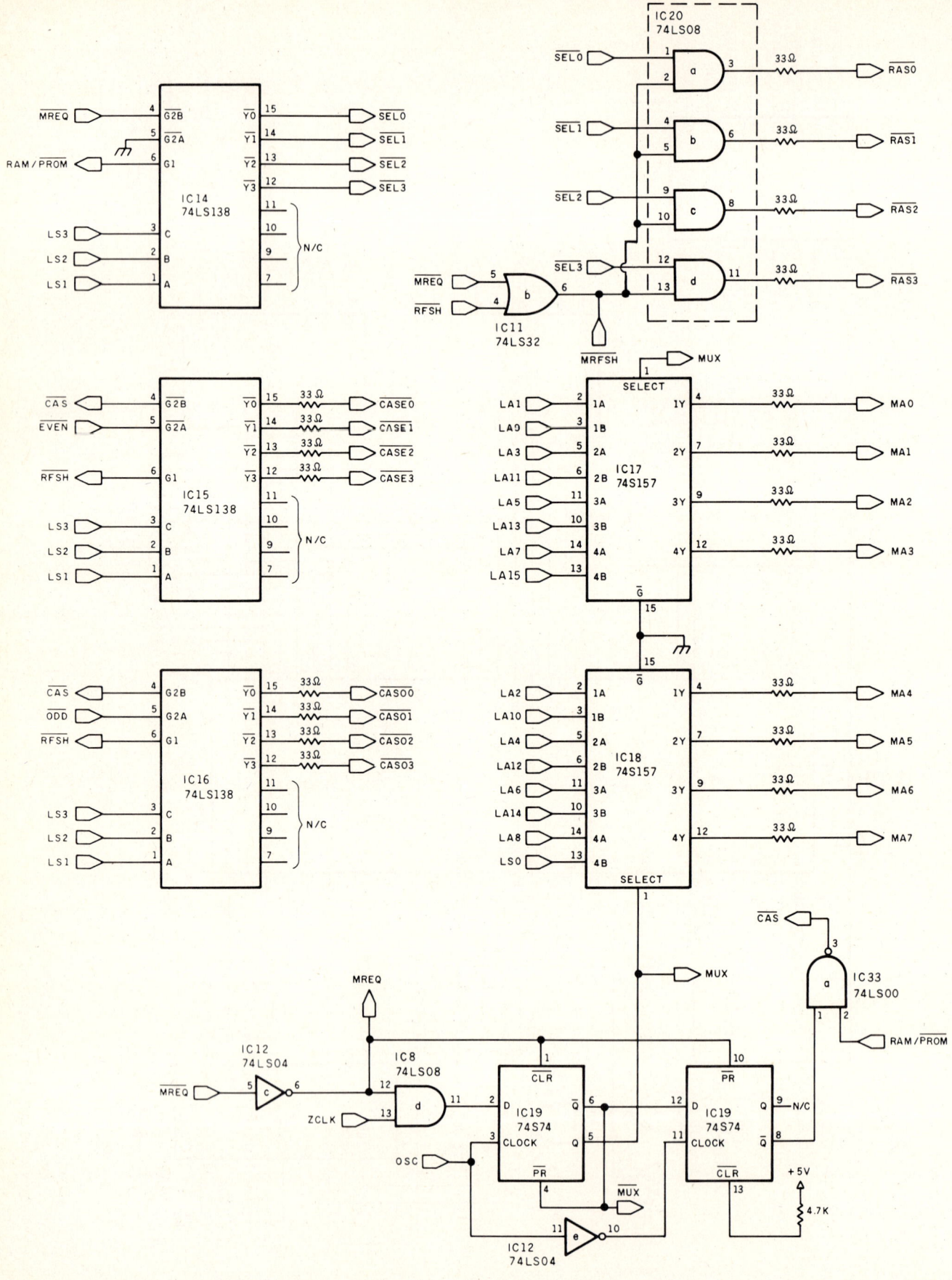

Figure 3 continued:

NOTE: IC's IN THIS SECTION ARE CONSECUTIVELY NUMBERED FROM IC36 UPPER LEFT CORNER TO IC99 LOWER RIGHT CORNER. IC's ARE ALL 4164's.

Figure 3 continued

Figure 3 continued:

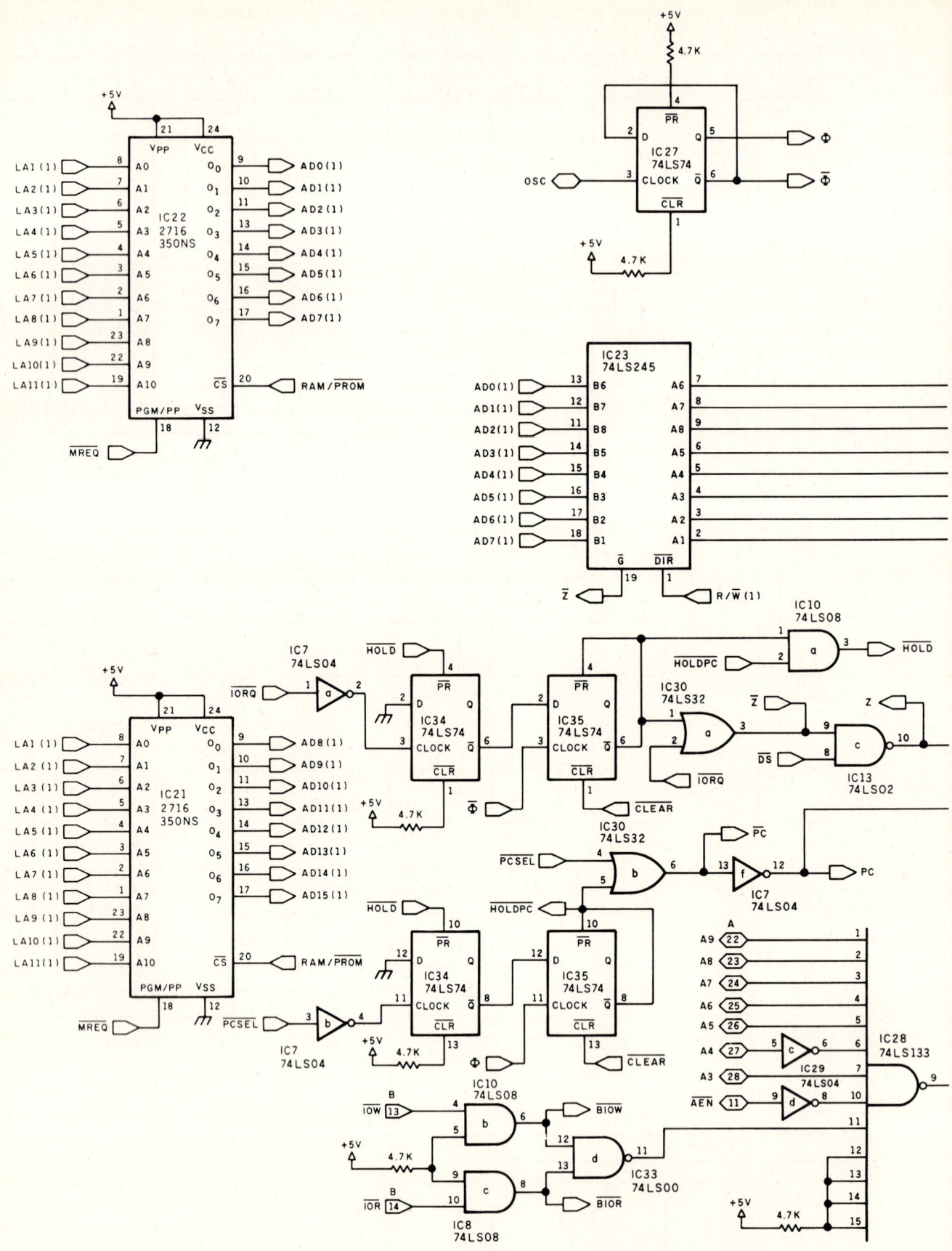

Figure 3 continued:

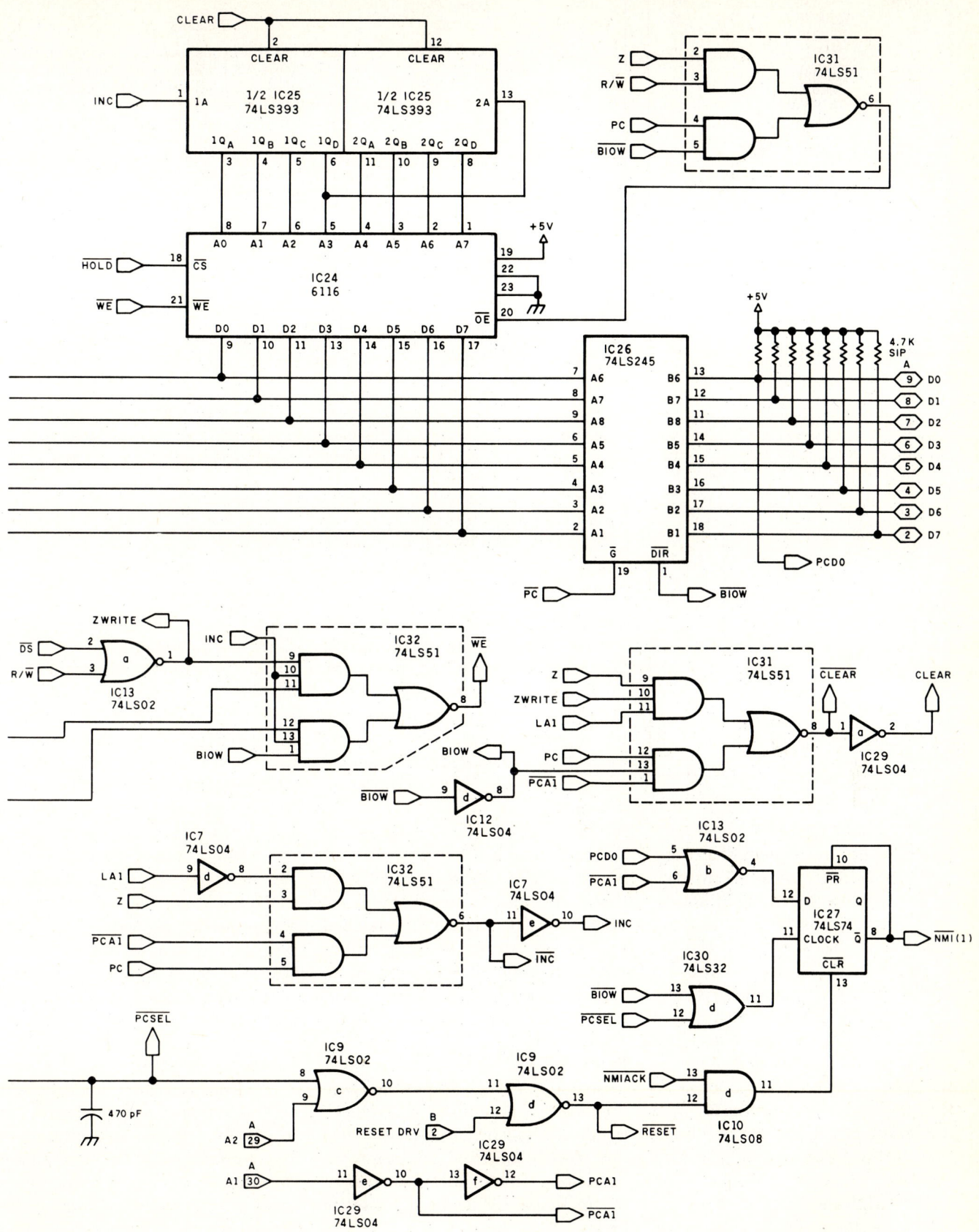

Number	Type	+5 V	GND
IC1	Z8001B	11	36
IC2	Z8581	5	14
IC3	74LS373	20	10
IC4	74S138	16	8
IC5	74LS373	20	10
IC6	74LS373	20	10
IC7	74LS04	14	7
IC8	74LS08	14	7
IC9	74LS02	14	7
IC10	74LS08	14	7
IC11	74LS32	14	7
IC12	74LS04	14	7
IC13	74LS02	14	7
IC14	74LS138	16	8
IC15	74LS138	16	8
IC16	74LS138	16	8
IC17	74157	16	8
IC18	74157	16	8
IC19	74S74	14	7
IC20	74LS08	14	7
IC21	2716	24	12
IC22	2716	24	12
IC23	74LS245	20	10
IC24	6116-3	24	12
IC25	74LS393	14	7
IC26	74LS245	20	10
IC27	74LS74	14	7
IC28	74LS133	16	8
IC29	74LS04	14	7
IC30	74LS32	14	7
IC31	74LS51	14	7
IC32	74LS51	14	7
IC33	74LS00	14	7
IC34	74LS74	14	7
IC35	74LS74	14	7
IC36	4164-15 (150 ns)	16	8
.			
IC99	4164-15	16	8

Power wiring table for figure 3.

PC Address	Trump Card Port	Function
03EE	3	A "write" to this port by the processor that has current use of the bucket will cause the 8-bit address counter, IC25, to be reset to 0. It will also release the bucket for use by the other processor. If bit 0 of the data bus is set to a 0 when this write is performed by the 8088 processor, a nonmaskable interrupt (NMI) is also issued to the Z8001. Reading this port allows either processor to see data at the current address of the counter without incrementing the counter. If the bucket is not available, a read operation to this port will return an FF.
03EC	1	A read operation to this port by the processor that has the bucket reserved will return the data at the current address of the counter and increment the counter at the end of the read operation. A read by the processor that does not have the bucket will return a value of hexadecimal FF and will not increment the counter. A write to this port by the processor that has reserved the bucket will enter data at the current address of the counter and increment the counter at the end of the operation. A write by the processor that does not have the bucket will not enter data and the counter will not be incremented.
03E8	x	This port is not used for data transfer by the Z8001. A write to this port by the 8088 will issue a reset to the Trump Card.

Table 2: *Communication between the Z8001 and the 8088 is through the "bucket," a FIFO buffer made from a type-6116 static-memory chip and support components. Shown here are the three basic bucket functions and the addresses and codes for each.*

which contain a bootstrap loader for cold-start-up and system-diagnostic routines. Address lines LA1 through LA11 are connected to the EPROMs, IC22 (even byte) and IC23 (odd byte). There is no need to use the $\overline{ODD}$ or $\overline{EVEN}$ bank-select lines since no data is ever written into the EPROMs. The signal RAM/$\overline{PROM}$ is connected to the $\overline{CS}$ pin on the 2716s. The $\overline{MREQ}$ (memory request) signal from the Z8001 is also connected to pin 18 ($\overline{OE}$ or output enable) of the 2716s, to inhibit the possibility of bus contention during I/O cycles.

Status Signals

Various status signals tell the rest of the system about the processor's condition and the type of information that is appearing on the address/data bus. The status signals are as follows:

Read/$\overline{Write}$: The R/$\overline{W}$ signal is used to indicate the direction of the current bus transaction. When high, the direction of data is toward the Z8001. Data is clocked into the processor at the occurrence of a positive-going pulse on $\overline{DS}$ (data strobe). When $\overline{DS}$ is low, data flows from the processor outward.

Normal/$\overline{System}$: The N/$\overline{S}$ signal indicates whether the processor is operating in the system (supervisory) mode or normal (user) mode of operation. This control line is used when there is a multitasking and/or multiuser type of environment to segregate system functions and memory. The line is unused in the Trump Card.

Byte/$\overline{Word}$: The B/$\overline{W}$ line is provided to enable the Z8001 to perform byte operations on memory. When high, it indicates that a byte operation is to take place; a low state indicates word operations. This signal is also used in the $\overline{ODD}$ or $\overline{EVEN}$ memory-select logic.

Status Lines: Lines ST0 through ST3 are utilized to define the exact type of transaction occurring on the bus. Only 4 of the 16 possible codes are required for operation of the Trump Card. The first status code, 0000 (Internal Operation), is decoded but unused. The second operation code, 0001 (Memory Refresh), is output by the internal Z8001 memory-refresh timer and is used in refreshing the on-board dynamic RAM. (This signal is ANDed with $\overline{MREQ}$ and is used as one of the two select signals in the row-address-strobe generation logic.) The third operation, 0010 (Standard I/O Reference), is used in the process of communicating with the host 8088 processor. The fourth operation code, 0011 (Special I/O), denotes I/O associated with the signal $\overline{SPIO}$ and is reserved for future expansion.

Clock Generation

The basic clock rate for the Z8001 on the Trump Card is provided by IC2, a Zilog Z8581 clock generator and controller (CGC). The Z8001's clock-input maximum voltage must come within a certain range of the power-supply potential (precisely $V_{cc} - 0.4$ V) and have a maximum rise and fall time of 10 nanoseconds (ns). Such requirements are difficult to meet with standard oscillators and

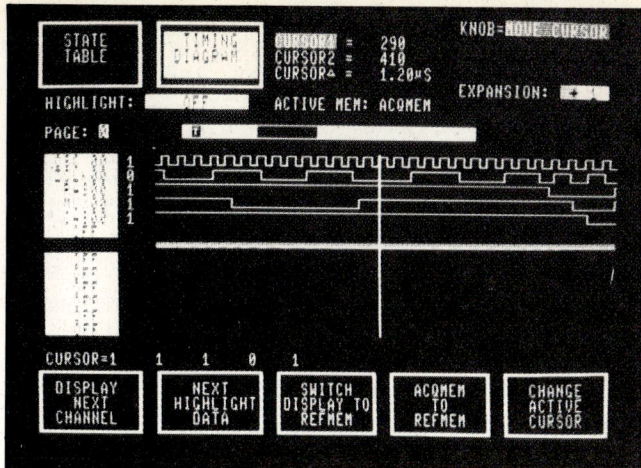

Photo 5: *A typical display on the Tektronix 1240 logic analyzer: the column-address-strobe/row-address-strobe timing of the Trump Card.*

Photo 6: *To aid in my initial development, a Zscan-8000 emulator is plugged by a ribbon cable into the Z8001 socket on the Trump Card. In emulation mode, the Zscan-8000 can run diagnostic programs and exercise all functions of the Trump Card at 4 MHz. Hardware debugging is greatly simplified because all sections of the hardware need not be working to use the emulator.*

TTL (transistor-transistor logic), but they are easily met by the CGC. The Z8581 also provides an easy and effective means of adjusting the processor's bus cycles to the speed of available memory devices.

The CGC is used on the Trump Card to stretch specific bus cycles. As used on the Trump Card, the Z8001 does three different basic categories of operations: internal operations, memory access, and input/output operations. The timing of the ZCLK signal emitted by the CGC depends on which of these bus activities is taking place. The Z8581 can be configured to add wait states that enable the use of 150- and 200-ns RAM chips.

Trump Card/Host Communication

The "bucket" is the communications interface between the PC and Trump Card. This FIFO (first-in/first-out)-type dual-port memory configuration consists of a 6116 static memory (IC24), an 8-bit address counter (IC25), two data-bus buffers (ICs 23 and 26), and the necessary control logic to arbitrate access. Programs and instructions are passed between the two computers via this FIFO circuitry. As far as the PC is concerned, the bucket appears as two I/O port addresses. A system of software handshaking between the computers determines which has reserved and is using the bucket. Table 2 shows the port addresses and their functions.

It is not possible for both processors to have use of the bucket at the same time. With the processors running asynchronously, arbitration is necessary. It is provided by four D-type flip-flops: two for access requests and two for access reservations. The two access-request flip-flops are clocked by the transition of an access-request signal from either processor ($\overline{IORQ}$ for the Z8001 and $\overline{PCSEL}$ for the 8088). The preset inputs of these flip-flops are connected to the $\overline{HOLD}$ signal, which is active whenever one of the processors has succeeded in reserving the use of the bucket. When $\overline{HOLD}$ is active, it prevents the other processor from gaining access.

The Z8001 communicates through the bucket for all its normal I/O by activating the $\overline{IORQ}$ line. The 8088 selects the bucket when it performs either an IOW (I/O write) or IOR (I/O read) in the range of the IBM's regular memory-address space from hexadecimal 03E8 to 03EE. Accesses to these addresses are decoded by IC28 to generate the Trump Card's $\overline{PCSEL}$ signal.

The two access-reserve flip-flops sample the output of the request flip-flops 180 degrees out of phase with each other. This is done to prohibit simultaneous requests from being honored. These flip-flops are cleared by a reset command issued from the reserving processor.

The $\overline{Q}$ outputs from these flip-flops are combined by a logical AND function with the processor request to form the active select states used by the bucket: $\overline{Z}$ ANDed with $\overline{DS}$ for the Z8001 and $\overline{PC}$ for the 8088. Whenever either request flip-flop is active, the $\overline{HOLD}$ signal is active and is used as the chip-select input on the 6116 memory. The FIFO memory, however, is written to by the Z8001 only when a "write bucket with increment" command is used.

The $\overline{WE}$ signal, connected to the write-enable input of the 6116 memory, is active during either a Z8001 I/O request (with R/$\overline{W}$ low and $\overline{DS}$ active) or an 8088-generated write to the bucket (with $\overline{PC}$ and INC active and $\overline{BIOR}$ inactive). The INC signal is active whenever the processor that has control of the bucket sets bit 1 of the address low. The CLEAR signal is active when bit 1 of the address generated by the selected processor is high and a write operation is occurring.

A nonmaskable interrupt to the Z8001 is generated when the 8088

Listing 2: *Bootstrap initialization program for the Trump Card written in Z8000 assembly language.*

```
0000              00   DW                    - Reserved control word
D000              02   DW                    - Flag and control word
0000              04   DW                    - Segment Register
0008              06   DW                    - Segment Offset
2100  9E01        08   LD  R0,#%9E01         - Set refresh freq and enable
1404  0003 0001   0C   LDL RR4,#%0003 0001   - Set port addresses
7D0B              12   LDCTL REFRESH,R0      - Load refresh value
3E40              14   OUTB @R4,RH0          - Set R4 as reset-bucket port
3C40              16   INB  RH0,@R4          - Read bucket without increment
A800              18   INCB RH0,#1           - Increment input value
E6FC              1A   JR 0 EQ,%0014         - Repeat if equal to 0
3C40              1C   INB  RH0,@R4          - Read bucket
8A80              1E   CPB  RH0,RL0          - Compare bucket value to 01
EFF9              20   JR NC UGE,%0014       - Do again if not > 01
C803              22   LDB  RL0,#%03         - Load R0 with bucket available #
3E58              24   OUTB @R5,RL0          - Load bucket with R0
8400              26   ORB  RH0,RH0          - Set zero flag if RH0 is 0
E6F5              28   JR Z EQ,%0014         - Restart boot, else continue
3C52              2A   INB  RH2,@R5          - Read bucket and save in register
3C53              2C   INB  RH3,@R5          - Read bucket and save in register
3C5B              2E   INB  RL3,@R5          - Read bucket and save in register
3C51              30   INB  RH1,@R5          - Read bucket and save in register
3C59              32   INB  RL1,@R5          - Read bucket and save in register
3A50  0120        34   INIRB @RR2,@R5 R1     - Read bucket into memory
F013              38   DBZNZ RH0             - Decr RH0 and at 0 goto 0014
3E40              3A   OUTB @R4,RH0          - Reset bucket
AB35              3C   DEC  R3,#6            - Decrement value in R3 six times
                                               to set up first addr of code
1E28              3E   JP  @RR2              - Jump to loc defined in RR2
```

performs a write operation to the bucket with address bit 1 and data bit 0 both low. This interrupt is latched in a D-type flip-flop and is not cleared until the Z8001 issues a Nonmaskable-Interrupt Acknowledge (status decode 5) or until the host computer resets the Trump Card. (See table 2 for more detail.)

Booting Trump Card

When you plug the Trump Card into a slot in the IBM PC and turn on the computer, the Trump Card automatically executes the bootstrap-loader routine contained on board in EPROM. The loader routine is only 31 words (62 bytes) long; its assembly code is shown in listing 2.

I used two 2716 EPROMs instead of bipolar PROMs to store the bootstrap loader because they are both cost-effective and easier to program than bipolar PROMs. Two byte-wide memory devices are required because the Z8001 is a processor with a 16-bit word length. Each machine-language instruction (expressed as four hexadecimal digits) is separated into high- and low-order bytes (or "even" and "odd," if you prefer); the high and low bytes are stored in separate EPROMs. When you examine a particular 16-bit memory location, you are actually viewing the information provided from two 8-bit sources.

Using Trump Card

Trump Card is transparent to normal PC operation. To start Trump Card, you run a program stored under PC-DOS called LDZSYS. This is the Trump Card communications software that runs on the 8088. If you always want Trump Card features available, you can add this program to your regular AUTOEXEC batch file. When LDZSYS has completed initialization, it returns to the PC-DOS A> prompt to wait further instructions.

At this point, I generally configure part of Trump Card's memory as a RAM disk, using a program called SETRMDSK. This is done as follows:

A> SETRMDSK 4
A>

SETRMDSK configures the additional C drive to your existing system under DOS 2.0. The number following SETRMDSK determines how many 64K-byte blocks you wish to reserve as a RAM disk. In this case, I set up a 256K-byte drive. The RAM-disk size can be 128K to 387K bytes, depending upon the amount of memory on the Trump Card board. (While you can set a 128K-byte RAM disk in a 256K-byte board, you might have problems running large BASIC or C programs concurrently.)

The memory that I've set as a RAM disk is completely separate and in addition to the regular IBM PC memory. Even if you have a 640K-byte PC, up to 387K bytes of Trump Card's memory would be available as additional RAM-disk configured storage space.

I used the RAM disk to speed up the process of writing these articles. Many word processors, like the Volkswriter I use, make extensive calls to the disk for help files and command-execution files. After a while, the noise and delay get aggravating. To remedy this situation, I run the SETRMDSK 4 sequence just described to create the 256K-byte RAM drive C and then add the following:

A> COPY *.* C:
A> C:
C> VW

This copies the entire contents of the Volkswriter distribution disk to the RAM disk, sets it as the default drive (C), and starts the word processor. When I now press a function key the action is instant and silent. To guard against power interruptions, drive B is designated as the hard-storage location and periodically I store the article file to it.

Trump Card's other features are equally simple to use. BASIC, C, Z80, and editing files can be stored on the same disk and executed with similar ease. While I'll explain it in greater detail next month, a possible sequence of Trump Card operations is shown in table 3.

Rewarding Diligence

I've been having a lot of fun with Trump Card. I haven't done much assembly-language or C programming yet, but it has renewed my faith in BASIC.

Trump Card is not an easy project to build. Compared to other Circuit Cellar projects, however, it's manageable. I was surprised at the number of readers who hand-wired the 121-chip MPX-16 PC-compatible computer that I presented last year. Their letters suggested that the motive was neither money nor masochism. In-

```
A> LDZSYS       (initialize Trump Card)
A>              (return to PC-DOS or use Trump Card)
A> G            (turn over PC operation to Trump Card)
:               (Trump Card prompt)
: EE filename   (edit a file)
    or
: Z80EM
  filename      (emulate CP/M-80 and run Z80 programs)
    or
: C filename    (compile and run a C program)
    or
: Y filename    (compile and run Z8000 assembly language)
    or
: BASIC filename (compile and run BASICA programs)
: //            (return to PC-DOS)
A>
```

Table 3: *A Trump Card operating sequence.*

stead, building these projects enabled them to experiment with digital circuitry yet be secure in the knowledge that their project would work. I hope this project elicits a similar response, and I'd like to reward such enthusiasm in advance.

Esoteric peripherals such as Trump Card depend a great deal on sophisticated software to fully exercise their capabilities. Unfortunately, when experimenters build rather than purchase boards, they often have to use great ingenuity to obtain software.

More than five man-years of development effort went into the present support packages for Trump Card. Some, like TBASIC and the RAM disk, were contracted by me, while others, like the C compiler and Y (a Z8000 assembler), were written by Zilog. Combined with the CP/M-80 emulator, Z8000 operating system, and telephone-book-size documentation, it is a formidable package that is difficult to independently price.

I want to encourage you to build your own Trump Card if that is your choice. If you send me a picture of the completed unit, I will send you a copy of the complete software and the documentation (provided it is for personal, noncommercial use) for the cost of duplication and shipping

---

The following items are available from

Sweet Micro Systems Inc.
50 Freeway Dr.
Cranston, RI 02910
(800) 341-8001 for orders
(401) 461-0530 for information

1. *Trump Card I, board-tested, expandable to 512KB RAM, Software and Documentation*
   TRC01 .... $955

2. *Trump Card II, board-tested, expandable to 2MB RAM,* Software and Documentation*
   TRC01 .... $1195

3. *128KB DRAM Set, 16-150nS 64K RAM chips (1 row), burned in and tested. (Use with TRC01 and TRC02.)*
   RAM01 .... $185

4. *512KB DRAM Set, 16-150nS 256K RAM chips (1 row), burned in and tested. (Use with TRC02 only.)*
   RAM02 .... $320

**All base units will accommodate four rows of RAM. The TRC02 unit will accommodate any combination of RAM01 & RAM02 above.*

($30). The software houses and other parties in this project have waived all royalties as a gesture of support for the Circuit Cellar.

Diagrams pertaining to the Z8000 are reprinted by permission of Zilog Corporation. Z8000 and Z80 are trademarks of Zilog.

References

1. Ciarcia, Steve. "Build the Circuit Cellar MPX-16 Computer System." Part 1, BYTE, November 1982, page 78. Part 2, BYTE, December 1982, page 42. Part 3, BYTE, January 1983, page 54.
2. Majundar, S., K. Kumar, and K. S. Raghunathan. "Interface Unites Z8000 with Other Families of Peripheral Devices." *Electronics*, July 28, 1981, page 156.
3. Rampil, Ira. "Preview of the Z8000." BYTE, March 1979, page 80.
4. Simington, R. B. "The Intel 8087 Numerics Processor Extension." BYTE, April 1983, page 154.
5. Shima, Masatoshi. "Genealogy of the Z8000." *Electronics*, December 21, 1978, page 83.
6. Williams, Gregg. "Benchmarking the Intel 8086 and 8088." BYTE, July 1983, page 147.
7. Zingale, Tony. "Intel's 80186: A 16-Bit Computer on a Chip." BYTE, April 1983, page 132.

Trump Card
Part 2: Software

TBASIC and C compilers and an assembler

Last month, we looked at the hardware of the Trump Card, a coprocessor board for use with the IBM Personal Computer (PC) or compatible computers. The presentation centered mainly on the Zilog Z8000's processor architecture, the support circuitry, and the interface between the Z8000 and the Intel 8088. But the power of the Trump Card can be unleashed only by the right software. This month, I'll describe the collection of software I've assembled for the Trump Card from several sources—most of it designed to support further program development. Let's first quickly review the features of the Trump Card.

WHAT IS THE TRUMP CARD?

The Trump Card (see photo 1) is a printed-circuit board that plugs into any I/O (input/output) expansion slot of an IBM PC, an IBM PC XT, or any computer compatible with them. It contains a Zilog Z8001 16-/16-bit microprocessor (the memory-segmented version of the Z8000) running at 10 MHz and up to 512K bytes of RAM (random-access read/write memory). The Trump Card communicates with the PC's built-in 8088 processor through a 256-byte FIFO (first-in/first-out) buffer.

A variety of software is available for the Trump Card. The most important, from my point of view, is the language system for its special version of BASIC. As you would expect, the Trump Card's TBASIC compiler excels at making user programs run fast, but it's also so easy to use that it makes some interpreted versions of BASIC look clumsy. The source language accepted by the TBASIC compiler is nearly identical with that of the IBM PC's Advanced BASIC interpreter (BASICA) and includes a few enhancements, such as compilation of programs larger than 64K bytes.

Other software included with the Trump Card follows:

- CP/M-80 emulator. The Trump Card can run programs designed to run under Digital Research's CP/M-80 DOS (disk operating system) by emulating the 8-bit Z80 instruction set and DOS calls. No special file headers or instruction-translation programs are required.
- C compiler. The source language accepted by this compiler follows that of Kernighan and Ritchie with a few minor differences (see reference 6).
- Screen editor. Incorporating many of the features normally found only in word-processing packages, the screen editor, called EE, enables you to write or examine ASCII (American National Standard Code for Information Interchange) text files for use either with the Trump Card or in the normal IBM PC environment.
- Y multilevel-language compiler. The unusual Y language system is essentially a structured assembler that enables Pascal-like control constructs and data types, arithmetic expressions with automatic or specified allocations of registers, and procedure calls with parameter passing
- Debugger. With the debugger, you can examine and replace the contents of memory and registers, set breakpoints, or single-step through programs. Intended to aid in program development, the debugger is an integral part of Y.
- Semiconductor disk emulator. Under versions of PC-DOS equal to or higher than 2.0, Trump Card can allocate 128K to 387K bytes of its on-board RAM to function as a RAM disk or disk emulator. This memory is separate from the memory already existing on the PC's motherboard or other expansion boards and resides in the Z8000's separate address space. The Trump Card can run another function concurrently with the disk emulator.

153

Photo 1: *The soldered prototype printed-circuit version of Trump Card. RAM sockets are at left, EPROMs are top center, and the Z8001 and support chips fill the remainder of the board.*

TBASIC is a new version of the BASIC language that looks like an interpreter and executes like a compiler.

Bringing the Trump Card Up

To initialize the Trump Card, run a program called LDZSYS.COM from PC-DOS. When it has completed setting up the Trump Card and installing the device driver needed by PC-DOS to communicate with it, LDZSYS returns control to PC-DOS and the host 8088 processor, with the Z8000 awaiting further instructions. Example 1 in the text box on page 156 contains examples of this and other typical user commands (in italics) and the system's response (in roman type). The operation of the Trump Card is transparent to programs running on the host 8088. (If you think that you will always want the Trump Card's capabilities available, you can add a line containing LDZSYS to your PC-DOS AUTOEXEC.BAT file.)

To begin using the Trump Card, execute the "go" program, G.COM (G). When the Z8000 has control of the system, it returns with a colon prompt, as the fourth line of example 1 shows, indicating that the Z8000 is ready to accept commands. The text box also shows the command format for editing and compiling files and programs, which may be stored on the same disk used to boot PC-DOS.

Interpreters versus Compilers

As I said last month, a chief cause for my building the Trump Card was a feeling of frustration with the slowness of BASIC interpreters. I had, of course, considered using an off-the-shelf BASIC compiler to speed up my programs, but I did not relish all the overhead operations required by the compilers I had seen, such as Microsoft's BASIC compiler.

The typical compiler requires three separate operations to run a BASIC program. First, the program source code must be written using an editor program. Next, the ASCII program text from the editor is compiled into object code and stored in a disk file, which often takes several minutes. Finally, the special BASIC run-time processor is loaded from the disk to supervise execution of the object program. At last, the program does its thing.

Interpreters, for all their inefficiency of execution, do have one important benefit: you quickly can add a line to your program and type RUN to see its effect. But if you want to change a line in a compiled program, it's back to the editor and all the way through the process again. So when you finally have your debugged, compiled program, it may indeed execute 100 times faster than under an interpreted one, but it may have taken you 10 times as long to get it running right. I think this is one reason BASIC compilers are not in wider use.

To counter this criticism, compiler manufacturers suggest developing code on an interpreted BASIC first and then compiling it. Such a suggestion, while valid, ignores the reason for a compiler in the first place. If a hundredfold increase in speed is necessary to achieve a program's objective, it hardly makes sense that to write and test the original program you must wait 100 times longer each time you must run it.

The answer seemed relatively trivial to me—simply write a version of BASIC that looks like an interpreter and executes like a compiler. The result is TBASIC.

The Trump Card's TBASIC language system is a BASIC compiler that offers significantly faster execution of BASIC programs than does a BASIC interpreter, while furnishing an operating environment much like that of an interpreter. TBASIC bridges the gap between traditional BASIC interpreters, which have built-in editors and are known for ease of use, and typical BASIC compilers, which produce rather efficient object code but can be difficult to work with. TBASIC's extremely fast compilation times and its capability for immediate-mode execution make working with it as easy as working with a friendly but slow interpreted BASIC, but the resulting programs run with the speed of a compiler. Unlike other compilers, the object code is not written into a disk file before execution (unless you request it). Therefore, no long delays are needed. When you load the file into the Trump Card, TBASIC compiles the program in a few tenths of a second.

Most programs that will run under the IBM PC's BASICA interpreter can be fed into TBASIC for compilation. You can use either the Trump Card's EE screen editor or the BASICA editor to write the programs. But if you then run the same program under both BASICA and TBASIC, depending upon the instructions you use, you will notice an increase in program performance by a factor of anywhere from 7 to 100. A listing of TBASIC's keywords is shown in table 1. TBASIC also supports most of BASICA's color and graphics commands (see photo 2).

Line numbers aren't required in the source code of programs written for TBASIC except where a line is to be referenced elsewhere in the program; for example, the destination of a GOTO or GOSUB statement would need a line number. Although not requiring them, TBASIC certainly allows line numbers on every line, so existing BASICA source code will run under TBASIC, to the extent that the program is compatible with TBASIC's syntax. Such programs can immediately benefit from the increase in performance provided by TBASIC.

The development of a program using a BASIC interpreter occurs in two modes: editing the program and running it. Developing a program with TBASIC involves three modes: editing, compiling, and running. Obviously, the only difference is compilation, which is invoked on the Trump Card by the DO command; once the program has been compiled, the familiar RUN command executes it.

Example 2 on page 156 shows some examples of the kind of interaction that occurs when you use TBASIC: how to enter a program using the EE editor, compile it, and run the compiled program. In the text box, input by the user is shown in italic type while the system's prompts and output are shown in roman characters.

During compilation of a program, error messages are issued each time an error is encountered. The line of the source file in which the error was detected is displayed; in some cases, an error message is also displayed. After an error is found and displayed, compilation continues and any other errors found also will be displayed. When the compilation has been completed, a list of any undefined symbols also may be output, in which case the program should not be run.

TBASIC Programs

Three methods can be used for entering program statements into the system for compilation under TBASIC. The first is to use the Trump Card's built-in EE screen editor, as mentioned previously (see photo 3). A second method is to enter the statements using TBASIC's direct-entry mode. The third choice is to enter and test the program using the computer's regular BASICA interpreter and then run it for effect using TBASIC. The three methods may be used interchangeably.

Example 3 shows an example of these functions with a minimally modified version of the Sieve of Eratosthenes program often used as a system benchmark (see references 4 and 5). A program called SIEVE.S was previously written in BASICA and stored as an ASCII file on the disk in drive B.

Suppose you want to run the program under both BASICA and TBASIC while recording how long it takes to be executed. You could use a stopwatch, but it's easier to add a few more program lines that record the starting and ending times automatically by calling the TIME$ function. It's possible to invoke the editor directly from TBASIC, as shown in example 3, to add two lines. And you can see that TBASIC took about 2 seconds to run the modified program as measured by the internal clock.

The program changes quickly were added and executed, and, when you left the editor with a QU command, the file SIEVE.S on drive B was updated to contain the TIME$-function statements. After running the slightly revised program under BASICA, you see that it takes 202 seconds, around 100 times as

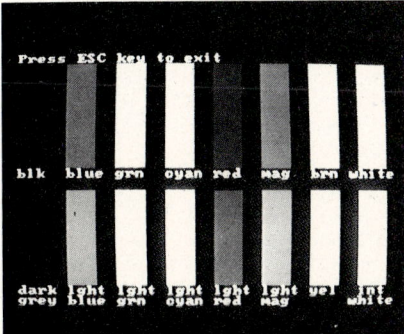 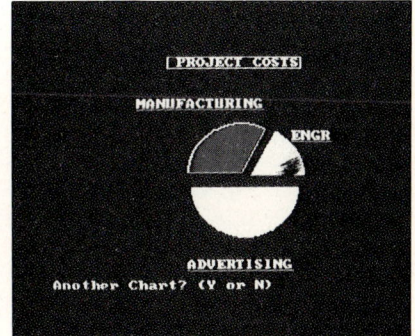

Photo 2: *Color (2a) and graphics (2b) tests demonstrate TBASIC's support of color/graphics commands normally associated with BASICA.*

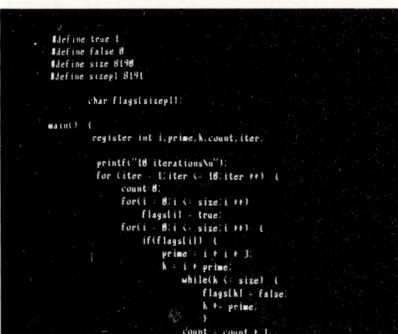 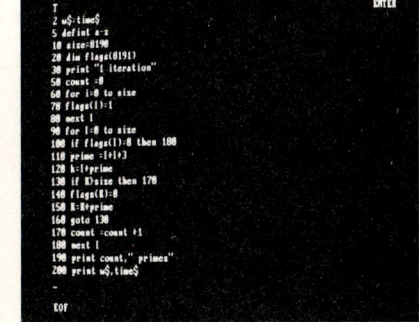

Photo 3: *Programs in BASICA (3a) and in C (3b) can be written for Trump Card or the PC by using Trump Card's built-in EE editor.*

long. Now consider the aggravation of making changes in programs that take this long to run and waiting for the results each time. Perhaps you now understand why I built the Trump Card. If you're interested in how fast some

TBASIC *speeds up development and debugging as well as execution.*

other computers and BASIC systems executed essentially the same program, see table 2. Another program that demonstrates how TBASIC speeds things up is the simple looping benchmark shown in listing 1. The results are shown in table 3.

Not all programs run a hundred times faster in TBASIC. The Sieve program purposely uses integer arithmetic and avoids difficult floating-point calculations. But we can get an idea of floating-point performance from the simple benchmark routine of listing 2. In this program, TBASIC takes 3.2 seconds while BASICA takes 24.2. This benchmark shows the wide variation in performance you can expect from a different mix of statements.

Of course, most other BASIC compilers for the IBM PC also can demonstrate dramatic speed increases over interpretive BASICA. But I believe that TBASIC is different because it speeds up development and debugging as well as execution.

(You might be wondering if the installation of an Intel 8087 Numeric Processor Extension in the IBM PC would help speed up execution of BASIC programs. Under BASICA, it would have no effect whatsoever because BASICA is not written to use it. I did a quick informal test using Morgan Professional BASIC, which uses the 8087. Morgan BASIC took 12.8 seconds to execute listing 2.)

TBASIC's EASE OF USE

TBASIC has many of the same convenience features for running programs that an interpreter has. You can use the commands RUN, RUN<*line number*>, GOTO<*line number*>, and GOSUB<*line number*> just as in BASICA. To stop a program from the console, you just hit

---

EXAMPLE 1

Computer Interaction	Comments
A > LDZSYS	Initialize Trump Card from PC-DOS.
A >	Control is returned to PC-DOS.
A > G	Turn control over to Trump Card.
:	(Trump Card's command prompt.)
: EE<filename>	Edit a file.
:	
: Z80EM<filename>	Emulate Z80 and run CP/M-80 programs.
:	
: C<filename>	Compile and run a C program.
:	
: Y<filename>	Compile and run Z8000 structured assembly language.
:	
: BASIC<filename>	Compile and run TBASIC programs.
:	
: //	Exit from Z8000 command interpreter.
A >	Control returns to PC-DOS.

EXAMPLE 2

Computer Interaction	Comments
A>B: (Return)	Set the PC-DOS default drive to B. TBASIC will also use this drive as its default drive.
B>G (Return)	Type G to "go to" the Z8000.
	The colon (:) is the Z8000 system command prompt, equivalent to the A> or B> prompt of PC-DOS.
:BASIC (Return)	Invoke TBASIC.
-	The hyphen (-) is the command prompt used by TBASIC; you may now invoke any TBASIC command.
-EDIT TESTFILE (Return)	Edit a new file using the EE editor.
T	You are now in the EE editor
EOF	in command mode.
E	Type "E" to enter text.
FOR I=1 TO 5 (Return)	Type in your BASIC program.
PRINT "Demo program" (Return)	
NEXT I (Return)	
(Escape)	Hit the Escape key to leave the Enter mode.
QU (Return)	Quit and save program on default disk B.
	The "-" prompt shows that you are now back in TBASIC.
-DO (Return)	Compile the program by using the DO command (takes about 0.1 second).
-	Your program is now compiled.
-RUN (Return)	Type RUN to execute the compiled program.
Demo program	Compiled program output.
Demo program	
Demo program	
Demo program	
Demo program	
-// (Return)	The // command exits TBASIC. (The SYSTEM command could be used instead.)
:DIR (Return)	Call for a disk directory from the command interpreter.
DIRECTORY OF DRIVE B:	
TESTFILE	There's the source file you created with the EE editor.
:// (Return)	The // command exits the Z8000's B> command mode and returns control to PC-DOS.

EXAMPLE 3

Computer Interaction	Comments
B> G (Return)	Go to the Z8000 operating system.
: BASIC SIEVE.S (Return)	Get SIEVE.S from disk and compile it in about 0.2 second.
- RUN (Return)	Execute program in TBASIC.
1 ITERATION	
1899 PRIMES	The program produces output and ends.
-	Awaiting next command.
-EDIT (Return)	Call the editor from TBASIC prompt.
T	T indicates display from top of file; the complete Sieve file is displayed, ready to edit.
5 DEFINT A¢Z	
10 SIZE = 8190	
20 DIM FLAGS(8191)	
30 PRINT "Only 1 iteration"	
50 COUNT = 0	

```
60 FOR I = 0 TO SIZE
70 FLAGS(I) = 1
80 NEXT I
90 FOR I = 0 TO SIZE
100 IF FLAGS(I) = 0 THEN 180
110 PRIME = I+I +3
120 K = I + PRIME
130 IF K > SIZE THEN 170
140 FLAGS(K) = 0
150 K = K + PRIME
160 GOTO 130
170 COUNT = COUNT + 1
180 NEXT I
190 PRINT COUNT;" PRIMES"
```

E (RETURN)	Enter mode, allows text entry.
2 J$ = TIME$	Two lines are added to print the time.
200 PRINT J$, TIME$	
(Escape, Return)	Type Escape key to exit Enter mode.
QU (Return)	Finished changes. Leave editor and return to TBASIC.
- DO (Return)	The file is recompiled with the DO command, taking about 0.2 second.
- RUN (Return)	The program is run again with changes.
1 ITERATION	The program produces output.
1899 PRIMES	
01:01:25 01:01:27	
-	The prompt returns after execution ends.
-// (Return)	Exit TBASIC.
:// (Return)	Exit the Trump Card system.
B>BASICA (Return)	Get BASICA and run SIEVE.S.
LOAD "SIEVE.S"	(SIEVE.S was stored in ASCII format.)
RUN	
1 ITERATION	The program produces output.
1899 PRIMES	
01:05:35 01:09:01	

EXAMPLE 4

Computer Interaction	Comments
B>G (Return)	Activate the Trump Card.
: BASIC (Return)	Enter TBASIC.
- /DIAG (Return)	Invoke subroutine-diagnostic mode.
- PRINT 2+3 (Return)	Directly add and print 2+3.
CExit;CImmxInit;Ki00000000; CPrtInit;Ki00000002;Ki00000003; b+;CPrtI;CPrtCR;R: 5	The listing shows the compiler subroutines that are executed to perform the function. CExit (call exit) jumps out of the console-input mode; CImmxInit calls for immediate execution with a flag integer-constant value of 0 set as Ki00000000. CPrtInit (call printer) directs printing to the console: the two integer values are expressed as Ki00000002 and Ki00000003, respectively; b+, calls a binary add routine; CPrtI prints the integer. CPrtCR finishes by sending a carriage return to the printer or console while R designates a return to the system. The computed value, 5, appears at the end.
- PRINT2.027+3.094 (Return)	Floating-point values produce a slightly different result.
CExit;CImmxInit;Ki00000000; CPrtInit;Kf01BA5E82;Kf46041982; CFltAdd;CPrtF;CPrtCR;R: 5.121	This time the constants are stored as floating-point numbers, and floating-point add and print routines are called instead.
-	

EXAMPLE 5

:	Back in command interpreter.
:C (Return)	Call C compiler, the "-" is
-	the C compiler prompt.
-/DO BASICIO.C (Return)	Compile I/O routines.
-/DO CDEMO.C (Return)	Compile CDEMO.C program (listing 3).
-/IMAGE CDEMO E=MAIN (Return)	Save memory image of compiled program in a disk file called CDEMO.
-// (Return)	Get out of C compiler.
:	Back in command interpreter.
:CDEMO (Return)	Run compiled program.
C language	
C language	The program produces output.
C language	
C language	
:	Back in command interpreter.
:// (Return)	Get out of interpreter.
B>	Back to IBM PC-DOS command prompt.

Control-C. If possible, TBASIC will display the statement label nearest the point in the program where the stop occurred. Programs may contain STOP statements and may be restarted by a CONT command.

TBASIC also can execute statements and commands in immediate mode. You simply type the program line without a line number. (If you precede a statement with a line number, it will be compiled into the existing program.) You can get results like

–PRINT SQR(2)
1.414214
–
–PRINT 2*3
6
–

You can print out variables or run specific program lines that contain line-identifier labels. Immediate-mode statements and commands also may be included in program files.

TBASIC also has some commands useful in debugging and problem diagnosis that you probably have not seen before. You can examine the actual compiled machine-language object code with commands like /DIAG. If you give the /DIAG command before a program is compiled, a complete list of compiler subroutine calls will be produced. This can be demonstrated in the direct-execution immediate mode, as shown in example 4 for both integer and floating-point values.

C COMPILER

For more ambitious program development, the Trump Card also supports a compiler for programs in the C language, as described by Kernighan and Ritchie (see reference 6). Programs need

> *The Trump Card also supports a compiler for programs written in the C language.*

only slight modifications for compilation. Developing and running a C program is a three-step operation similar to the process used in TBASIC: editing, compiling, and running.

C compilers expect to find input and output routines in a subroutine library separate from the compiler. Kernighan and Ritchie describe a file called "stdio.h" that contains the I/O facilities. The Trump Card's C compiler uses a file of I/O routines called "basicio.c", which includes the following routines: "getchar", "putchar", "open", "close", "read", "write", "printf", "scanf", "lseek", and "creat".

The implementation of "scanf" and "printf" in the Trump Card's version of C differs slightly from that of Kernighan and Ritchie. In their implementation, the conversion characters "d" and "x" may each be preceded by an "l" to indicate a pointer to a "long" value rather than a pointer to an "int" value appears in the argument list. In this implementation, the uppercase conversion characters "D" and "X" are used for the same purpose. The conversion character "f" is used for floating point. The "scanf" routine assumes that the input values are separated by Space or Tab characters and that a Return character ends an input sequence.

The Trump Card's C compiler was designed with a user interface similar to that of TBASIC, and it's just as easy to use. Listing 3 shows a C program that is entered into the system using the EE editor in a manner such as that used for TBASIC. Example 5 shows how the program is compiled and run. Should you care to try the Sieve program in C, it is shown in listing 4 set up for 10 iterations. It runs in 3.2 seconds on the Trump Card, which compares quite favorably with versions of C running on 8-MHz MC68000 processors and with assembly-language versions on the IBM's 4.77-MHz 8088.

Y MULTILEVEL LANGUAGE

The Y language system compiles a multilevel language that can be best described as structured assembler code. It allows you to write programs using a mixture of Z8000 assembly language (in Zilog mnemonics), Pascal-like control structures, data types, arithmetic expressions with automatic or specified allocation of registers, procedure calls with parameter passing, and a descriptive compiler language. The different levels of constructs may, for the most part, be freely mixed.

The Y compiler generates code directly into memory with one pass and supports immediate execution of statements, conditional compilation, user-defined extensions to the language, and symbolic debugging. Most of the Z8000

Table 1: Keywords for statements and functions available in the TBASIC compiler for the Trump Card. An asterisk indicates a new feature.

Function	Statement	Command	Variable
ABS	BEEP	ALLOCATE*	CSRLIN
ASC	CALL	BLOAD	DATE$
ATN	CLOSE	BSAVE	ERR
CALLINTS*	CIRCLE	CONT	INKEY$
CDBL	CLS	DIAG*	TIME$
CHR$	COLOR	DISP*	
CINT	DATA	DO*	
COS	DATE$	EDIT	
CSNG	DEF FN	KILL	
CVI	DEF SEG	LIST	
CVS	DEFtype	MAP*	
CVD	DIM	NAME	
EOF	END	NEW	
EXP	FIELD	REGIONS*	
FIX	FOR...NEXT	REGS*	
HEX$	GET	RESET	
INP	GOSUB	RUN	
INPUT$	GOTO	SAVE	
INSTR	IF	SYSTEM	
INT	INPUT		
LEFT$	INPUT#		
LEN	LSET		
LOC	LET		
LOF	LINE		
LOG	LINE INPUT		
LPOS	LINE INPUT#		
MID$	LOCATE		
MKI$	LPRINT		
MKS$	LPRINT USING		
MKD$	ON ERROR		
OCT$	ON GOSUB		
PEEK	ON GOTO		
POINT	OPEN		
POS	OUT		
RIGHT$	PAINT		
RND	POKE		
SCREEN	PRINT		
SGN	PRINT USING		
SIN	PRINT#		
SPACE	PRINT# USING		
SPC	PSET		
SQR	PUT		
STR$	PRESET		
STRING$	RANDOMIZE		
TAB	READ		
TAN	REM		
VAL	RESTORE		
	RESUME		
	RETURN		
	RSET		
	SCREEN		
	SEEK*		
	SOUND		
	STOP		
	TIME$		
	WAIT		
	WHILE...WEND		
	WIDTH		
	WRITE		
	WRITE#		

Table 2: Comparison of Sieve benchmark results (one iteration) on other computers running Microsoft-derived BASIC interpreters (times measured in seconds).

Apple II	Apple III	TRS-80 Model II	IBM PC (BASICA)	IBM PC (TBASIC with Trump Card)
224	222	189	206	2.4

Table 3: Execution time in seconds for the looping program of listing 1 on several interpreters.

Apple II	IBM PC (CBASIC-86)	IBM PC (BASICA)	IBM PC (TBASIC with Trump Card)
101	275	80	0.9

Table 4: A listing of the standard CP/M-80 2.2 functions. Those marked with an asterisk are supported by the Trump Card Z80 emulator.

Function	Supported?
0 System Reset	*
1 Console Input	*
2 Console Output	*
3 Reader Input	
4 Punch Output	
5 List Output	*
6 Dir Console I/O	*
7 Get I/O Byte	
8 Set I/O Byte	
9 Print String	*
10 Read Con Buffer	*
11 Console Status	*
12 Version Number	*
13 Reset Disk Sys	*
14 Select Disk	*
15 Open File	*
16 Close File	*
17 Search For 1st	*
18 Search For Next	*
19 Delete File	*
20 Read Sequential	*
21 Write Sequential	*
22 Make File	*
23 Rename File	*
24 Login Vector	*
25 Current Disk	*
26 Set DMA Address	*
27 Get Alloc Addr	
28 Write Protect	
29 Get R/O Vector	*
30 File Attributes	
31 Disk Params Addr	
32 User Codes	
33 Read Random	*
34 Write Random	*
35 Comp File Size	*
36 Set Random Rec	*

mnemonics are implemented; those that are not can be used via the WORD pseudo-operation, as in the following: LDCTL REFRESH,R3 = WORD 07D3B.

The TBASIC and C compilers are written in Y. Each of the compiler subroutines is a Y file that has been compiled into assembly-language code. A full explanation of Y is beyond the scope of this article, but listing 5 shows some Y code for your inspection. Y is an advanced tool for the experienced programmer.

CP/M-80 Emulator

The Trump Card supports a software emulator for CP/M-80 version 2.2, which allows the Trump Card to execute assembly-language programs for the 8-bit Z80 microprocessor.

The Z80 program must be transferred to a PC-DOS (or MS-DOS) floppy disk. (This can be done by linking a Z80-based computer and an IBM PC through a serial RS-232C connection, either through a direct cable or through a modem.) Once the Z80 program is on the IBM-format disk, its filename extension must be changed from ".COM" to ".CMD", which is consistent with the CP/M-86 convention and avoids the problem of trying to run a Z80 program under IBM PC-DOS.

The emulator normally resides on a disk in drive B and is used in a manner very much like that of the other Trump Card software we've looked at. Nearly all the normal CP/M-80 system calls are supported by the emulator, with a few exceptions as shown in table 4. The standard CP/M-80 BIOS (basic input/output system) calls dealing with the disk, punch, and reader devices are not supported by the Z80 emulator; the remaining BIOS calls are supported.

In Conclusion

The Trump Card is a board-level hardware approach to upgrading the performance of your IBM PC (or a compatible system). Aside from its function as a

Listing 1: A simple FOR...NEXT loop benchmark program in BASIC.

```
100 FOR A = 1 TO 10
115 FOR J = 1 TO 10
120 FOR T = 0 TO 200
130 GOSUB 200
140 B = 1
150 NEXT T
155 NEXT J
160 NEXT A
170 PRINT "DONE"
200 RETURN
```

Listing 2: A simple BASIC benchmark program for floating-point division.

```
60  A = 2.71828
80  B = 3.14159
100 FOR I = 1 TO 5000
120 C = A/B
320 NEXT I
```

Listing 3: A demonstration program for the C compiler.

```
main()
{
    int count,step;
    count = 1;
    step = 1;
    while (count <= 5)
        {
        printf(" C language\n");
        count = count + step;
        }
}
```

Z8000 development system, it provides many popular system enhancements in a single package: add-on memory, execution of Z80 programs, a separate editor, and language compilers. It was designed to solve my specific personal problem—I wanted a better BASIC that wasn't slow or cumbersome—and to support the PC in other ways: as a language and RAM-disk peripheral. If you're like me, these characteristics will be the most important ones to you.

In the process of building the Trump Card, however, I've found that it has potential I never imagined. Besides the software I've described, I expect that object-code translators for Z80-to-Z8000 and 8088-to-Z8000 conversions will soon be available, along with other utilities such as a print spooler. You also eventually will see Bell Laboratories' UNIX operating system for the Trump Card.

Z8000 and Z80 are trademarks of Zilog Corporation, a subsidiary of Exxon. CP/M-80 is a trademark of Digital Research.

Listing 4: *The Sieve of Eratosthenes benchmark in C.*

```c
#define true 1
#define false 0
#define size 8190
#define sizepl 8191
    char flags[sizepl];
main(){
    register int i,prime,k,count,iter;
    printf("10 iterations\n");
    for (iter = 1;iter <= 10;iter ++){
        count = 0;
        for(i = 0;i <= size;i ++)
            flags[i] = true;
        for(i = 0;i <= size;i ++){
            if(flags[i]){
                prime = i + i + 3;
                k = i + prime;
                while(k <= size){
                    flags[k] = false;
                    k += prime;
                }
                count = count + 1;
            }
        }
    }
    printf("\n%d primes",count);
}
```

Listing 5: *TBASIC subroutines written on the Y multilevel-language compiler.*

```
[5a]
if SWITCH=0 or CNT>100 then begin
SWITCH: = 1; GODOIT(2, VAL&0F)
end
else begin
    R3: = ^ABC; R5: =@R9|2|; R1: =CNT/2
    LDIR @R3,@R5,R1
end

[5b]
COLOR: PROC ...passed flag, then other params
depending on flag
    ...if flag bit 2 = 1, then set border color (if text mode)
    ...if bit 1 = 1, set background color (text) or palette (graphics)
    ...if bit 0 = 1, set foreground color (text) or background color (graphics)
    save R6,R7
    POPL RR6,@RR12
    if BIT R7.2 not zero then begin
        POPL RR2,@RR12
        if SCRMODE<=1 then SETBORDER(R3)
    end
    if BIT R7.1 not zero then begin
        POPL RR2,@RR12
        if R0: =SCRMODE<=1 then SETBG(R3) else
    if R0=2 then
SETPALET(R3)
    end
    if BIT R7.0 not zero then begin
        POPL RR2,@RR12
        if R0: =SCRMODE<=1 then SETFG(R3) else
    if R0=2 then
```

```
SETGRAPHBG(R3)
end
restore R6,R7
RET
SOUND: PROC ...passed duration
(in 1/18.2 secs) and frequency
    ...make sound
POPL RR4,@RR12     ...duration
POPL RR2,@RR12     ...frequency
EXB RL3,RH3; EXB RL5,RH5
R3: − >BX; R5: − >CX
AH: = 4      ...sound
EXTCALL(SPSCRINT)
RET
```

REFERENCES

1. Brown, Peter J. *Writing Interactive Compilers and Interpreters*. New York: John Wiley & Sons, 1979.
2. Ciarcia, Steve. "Trump Card, Part 1: Hardware." BYTE, May 1984, page 40.
3. George, Donald P. "Professional BASIC." BYTE, April 1984, page 334.
4. Gilbreath, Jim. "A High-Level Language Benchmark." BYTE, September 1981, page 180.
5. Gilbreath, Jim, and Gary Gilbreath. "Eratosthenes Revisited: Once More through the Sieve." BYTE, January 1983, page 283.
6. Kernighan, Brian W., and Dennis M. Ritchie. *The C Programming Language*. New York: Prentice-Hall, 1978.
7. Lee, J. A. N. *The Anatomy of a Compiler*, 2nd ed. New York: Van Nostrand Reinhold, 1974.
8. Mello-Grand, Sergio. "The Docutel/Olivetti M20: A Sleek Import." BYTE, June 1983, page 188.

The following items are available from

Sweet Micro Systems Inc.
50 Freeway Dr.
Cranston, RI 02910
(800) 341-8001 for orders
(401) 461-0530 for information

1. Trump Card I, board-tested, expandable to 512KB RAM, Software and Documentation
TRC01 . . . . $955
2. Trump Card II, board-tested, expandable to 2MB RAM,* Software and Documentation
TRC01 . . . . $1195
3. 128KB DRAM Set, 16-150nS 64K RAM chips (1 row), burned in and tested. (Use with TRC01 and TRC02.)
RAM01 . . . . $185
4. 512KB DRAM Set, 16-150nS 256K RAM chips (1 row), burned in and tested. (Use with TRC02 only.)
RAM02 . . . . $320

*All base units will accommodate four rows of RAM. The TRC02 unit will accommodate any combination of RAM01 & RAM02 above.

Editor's Note: Steve often refers to previous Circuit Cellar articles. Most of these are available in reprint books from BYTE Books, McGraw-Hill Book Company, POB 400, Hightstown, NJ 08250.

Ciarcia's Circuit Cellar, Volume I covers articles that appeared in BYTE from September 1977 through November 1978. *Ciarcia's Circuit Cellar, Volume* II contains articles from December 1978 through June 1980. *Ciarcia's Circuit Cellar, Volume* III contains articles from July 1980 through December 1981. *Ciarcia's Circuit Cellar, Volume* IV contains articles from January 1982 through June 1983.

A Musical Telephone Bell

Personalizing the sound of your telephone

About a month ago, I was visiting an IC (integrated circuit) manufacturer that produces a line of communication and voice-synthesis chips, among others. I was there to get some firsthand information on some new chips that will be the heart of a low-cost Circuit Cellar voice-recognition project coming later this year.

The visit started like most business meetings I'm used to: it included a tour of the facility, discussions with the technical staff, and lunch. After returning from lunch, a group of us were standing in the corridor adjacent to a large divided office area deciding who should be part of the next meeting when a phone rang in the middle of the room. Everyone immediately stopped talking and looked toward their section of the office as the phone rang again.

They glanced at each other again as one said, "Is that my phone or yours?"

The consensus was, "It must be yours, it doesn't sound like mine."

I watched as all four started walking toward their offices. About halfway there, each one stopped dead and turned around. With a somewhat exasperated sigh, one of them said, "It was George's phone."

"It gets so frustrating. All the bells sound alike, and with the acoustics in here it's impossible to tell whose phone it is unless you're within 10 feet of it."

The others shared the same expression of annoyance as the discussion shifted from electronics-related topics to an area of more immediate concern: Why can't telephone manufacturers make different-sounding bells for phones? It would seem that in a free-enterprise system, adept at producing pet rocks and Cabbage Patch dolls, a custom telephone bell would be trivial.

While I didn't make any pledge to solve this problem, I recognized legitimate concerns and decided to intervene electronically. Therefore, projects on talking robots, automatic houses, and rainmaking machines will have to be put off for another month as I try to build a better mousetrap.

I hope that this month's project will solve the auditory confusion in an office or at least add a little spice to an otherwise boring telephone. Rather than just ring, the Circuit Cellar Whimsi-Bell plays the first few bars of 25 preselected tunes. Instead of hearing an annoying metallic clamor, you can be greeted by the theme to *Star Wars*, or perhaps you would prefer the "William Tell Overture."

Cleverly disguised in this whimsical project is a discussion of the telephone system. My intention is to help you understand characteristics and specifications that govern the telephone products you purchase or the telephone interfaces you might build. It also sets the stage for future Circuit Cellar projects dealing with telephone lines.

Reading between the lines, however, you'll soon realize that the central theme is not musical phones but rather ring detection and auto-answering the phone by computer. It may seem like a trivial consideration, but the environment is hostile and connection restrictions abound. First, a little about the phone system.

THE STATUS QUO TELEPHONE

Essentially, the characteristics of the telephone have been unchanged for 90 years. It originally used a carbon microphone and electromagnetic earphone with a capacitively coupled electromagnetic ringer triggered from a hand-turned magneto. Today's Western Electric phones incorporate many of the same materials, and new electronic phones merely simulate their archaic predecessors. For example, the characteristic impedance of the early Edison phones was 600 ohms, and today's electronic units must still abide by this specification.

The design and use of telephone equipment are dominated by line resistance. When you wish to answer or initiate a telephone call, the only requirement is to place a load across the phone lines (between tip and ring). The handset, or the data-access arrangement (DAA) in either your modem or your auto-dialer, will cause a DC current flow of approximately 25 to 30 milliamperes (mA). A current-sensing relay at the telephone company then signals the system that you are "off hook." If it is an outgoing call, you will receive a dial tone; if it is an incoming call, the ringing will stop and you will be connected to the incoming party.

The on-hook voltage between tip and ring of the telephone line is approximately 48 volts (V) DC. It is generally supplied by a battery from the telephone office and can range from 42.75 to 52.5 V. Tip and ring have nothing to do with the telephone ring itself; they refer to the plugs that the operators used to connect callers many years before automatic dial exchanges. The original system had large arrays of connection jacks with operators who would physically insert patch-cord jumpers between initiating and receiving calls. The two conductor patch cords made their electrical connection to the tip-and-ring portion of the plug. The tip connection was usually ground. (Anything designed for connection to the telephone line should not be polarity sensitive. Polarity is sometimes reversed, and 200-V test voltages are sometimes placed on the line.)

Once the line is captured, an off-hook situation exists and a dial tone will be heard in approximately a second. The dial tone actually consists of two tones: 350 Hz and 440 Hz. If you are contemplating building a computer-activated automatic dialer, a tone decoder should be incorporated to signal the computer that dialing can commence. It will also signal you when the call has terminated if it was not initiated at your end. (If you want to get elaborate, the awful sound that the phone company blares at you to attract your attention when you forget to hang up a phone consists of four tones: 1400 Hz, 2060 Hz, 2450 Hz, and 2600 Hz. Pulsed at a rate of 5 Hz, it is called the receiver off-hook tone.)

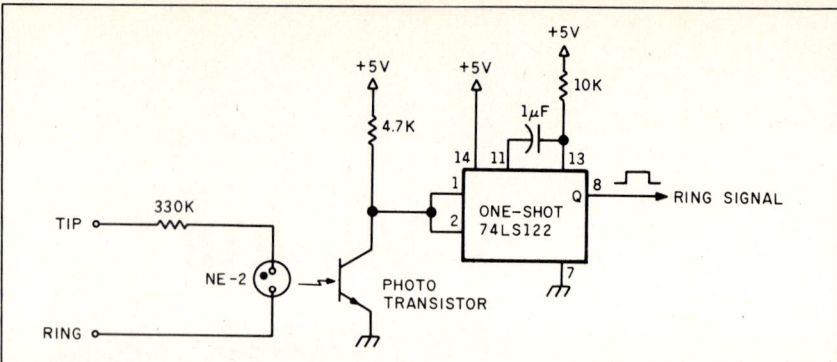

Figure 1: *A simple ring-detector circuit.*

Dialing

Most residential telephones in this country still use mechanical pulse dialers. When you turn and release the telephone dial, the current in the telephone line through the off-hook relay is interrupted by the number of times indicated on the dial position. These dial pulses are issued at approximately 10 pulses per second (pps)—some exchanges will accept up to 21 pps—with about a 60/40 percent on-/off-hook duty cycle. Nine breaks in succession are interpreted as a "9" digit, and three breaks define a "3" digit.

The separation of digits is determined by the *interdigit time*. Any succession of pulses occurring more than 750 milliseconds (ms) after the last pulse are considered part of the next digit. For example, if you send five pulses at 10 pps, wait 200 ms and send three more pulses, the telephone company will interpret this as the single digit "8." If you had waited 750 ms between transmissions, they would have been interpreted as the two digits "5" and "3."

The latest innovation in the telephone system is Touch-Tone, the registered trademark for AT&T's dual-tone multifrequency (DTMF) communication. The pulses go only as far as the local exchange office. From there, and throughout the rest of the telephone system, DTMF tones are used to direct calls.

The advantages of DTMF are increased dialing speed and, more important, the ability to transmit data. DTMF tones are sent with a minimum duration of 50 ms and an interdigit time of 45 ms. A seven-digit number on a pulse-dial system would take 11 seconds versus less than 1 second for DTMF.

I won't dwell on DTMF because it is covered in detail in my "Build a Touch Tone Decoder for Remote Control" on page 42 of the December 1981 BYTE.

The Busy and Ring Signals

Again, if you intend to build an auto-dialer at some point, you should incorporate some means to recognize a busy signal. The busy signal consists of two tones, 440 Hz and 620 Hz, that are on for 0.5 second, then off for 0.5 second. Either the tones themselves can be recognized, or the unique 50/50 0.5-second duty cycle can be monitored.

When your telephone rings, it is the result of a high AC voltage being ap-

Photo 1: *Whimsi-Bell attaches to the tip and ring wires in parallel with your phone.*

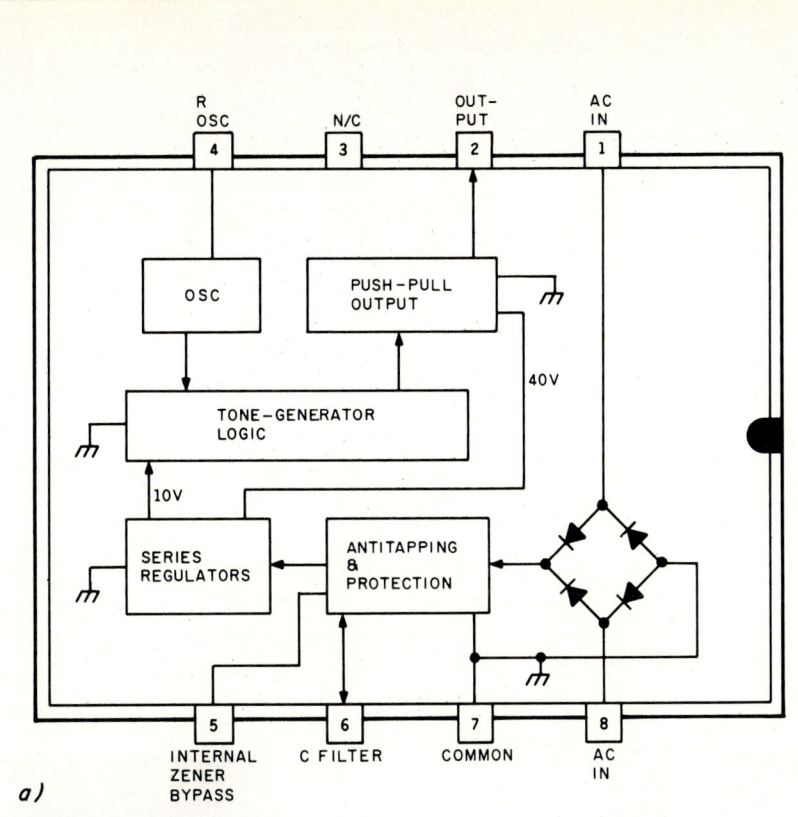

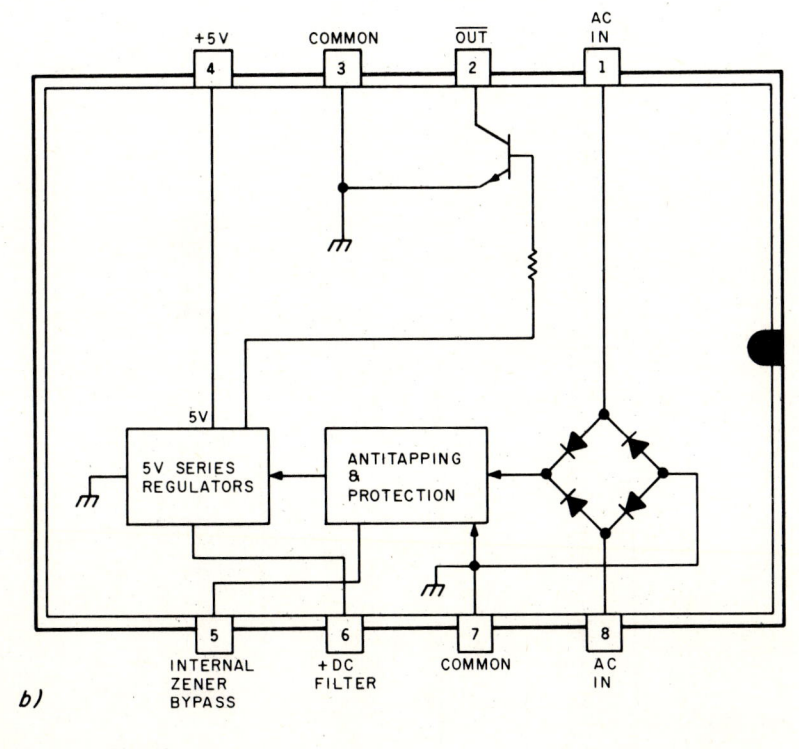

Figure 2: *Texas Instruments CMOS ring-detector chips. (2a) Block diagram of the TCM1501A, TCM1505A, TCM1506A, and TCM1512A versions. (2b) Block diagram of the TCM1520A version.*

If you are building a computer-activated automatic dialer, use a tone decoder to signal the computer.

plied to the telephone lines. The voltage is capacitively coupled to the electromagnetic bell. When you are initiating a call and you hear a ringing in the earpiece, you are not hearing the ringing voltage. Instead, you are hearing a pair of tones, 440 Hz and 480 Hz, used by the phone company for signaling. The on/off duty cycle depends on the exchange being dialed. When the ringing stops, it usually means the call has been completed or that you are irrevocably lost among the trunk lines.

Ring Detection

The solution to the problem described earlier is a ring detector. Rather than try to change all the bells in an office, it is a relatively simple matter to monitor the telephone line in parallel with the existing phone (with its bell turned low or off) and generate a different sound corresponding to the incoming ring signal. This new sound can be a buzzer, a slightly different bell, or an entirely new electronic signaling device. The actual alerting mechanism is secondary. It is all triggered by a circuit called a ring detector.

The incoming ring is the highest non-test voltage encountered on the telephone line. The normal on-hook condition is a high-impedance state with approximately 50 V DC between tip and ring. When the phone rings (usually 2 seconds on and 4 seconds off), it is because an additional 86 V AC of ringing voltage has been superimposed on the line. This 20 Hz ±3 Hz signal is passed through a capacitor to the telephone bell, causing it to ring. While 86 V AC is nominal, the ringing voltage can vary from 65 to 130 V AC, and the DC component can appear as much as 70 V negative.

Because the ringing voltage is so different from other telephone signals, a ring detector is simple in theory to construct. The simplest ring-detector circuit (shown in figure 1) consists of a neon lamp and a phototransistor. Neon lamps such as the NE-2 have a turn-on threshold of about 65 V and therefore would respond only to the higher ringing voltages. When the neon lamp lights, it in turn causes the transistor to conduct and triggers the one-shot.

Variations on this circuit employing LED (light-emitting diode) optoisolators and level-detection circuits are available, but a price is paid for simplicity. The telephone line is not an ideal environment and contains many aberrations that can lead to false triggering by crude ring detectors. Just the action of going on hook or off hook (also called tapping) generates local line transients that are sufficient to cause a neon bulb or LED to fire. We can add more components or compensate for these peculiarities in our communications software, but, fortunately, alternatives exist.

SPECIALLY DESIGNED RING-DETECTOR CHIPS

Texas Instruments produces a line of CMOS (complementary metal-oxide semiconductor) ring-detector chips that offer all the necessary features. (See figures 2a and 2b.)

The normal installation of the ring-detector chip uses a capacitor and a 2.2K-ohm resistor connected between the detector and the line. The network formed by the DC-blocking capacitor, current-limiting resistor, and full-wave bridge rectifier supplies power to the IC from the phone lines. The rectified AC signal is filtered by a 10-microfarad (μF) capacitor attached between pins 6 and 7. The value of this capacitor determines the minimum input voltage and turn-on time of the ring detector. It is also used to suppress response to dial tapping (tapping is a false triggering of the bell due to transient-producing pulses on the phone line, usually from rotary-dial phones).

In use, the ring-detector IC stays in standby mode until the incoming

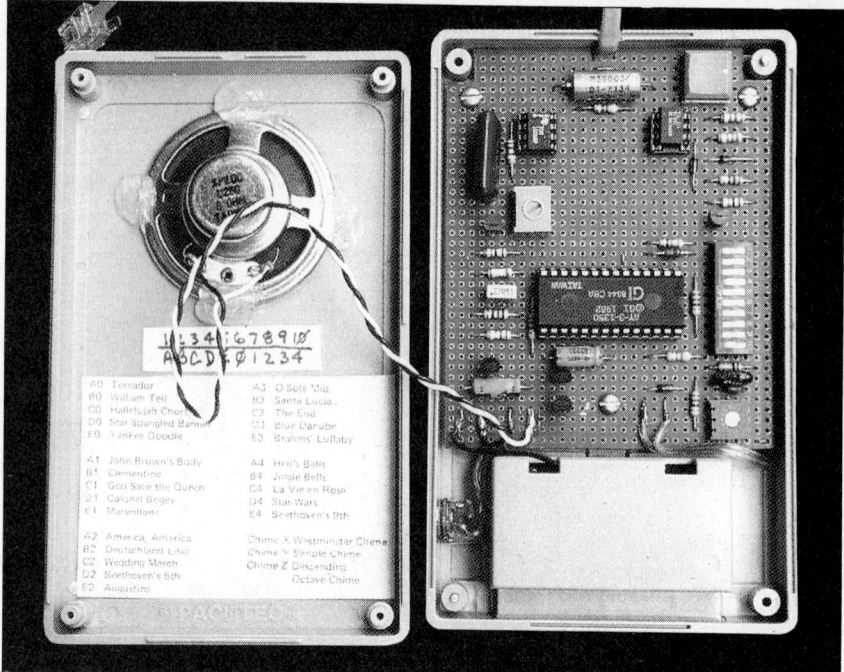

Photo 2: *A view inside the Whimsi-Bell prototype. The doorbell synthesizer chip can play 25 different tunes.*

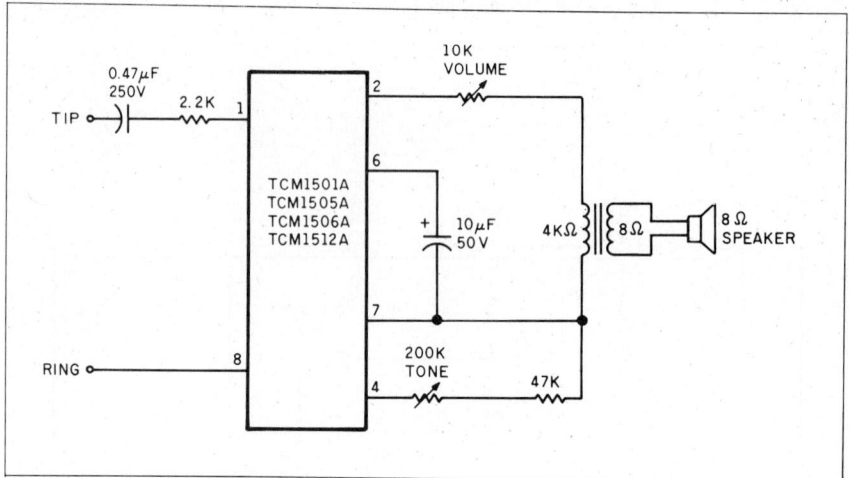

Figure 3: *A typical ring detector using the TI ring-detector chip.*

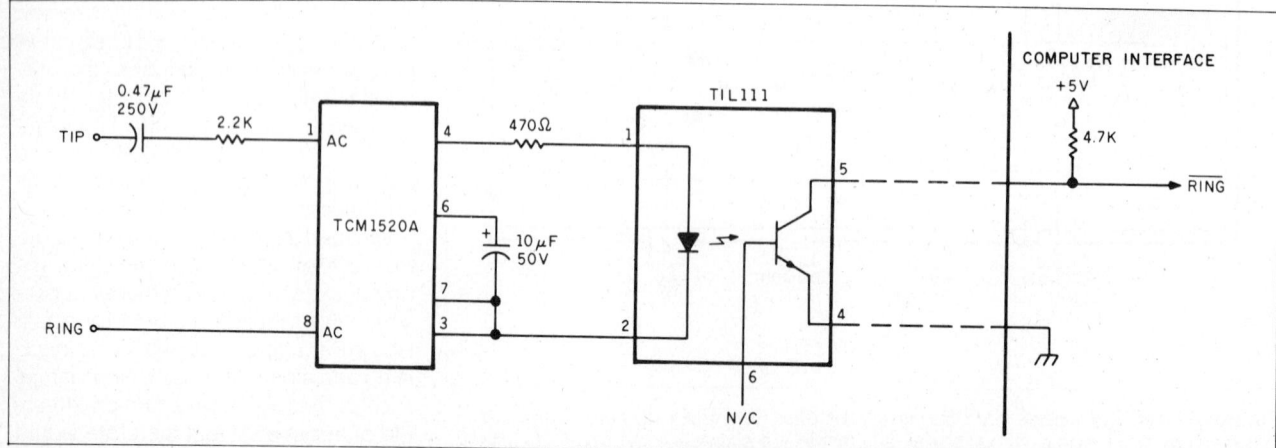

Figure 4: *A two-chip ring-detection circuit that provides an optoisolated ring-detector signal for your computer.*

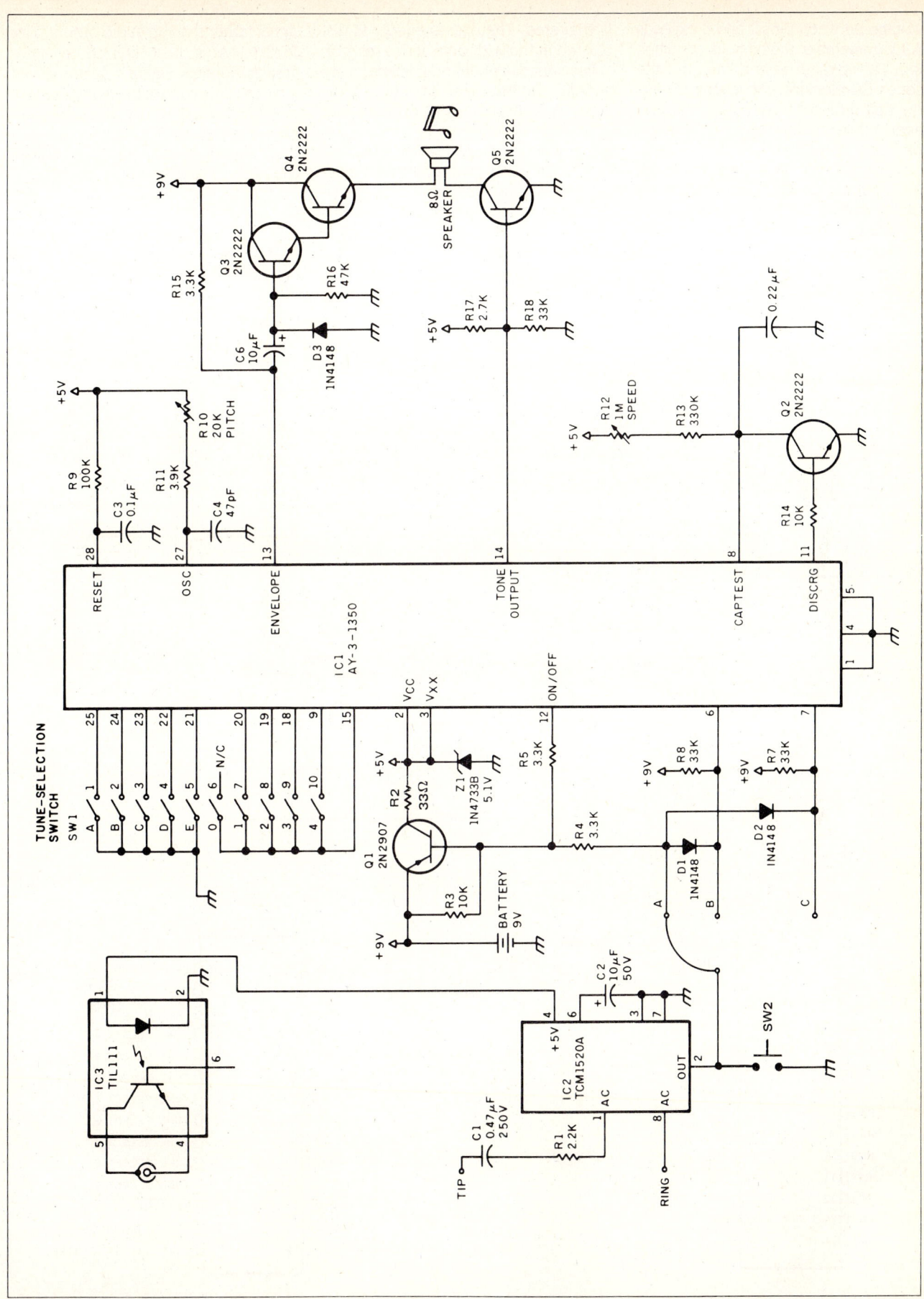

Figure 5: *The schematic of the Whimsi-Bell.*

voltage between pins 1 and 8 exceeds 8.9 V (remember, the off-hook condition is a DC voltage on the line, and the series DC-blocking capacitor prevents the ring detector from seeing this voltage). While in standby mode, the impedance is approximately 1 megohm.

When the voltage exceeds 8.9 V, the IC begins to conduct. This energy is not transferred to the load, however, until the input reaches 17 V. Should the input voltage continue to rise beyond a predetermined limit (a transient instead of a true bell signal), an internal high-current SCR (silicon-controlled rectifier) is triggered. The excess energy is dissipated in the 2.2K-ohm series resistor.

Two versions of the ring detectors are supplied by Texas Instruments. Models TCM1501A, TCM1505A, TCM1506A, and TCM1512A are detector tone drivers intended for use as electronic alternatives to the standard electromagnetic telephone bell. These chips incorporate an oscillator and a power audio-generator section to drive a piezoelectric transducer or speaker. Figure 3 demonstrates a typical circuit.

The TCM1520A is a ring detector only. It has a +5-V output signifying the ring signal rather than an audio output. The TCM1520A is best utilized in auto-answer modems. Figure 4 demonstrates a two-chip circuit that provides an optoisolated ring-detection signal that can be connected to a computer. When a ring signal is applied, the 10-μF capacitor charges until it passes the 17-V threshold. At that point, pin 4 outputs +5 V, which in turn drives the LED side of an optocoupler. The illuminated LED causes the transistor portion of the optocoupler to saturate, providing a low true signal to the computer. The computer, on recognizing a valid ring, gives

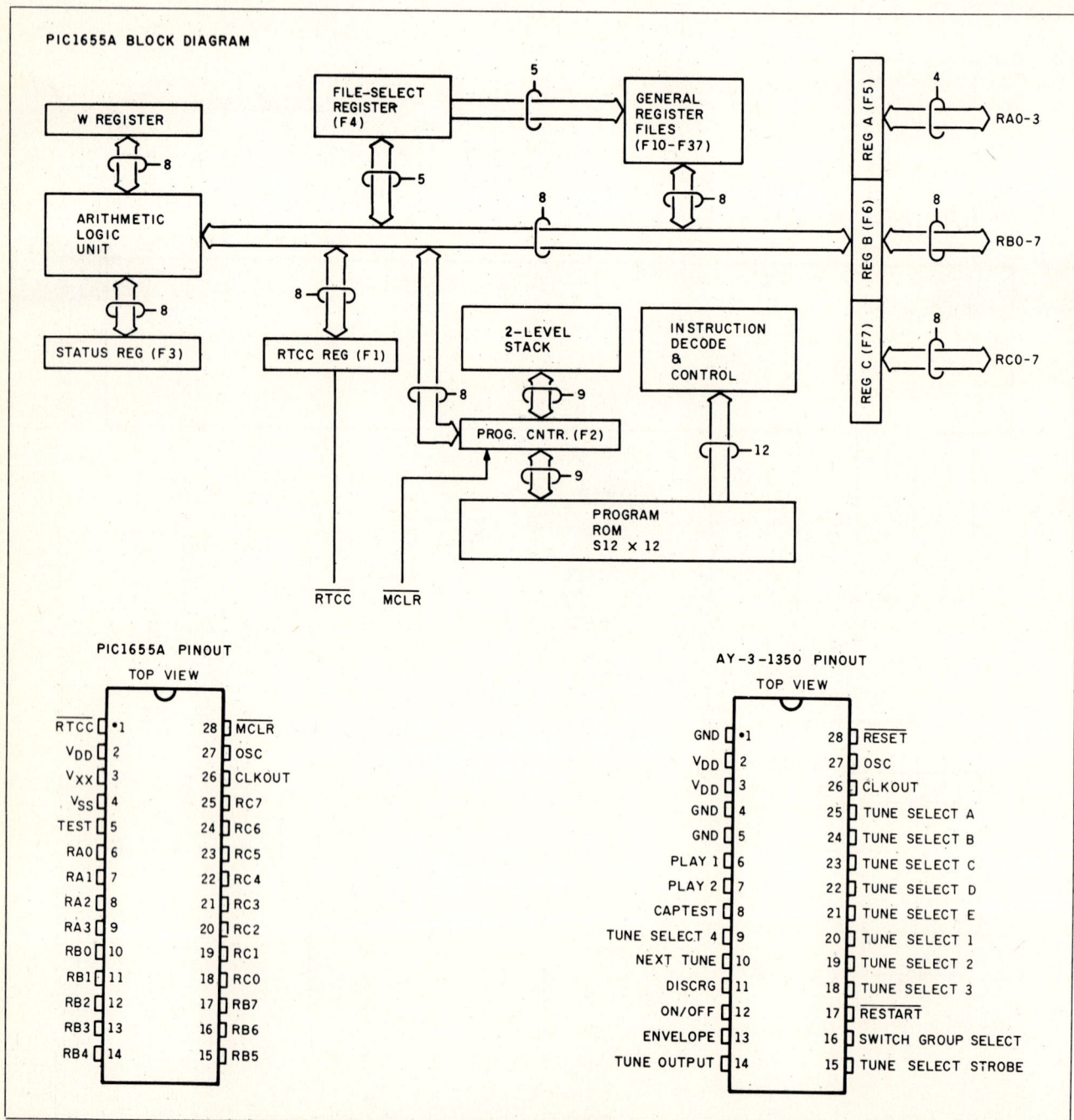

Figure 6: *Block diagram of the General Instrument AY-3-1350 tone synthesizer.*

an off-hook signal to the DAA or the modem to answer the phone.

BACK TO THE INITIAL PROBLEM

The office situation described earlier can be easily solved for one or two individuals with the one-chip circuit shown in figure 3. Here, a TCM1512A is used to drive a speaker. The circuit is connected in parallel to an existing phone that has had the bell disconnected or turned down. R1 sets the frequency of the "bell"; R2 sets its volume. When the phone rings, a different kind of bell is heard.

Such a circuit might be useful in some applications. Beyond a few installed in a large room, however, the same confusion would arise. Instead, a more varied signaling device must be employed.

A variety of alternatives comes to mind. Once you've detected the ring, virtually any triggerable event can occur. Initially, I thought it might be cute to attach a voice synthesizer or other equally nonstandard signaling device. However, I'd rather reserve voice synthesizers for more serious purposes. Instead, in the name of cost-effectiveness and mass production, I opted for Whimsi-Bell.

WHIMSI-BELL

The Circuit Cellar Whimsi-Bell, shown in photos 1 and 2 and schematically diagramed in figure 5, combines a ring detector and a sophisticated microprocessor-controlled doorbell synthesizer. The phone is connected to the Whimsi-Bell through a Y-junction phone-line connector. When the phone rings, it plays the first few bars of one of 25 tunes or three chimes. The switch-selectable repertoire is listed in table 1. The circuit also includes an optoisolated output so you can connect the device to a computer as well.

The heart of Whimsi-Bell is the General Instrument AY-3-1350 tone synthesizer, which is actually a PIC 1655A 8-bit microcomputer that has been specifically mask-programmed as an electronic doorbell chime. Block-diagramed in figure 6, the PIC 1655A includes thirty-two 8-bit RAM (random-access read/write memory) registers, 512 by 12 bits of program ROM (read-only memory), an on-board oscillator and real-time clock, and 20 I/O (input/output) lines on a single chip. The chip runs on +5 V.

The Whimsi-Bell is powered by a 9-V battery and consumes virtually no power until it is triggered by a ring-detect signal from IC2. This signal turns power on to IC1 through Q1. Q1 is sustained for the musical period regardless of the length of the ring by the on/off signal, IC1 pin 12. The chip shuts itself off when it has concluded playing.

The ring-detect signal is attached to one of three points in the circuit (A, B, or C), depending upon the combination of tones you wish. All 25 tunes are accessible if the connection is to point A. The chimes are selected by opening all the selection switches and connecting the ring-detect input to one of the three input terminals. If A, it will play the Westminster chime; if B, it will play the simple chime; if C, it will play the descending-octave chime.

With the ring-detector connection at point A, where all 25 tunes are available, a particular tune is selected on the 10-position DIP (dual-inline package) switch SW1. SW1 positions 1 through 5 are assigned to selection codes A through E; SW1 positions 6 through 10 are assigned to selection codes 0 through 4. To select the *Star Wars* theme, D4, positions 4 and 10 would be closed (all others open). Every time the phone rings, that tune will be played.

The 25 tunes average nine notes each. The tunes are stored internally as a series of notes with a stop code following the last note in each tune. Each 8-bit note comprises a 5-bit pitch value and 3-bit duration. Two-and-a-half octaves of notes can be accommodated, with the durations ranging from a sixteenth note to a whole note. (I mention this only because the AY-3-1350 can be configured for external ROM should you care to play the entire "Star-Spangled Banner" when the phone rings. Contact General Instrument for an application note describing this procedure.)

There are two potentiometer adjustments. R10 varies the processor clock speed about 1 MHz to set the pitch. The other adjustment potentiometer, R12, varies the charging time on C7 to set the speed at which the notes are played.

The output of IC1 is a combination of signals. The actual tune output comes from pin 14, while pin 13 serves as an envelope generator to control the volume. A three-transistor amplifier directly drives an 8-ohm speaker.

IN CONCLUSION

This month's project is a little less taxing than the 99-chip Trump Card from the last two months. Nonetheless, ring detection and the telephone system are important subjects for discussion.

While I don't think everyone is going to want a Whimsi-Bell attached to their phone, all the people who have heard my prototype have been amused enough to want one, if only as a unique conversation piece. More important, I'm encouraged by all the experimenters who have built one of the Circuit Cellar modem projects and see this as a means to add auto-answering.

Table 1: *Tunes and switch settings for the Whimsi-Bell.*

Switch	Tune	Switch	Tune
A0	Toreador	A3	O Sole Mio
B0	William Tell	B3	Santa Lucia
C0	Hallelujah Chorus	C3	The End
D0	Star-Spangled Banner	D3	Blue Danube
		E3	Brahm's Lullaby
E0	Yankee Doodle		
A1	John Brown's Body	A4	Hell's Bells
B1	Clementine	B4	Jingle Bells
C1	God Save the Queen	C4	La Vie en Rose
		D4	Star Wars
D1	Colonel Bogey	E4	Beethoven's Ninth
E1	Marseillaise		
A2	America, America		
B2	Deutschland Leid		
C2	Wedding March		
D2	Beethoven's Fifth		
E2	Augustine		

The following item is available from
The Micromint, Inc.
25 Terrace Drive
Vernon, CT 06266
For orders: (800) 635-3355
For information: (203) 871-6170

Whimsi-Bell kit. Includes printed-circuit board, ICs, and all discrete components, as described in figure 5. Case and battery are not included.....................$49

Please include $3 for shipping and handling in the continental United States, $10 elsewhere. New York residents please include 7 percent sales tax.

BUILD THE AC POWER MONITOR

Your computer can measure electrical power consumption

I recently saw a commercial on television for the local power utility. I don't remember the exact details, but it compared the cost of certain commodities in the early 1900s with today's cost. Almost without exception, costs have increased. The only continuing bargain mentioned was electric power. A kilowatt-hour of electricity is still about a dime after 70 years of inflation. Unfortunately, this is little consolation when all these cheap kilowatt-hours add up to a $300 monthly electric bill.

While discussing utility costs with a friend, she mentioned that her electric bill averaged $56 a month. We each have electric stoves and oil hot-water heating. I assumed that her extra use of the stove, dishwasher, and washing machine, for a family of four, would more than offset all my gadgets in a house with only two occupants. Thus, my average bill should be about the same.

Was I wrong! I looked at my electric bills and determined that my power usage during the same period averaged $175 a month. Was the difference in our life-styles so significant, or could there be some hidden power-consuming devices that I didn't normally consider? It was time for a little analysis of the situation.

Lighting is the most obvious power usage in the home, and this was my first consideration. Unlike most homes, I always have a few lights on 24 hours a day. I hate searching in the dark for light switches, so I leave lights on in strategic locations. Together, they amount to perhaps 200 watts (W).

My office lighting is also a consideration. Generally speaking, if I'm in the house, I'm either asleep or working in the Circuit Cellar. Most of my power consumption occurs there: 12 ceiling lights (@ 75 W), six fluorescent tubes (@ 40 W), two table lamps (@ 150 W), and one floor lamp (five @ 40 W). That's 1640 W of lights. But that's only part of the story. With my computer equipment, copier, and monitors, I must be using 3000 to 5000 W.

The first reaction might be to pull the plugs. Fortunately, most of my real power consumption is not constant duty cycle. Logic suggests that the copier, for instance, will consume less power in standby mode between copies than when the heaters and drying fans are on as it makes copies. Similarly, while I have an oil-heating system like my friend's, mine has four zones instead of one, and they do not necessarily operate at the same time.

Given the intermittent but heavy power demand of the heating system; the outside floodlights; the copier; and about 10 other pumps, heaters, and motors around the house, it is no wonder that my electric bill is higher than hers. Unfortunately, there is no easy way to determine power consumption of specific electrical appliances, especially those that exhibit dynamic load changes.

Hundreds of articles have been written using the words energy management and conservation, but rarely do they go into the actual measurement of the kilowatt-hours used by an appliance. Perhaps power monitoring is such a trivial subject that no one pays much attention to it. In those cases where such information is required, the expensive data-acquisition equipment required is considered a necessary evil.

Have you ever asked yourself, "I wonder how much this costs to run?" This can be in reference to an air con-

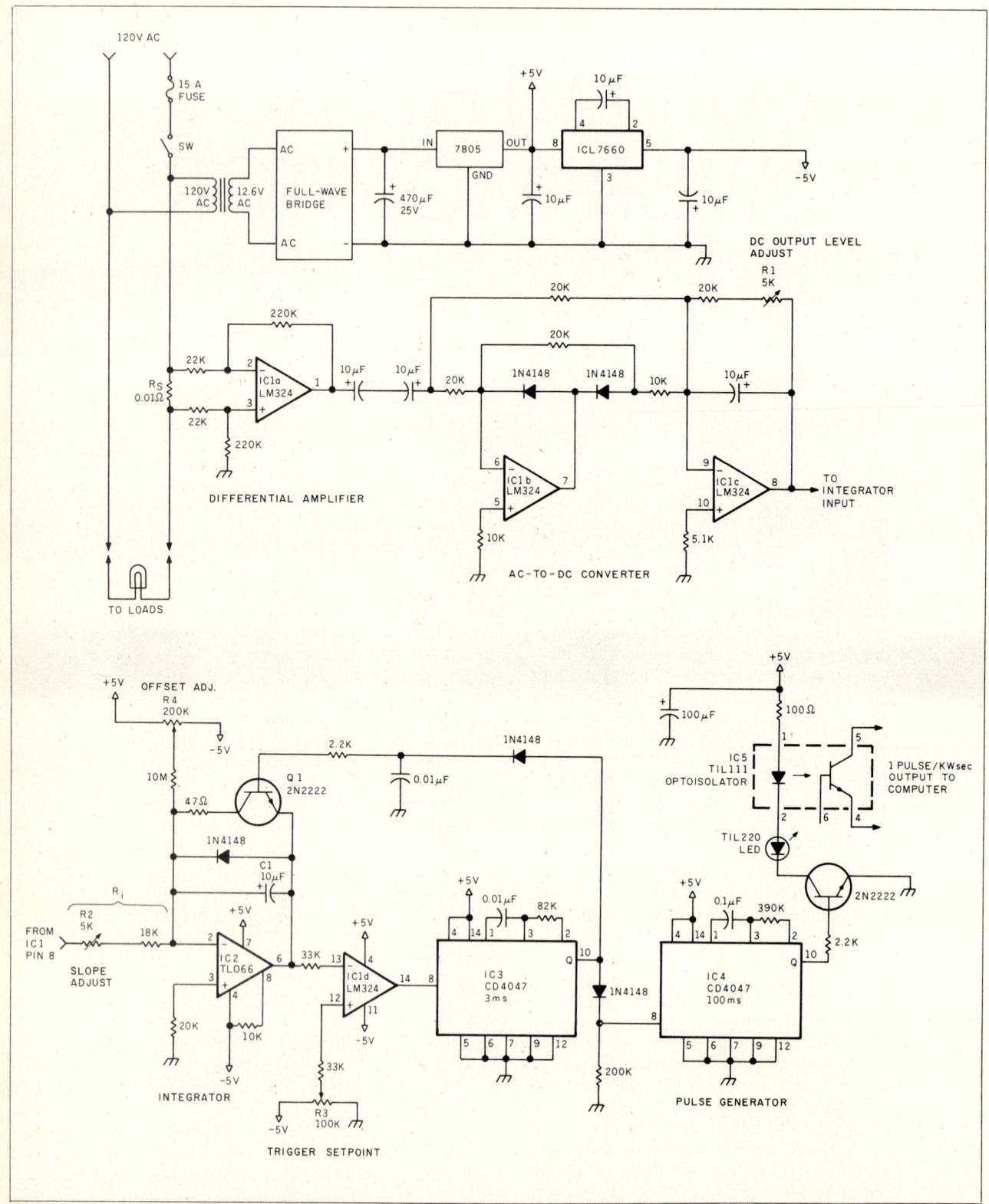

Figure 1: *A schematic diagram of the Circuit Cellar AC power monitor.*

ditioner, the pumps on the solar-heating system you are designing, or even your computer system. Have you heard that running a large air conditioner for a short period of time is supposed to be more efficient than a small one for a long time and wanted to prove it?

I decided to find a solution.

How Is Power Monitored?

Power is measured in watts. Disregarding power factors and dealing in general terms, watts equal voltage times current in most situations. If you have an electric heater that lists its power rating at 1320 W, it would draw 11 amperes (A) of current at 120 volts (V) AC. (If the voltage were lowered to 110 V AC, the same resistive element would consume only 1109 W. This is why power companies reduce voltage during periods of heavy usage.) Similarly, at 120 V AC, a 100-W bulb corresponds to a current of 0.8333 A.

The power company bills you for power usage based on the amount of power (kilowatts or kW) that you use over a given period of time (hours). Collectively, the expression is kilowatt-hour (kWhr), which means 1000 W of power used for one hour. In the case of a 100-W light bulb on for one hour, this would be 0.1 kWhr. If the light were on for a half hour, it would be 0.05 kWhr.

The power company charges by the kilowatt-hour. A typical cost is about 10 cents per kWhr (in central Connecticut, the cost is $.08902). The 100-W bulb on for one hour would therefore cost about a cent. More power hungry devices such as the heater would take 13.2 cents an hour to run.

Unfortunately, it's difficult to calculate the true operating costs when you factor duty cycle into the calculation. Once the heater is up to temperature, a thermostat causes it to cycle on and off to maintain the room at a constant temperature. Since this is less than a 100 percent duty cycle, it costs less than 13.2 cents per hour to run. But how much less?

One way to measure it is to sit there and hit a stopwatch or in some way automatically record every time the heater goes on and off and total these periods into a kilowatt-hour value. This is much the same way that the kilowatt-hour meter on the side of your house works. Basically, it is nothing more than a motor whose speed is proportional to the power passing through it. Its shaft is connected to a series of gears and dial indicators that register how many revolutions the motor has made. It turns very slowly when only a few lights are on and very quickly when heavy-current items such as electric stoves are turned on. In my case, I think it's been turning pretty fast.

The one advantage of the mechanical kilowatt-hour meter is that it is a true integrating indicator. Assuming that the voltage remains constant, if the current to a load doubles (such as a second element in the heater turning on concurrently with the first), the meter turns twice as fast. To equal this using our stopwatch and clipboard, we would have to record current value as well as time.

Most commercial electronic kilowatt-hour monitors measure time, voltage, and current. Many also employ a computing capability to derive the results. Conceivably, we could strive to duplicate these devices. Since most of the people reading BYTE already have a computer that could record time and perform the calculations, we only need to measure voltage and current. It must be remembered that voltage and current are analog parameters that must be converted to digital values if a computer is involved. Typically, acquiring this data involves using a 10- to 12-bit analog-to-digital (A/D) converter. It must have enough dynamic range to read the high voltages across the load as well as the millivolt shunt indications in series with the load.

The expense associated with such an interface has generally limited discussion of power-consumption monitors. For me to produce a project that is both cost-effective and universally applicable, I have to consider alternative means of measuring time, current, and voltage to compute energy use.

Computing Power Consumption

The first consideration is voltage. Nominally 120 V AC, the AC power-line voltage averages between 115 and 125 V AC in most places. If I arbitrarily assume that the voltage is a constant 120 V AC, I would have a worst-case error of only ±4 percent should it vary within those limits.

With that as a prerequisite, determining the power consumed is merely a matter of integrating the load current over a given period of time and calculating watts versus time. Such refined data, however more easily obtained, is no less easily conveyed to a computer than the original data variables themselves. Communicating 0.0830 kWhr to the computer through a serial or parallel interface can be just as involved as building an A/D converter.

The Circuit Cellar power monitor fits in a power strip and has a single-bit optoisolated output that can be attached to your computer.

The alternative is to make time one of the variables in the calculation. Rather than measuring the power consumed between fixed starting and ending points and totaling and reducing the data to a value such as 0.765 kWhr, I'll describe the amount of power used in smaller, more easily defined increments.

For example, 1 kWhr is the same as 60 kilowatt-minutes (kWmin) or 3600 kilowatt-seconds (kWsec). A 100-W bulb on for 6 minutes consumes 0.01 kWhr, 0.6 kWmin, or 36 kWsec, depending on how you want to record it. The 0.765-kWhr value mentioned earlier is the same as 2754 kWsec.

To input the number 0.765 into the computer requires many parallel lines or a serial interface. Communicating 2754 kWsec, however, requires only one wire (plus ground) between the power monitor and the computer if these smaller increments are communicated in real time. We can let the computer count the total kilowatt-seconds and do the calculations.

The cost-effective solution is to design a power monitor that generates a pulse every time the load uses 1 kWsec. During the 6 minutes a 100-W bulb is on, pulses will be occurring at a rate of one every 10 seconds (a total of 36 pulses). Total power cost is then computed by dividing the number of accumulated pulses by 3600 to get kilowatt-hours and then multiplying the result by the cost per kilowatt-hour in your area.

The AC Power Monitor

The Circuit Cellar power monitor is designed to perform as I've described. It fits inside a power strip and has a single-bit optoisolated output that can be attached to any available input bit on your computer. A convenient choice is a bit on the parallel printer port. Output from the power monitor is a 100-millisecond (ms) pulse every kilowatt-second consumed by the load. For a 250-W load, the pulse rate is once every 4 seconds. At 1000 W, pulses occur once every sec-

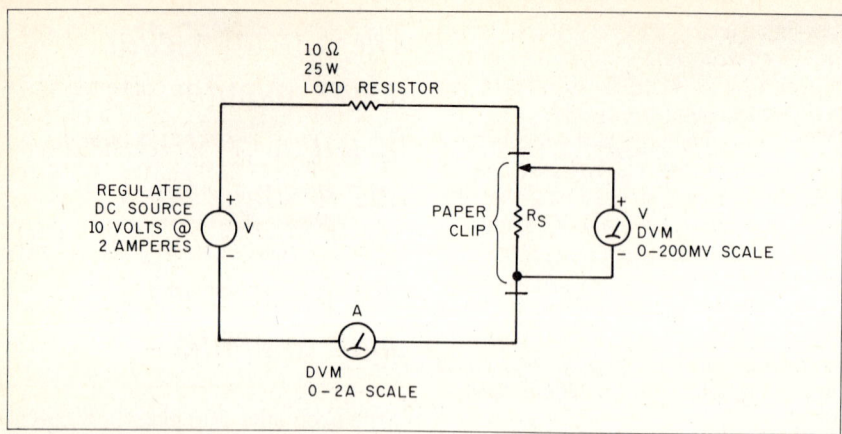

Figure 2: *Calibrating the paper clip.*

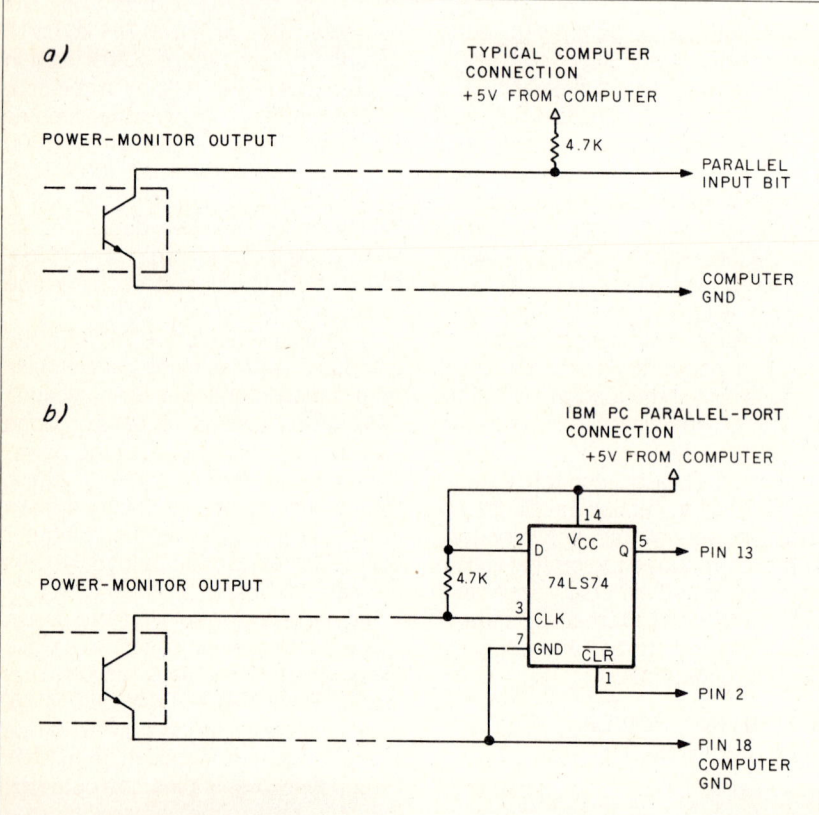

Figure 3: *The connection between the computer and the power monitor: (a) shows the usual connecting circuit; (b) shows the circuit needed to connect the monitor to an IBM PC.*

ond. As the load changes (motors and heaters cycling on and off), the pulse rate increases or decreases proportionally.

The monitoring range of the power monitor is 30 to 1800 W in the prototype configuration described. The maximum power that can be monitored is limited by the 15-A circuit breaker on the power strip rather than by the circuit design. In my opinion, the circuit as shown should be equally accurate to loads of 3000 W or more. I've included host-computer software that totals and displays the data generated by the power monitor.

Please remember that part of the Circuit Cellar power monitor attaches directly to the power line, which means a shock hazard exists. Only experienced individuals should attempt to build this device. Use extreme care while testing the circuit on line. If any connection is made to a computer, it should be made only through the optoisolated output provided. Do not connect the power monitor's +5-V supply to the 4.7k-ohm pull-up resistor on the monitor's output to the computer!

The circuitry of the power monitor (see figure 1) is relatively straightforward. It is basically an AC measuring device with the addition of integrating and pulse-generating sections.

While not generally of major concern, the power supply for the power monitor should be carefully considered. The power monitor requires both a positive and negative 5-V supply to accommodate the bipolar levels in the circuit. While the circuit itself takes only 4 milliamperes (another 20 when the LEDs [light-emitting diodes] fire for 100 ms) to run, the voltages must be regulated.

The power-supply circuit I have illustrated uses a single 12.6-V AC transformer winding to produce a regulated +5 V. An ICL7660 provides the −5 V. While other inverter circuits abound, if you don't have an ICL7660, don't substitute an alternate inverter circuit because it may not have sufficient regulation. Instead, substitute a 24-V AC transformer and construct a dual-polarity supply incorporating two regulators such as the 7805 and 7905. The approach I took, using a small transformer from Radio Shack (catalog #273-1385) and a DC-to-DC converter, was chosen merely to fit into the power-strip enclosure. Since many of you might want to build it the same way, that is the way I've documented.

Measuring Current

The most critical aspect of the monitor is the current-measuring section: Rs and IC1a. A precisely calibrated current-sensing resistor (also called a current-measuring shunt resistor), Rs, is placed in series between the voltage source and the load. As current flows through the wires, a voltage is produced across Rs proportional to the current flowing through it, according to the relation E=I×R, that is, voltage equals current times resistance.

The shunt resistance, in this case 0.01 ohm, is low to reduce internal power dissipation, which would change the temperature and cause errors. A current through Rs of 1 A produces 10 millivolts (mV), while a 10-A current produces 0.1 V.

Before you say that you don't have a 0.01-ohm shunt lying around or a meter that can measure 0.01 ohm, let's derive the proper resistance using more easily measured values such as voltage and current.

The shunt resistor is most conveniently incorporated as a piece of connecting wire in the path between source and

Part of the power monitor attaches directly to the power line, which means a shock hazard exists. Only experienced individuals should attempt to build it and extreme care should be used.

load. Since it has to carry 10 to 15 A, it must be fairly thick wire. The #12 copper wire, which is used to connect the receptacles and the wall outlet, has too little resistance. It would take about a foot of copper wire to equal 0.01 ohm. An easily obtained high-resistance material is steel, and a convenient source in the proper form is a jumbo paper clip. (Would I lie to you?) About 2 inches of a paper clip is 0.01 ohm.

Proper use of the paper clip, however, requires calibration using the circuit shown in figure 2. First, a load resistor and the paper clip are connected in series to a regulated power supply. A piece of #22-gauge wire is soldered near one end of the paper clip. With an ammeter—preferably a DVM (digital voltmeter)—connected in series with these items, adjust the power-supply voltage to approximately 1 A. The absolute value is unimportant, but record the meter reading exactly (perhaps it was 1.104 A).

Next, without changing the voltage setting, remove the meter and reconnect the circuit so that the same current is flowing through the shunt. With the DVM set on the DC millivolt scale and one probe connected to the wire previously soldered to the paper clip, move the other probe along the paper clip until you read a voltage corresponding to the current you recorded, times 0.01 (about 10 mV). If that number were 1.104 A, you would be looking for 11.04 mV. At the point on the paper clip where you get the correct value, carefully solder the second sensing wire, as shown in photo 1. Let it cool and recheck your calibration. You may have to move the wire a millimeter or two in either direction.

When the sensing resistor is installed, the two sensing wires are connected to IC1a. This op amp is configured as a times 10 gain, inverting, differential amplifier. With a 100-W load plugged into the power monitor, the output of IC1a should be 0.8333 × 0.01 × 10, or 83.33 mV AC.

Next, the AC output from IC1a is converted to DC. IC1b and 1c are configured as an AC-to-DC converter that, unlike diodes alone, will work in the millivolt range. The R1 adjustment potentiometer is used to set the output of the converter to a DC value equal to the AC RMS (root mean square) voltage input to the converter. If the input is 0.973 V AC from IC1a, the output of IC1c is set at 0.973 V DC.

The output of the DC converter is connected to an integrator, IC2, which converts current and time to kilowatt-seconds. IC2 is a TL066-type low-power, high-impedance op amp. (Only this or a similarly rated op amp, such as an ICL7611, should be used if you want accuracy while measuring low-wattage

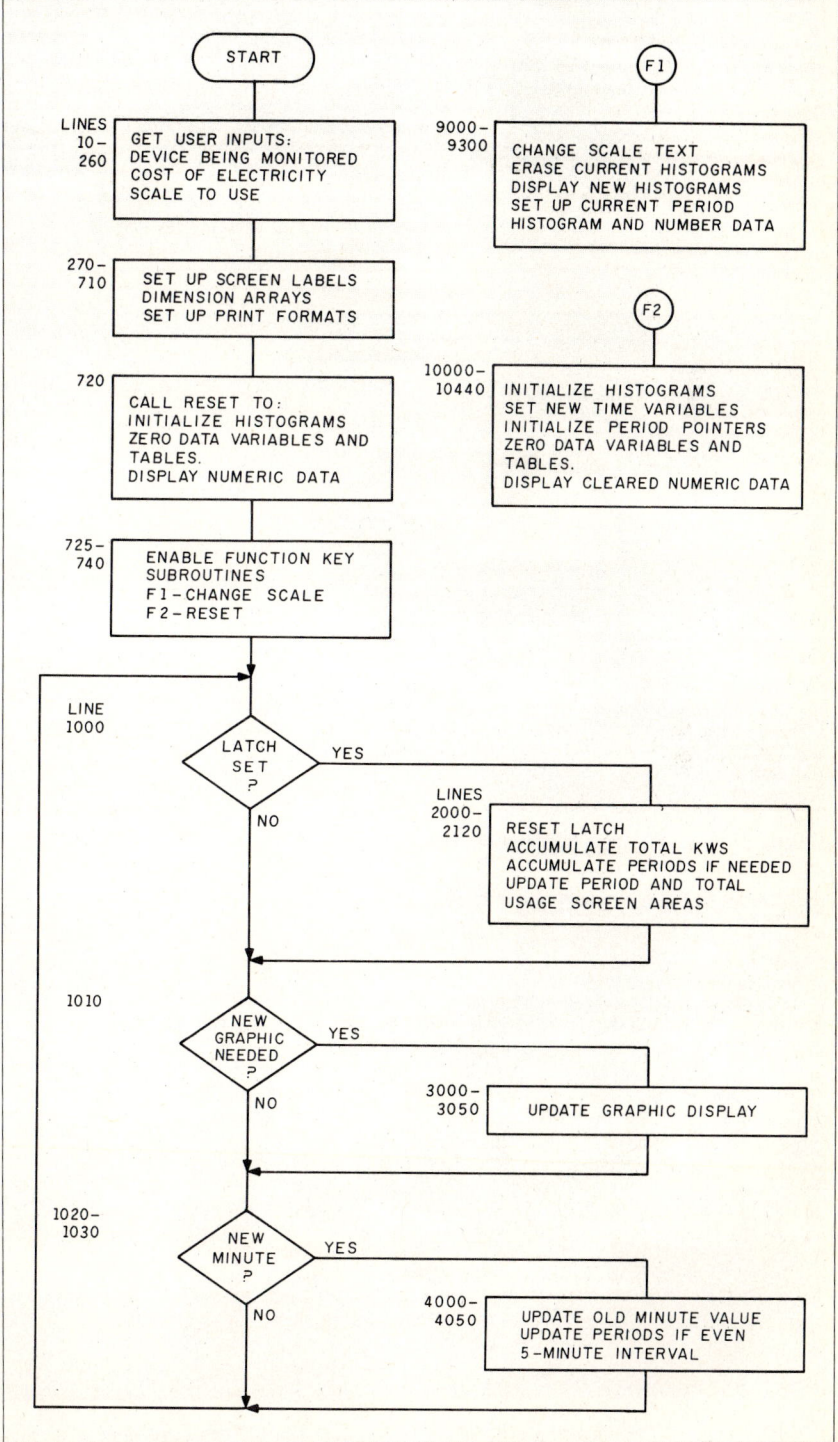

Figure 4: *A flowchart of listing 1.*

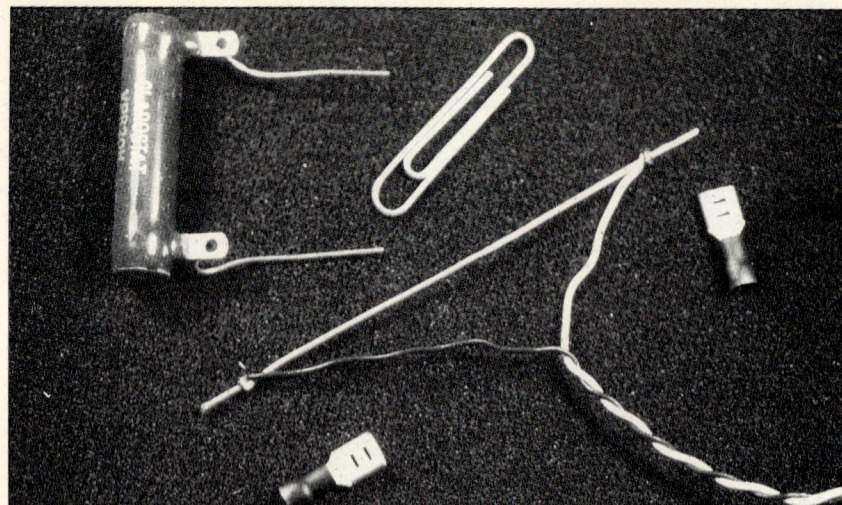

Photo 1: *A 0.01-ohm sensing resistor is created from a jumbo paper clip. A load resistor combined with a regulated DC power source is used to calibrate the sensing resistor. Two wires are attached at the 0.01-ohm resistance points.*

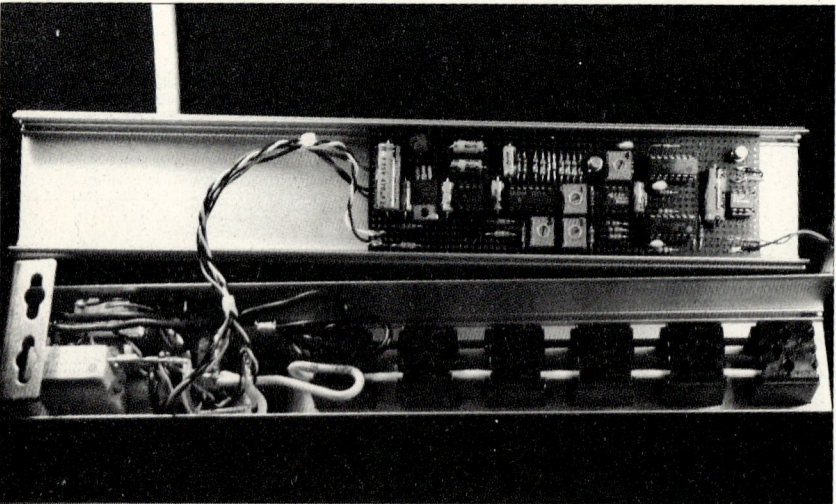

Photo 2: *A completed prototype power monitor.*

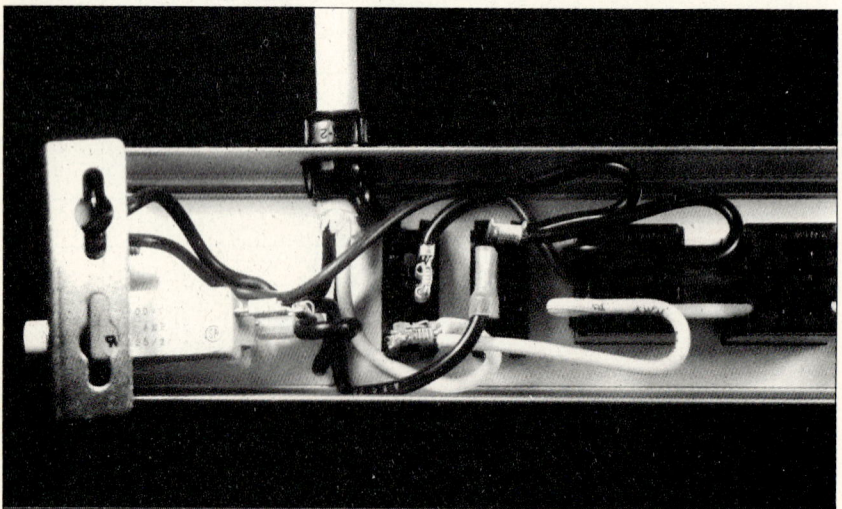

Photo 3: *An internal view of the Radio Shack six-outlet power strip. Power enters through the white cord and travels through the switch and a 15-A circuit breaker to the outlets.*

loads.) It is configured as an integrator whose output level is defined as

$$Vo = -(Vi)(T)/(Ri)(C)$$

where Vi is the voltage from IC1 pin 8, T is time, Ri is R2 plus 18 kilohms, and C is essentially C1.

The integration cycle starts after discharge of the capacitor. Then, depending upon the positive level applied to the integrator, it ramps toward negative saturation. A 1000-W load provides an input voltage that causes the integrator to reach −4 V in exactly 1 second. IC1d, configured as a voltage comparator, senses when it reaches −4 V and triggers one-shot IC3 when it occurs. IC3's output, set for 3 ms, turns on transistor Q1, discharging the integrating capacitor C1 through a 47-ohm resistor. It also fires a second, longer one-shot (IC4) that signals the computer through an optoisolated connection that 1 kWsec has been consumed. At lower currents, the input to the integrator is less, and it takes longer for it to reach −4 V (a 100-W load takes 10 seconds, for example).

The integrator can also be used with intermittent loads such as a thermostatically controlled heater or a level-activated pump. When a load is seen by the power monitor, the integrator starts to climb toward the trigger value. If the load is suddenly removed, the integration voltage level remains constant for a relatively long period of time. If the load is resumed, the voltage level starts from the same point at which it left off. If the load is off for more than a few minutes, the charge gradually leaks off C1, and some portion of a kilowatt-second is lost. The purpose of offset adjustment R4 is to minimize this drift when no load is applied and to make sure that the drift is in a direction (away from −4 V) such that a false trigger will not occur.

The connection to the computer is through an LED optoisolator. Every time a kilowatt-second is detected by IC1d, IC4 is fired for 100 ms. Two LEDs are connected to its output: the optoisolator and one that serves as a visual indicator of power consumption. The collector and emitter leads of the optoisolator, IC5, are connected to the computer. The emitter is connected to ground while the collector is tied to +5 V (supplied from the computer) through a 4.7k-ohm resistor. The output at the collector is then tied directly to any available TTL (transistor-transistor logic) input line (an alternative approach using an IBM Personal Computer (PC) will be described later).

Construction and Calibration

Photo 2 shows the prototype power monitor that I built. The enclosure is a Radio Shack six-outlet power strip (catalog #61-2619A). It incorporates a 15-A circuit breaker, pilot light, power switch, and six receptacles.

Conversion to a power monitor is relatively straightforward. Following disassembly of the power strip, remove the black wire between the circuit breaker and the power switch; the wire is shown in photo 3. Replace it with the paper clip sensing resistor, as shown in photo 4. The paper clip should be insulated (heat-shrink tubing is good) to prevent short circuits.

I glued the power transformer next to the circuit breaker, as shown in photo 5, and installed the monitor LED next to the existing neon indicator. The rest of the circuit, shown in photo 6, was built on a 2- by 6-inch perforated board that slides into the rear cover. The entire unit snaps together to provide a convenient and safe prototype enclosure.

You can test and calibrate your power monitor in two ways: with a simulated load and with an actual load. Which method you choose depends on how accurately you have constructed your 0.01-ohm sensing resistor. If you are fairly confident that it is correct, simulated conditions will suffice. The alternative is to use some form of calibrated load that allows you to calibrate the entire power monitor, including the sensing resistor. Ultimately, you will need 83.3333 mV presented to the input of IC1a. Under actual operating conditions, this requires a 1000-W load.

Simulating this load is much easier. It is accomplished by disconnecting the sensing resistor and replacing it with a larger-value resistor, such as 47 ohms. With an additional series resistor, the combination is attached to the output of a low-voltage transformer. I took the same 12.6-V output transformer used in the power supply and varied the input to it with a Variac autotransformer. Using a DVM set for AC millivolts, the autotransformer is adjusted to produce 83.33 mV across the input sense points of IC1a.

Next, with the meter set for DC volts connected to IC1 pin 14 and ground, R1 is adjusted to get a reading of 0.8333 V DC.

An oscilloscope is needed for the next phase. The point of observation is IC2 pin 6 while the trigger point is either

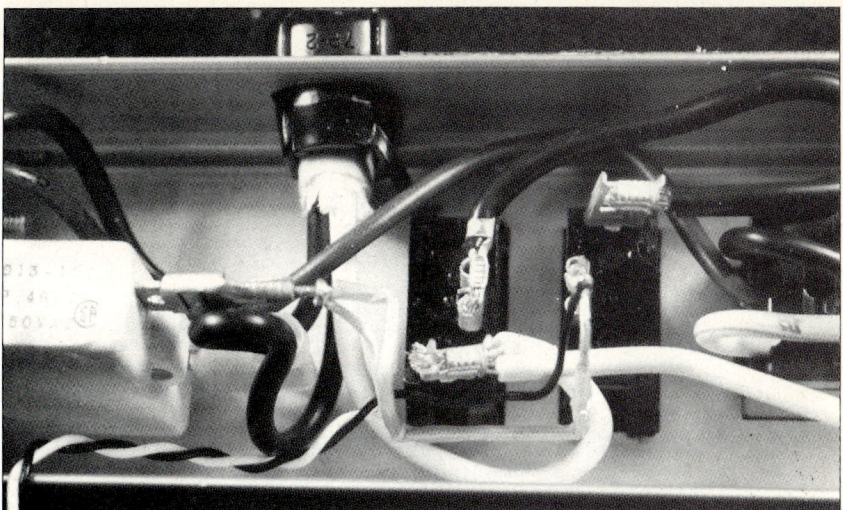

Photo 4: *The wire from the hot side of the breaker (black wire) to the switch is replaced with the paper clip sensing resistor. Notice how it is bent to fit the enclosure. It is insulated with heat-shrink tubing. The sensing wires (#22 black-and-white) are connected to the power-monitor electronics.*

Photo 5: *A 12.6-V AC, 300-mA transformer is inserted and secured within the case next to the circuit breaker. Its output is also connected to the power-monitor electronics.*

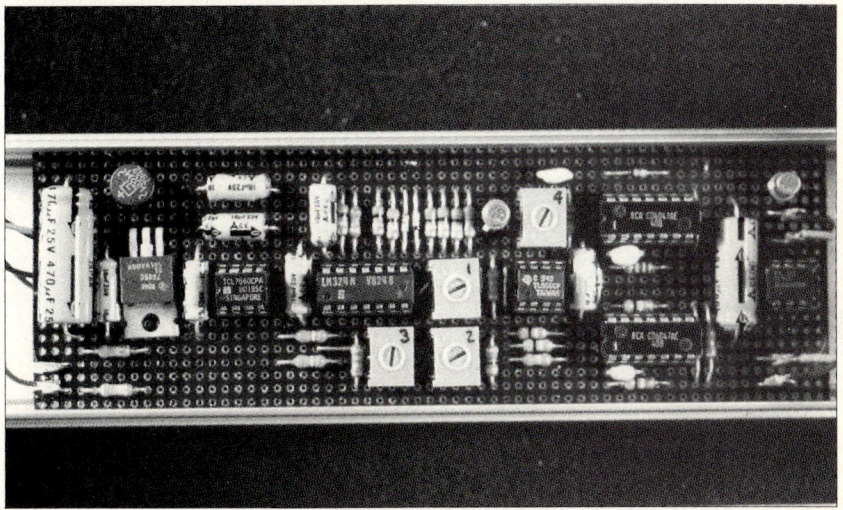

Photo 6: *A close-up of the power-monitor electronics.*

IC1 pin 7 or IC3 pin 10. The integrator output (IC2 pin 6) should either appear as a sawtooth waveform or be sitting somewhere between −4 and −5 V.

Using a clip lead, momentarily short out C1. The integrator output will return to 0 V and slowly ramp toward −5 V. If this occurs, adjust the setting of R3 until the integrator automatically resets and produces the sawtooth. Failure to achieve this activity suggests that Q1 is installed backward, IC1d is wired wrong, or one-shot IC3 is not firing (IC3 runs on + or −5 V while IC4 uses only +5 V and ground).

Once you have the integrator functioning, the time constant must be set. Using the oscilloscope, adjust the settings of R2 and R3 so that the output at IC2 pin 6 goes from 0 to −4 V in exactly a second each time it's reset.

Finally, with the DVM on the integrator output, switch off the load or simulated input source and note the reading. It should be stationary and drop only a millivolt or so a second. Adjust R4, if necessary, to achieve this. If it continues to drift after adjustment, even if only a small amount, set R4 so that the drift is in a positive direction away from −4 V.

Turn the load back on. If everything is operating correctly, the monitor LED attached to Q2 should be blinking once per second. If not, check the one-shot to see if one of the LEDs is backward.

Provided that you get the blinking indicator, the testing is completed. You can test your power monitor by simply plugging various loads into it and watching the LED. Try a table lamp, then a toaster.

COMPUTER MONITORING

The whole purpose for the particular design employed in the Circuit Cellar power monitor is to make it easily attachable to any computer. Once connected, the computer adds the kilowatt-second pulses, plots a power-usage-versus-time display, and calculates the total kilowatt-hour cost during the usage period.

I consider the main emphasis of this article the production of the monitor, but the software is a significant component. To best illustrate how the functions are performed, I wrote the program entirely in BASIC, which is relatively easy to understand and more transportable between systems. I chose to use the IBM PC simply for convenience.

By using only BASIC and no machine-specific assembly language, I have had to incorporate one modification to the power monitor/computer interface that would not normally be required if I had used assembly language. Since BASIC is relatively slow, I can't be absolutely sure that the program will be through with its housekeeping chores and not miss a pulse if connected to a high-wattage load. (While I could have used TBASIC and the Trump Card described in the May and June Circuit Cellar articles to speed things up, there is some merit in describing less costly alternative solutions.) To eliminate the concern, a flip-flop is installed at the computer input. When the pulse occurs, the flip-flop is set and remains set after the pulse is gone. When the program reads the flip-flop, it resets it.

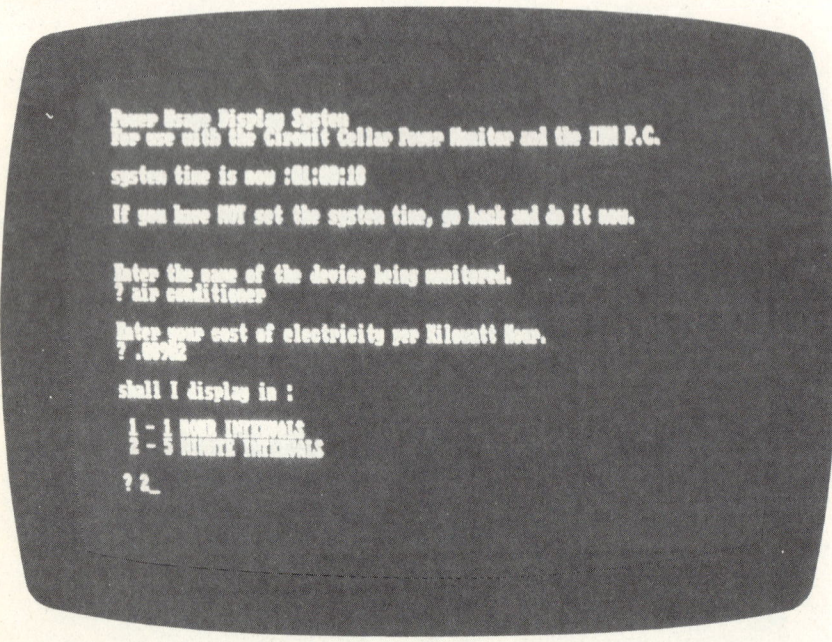

Photo 7: *Power-monitor software initialization.*

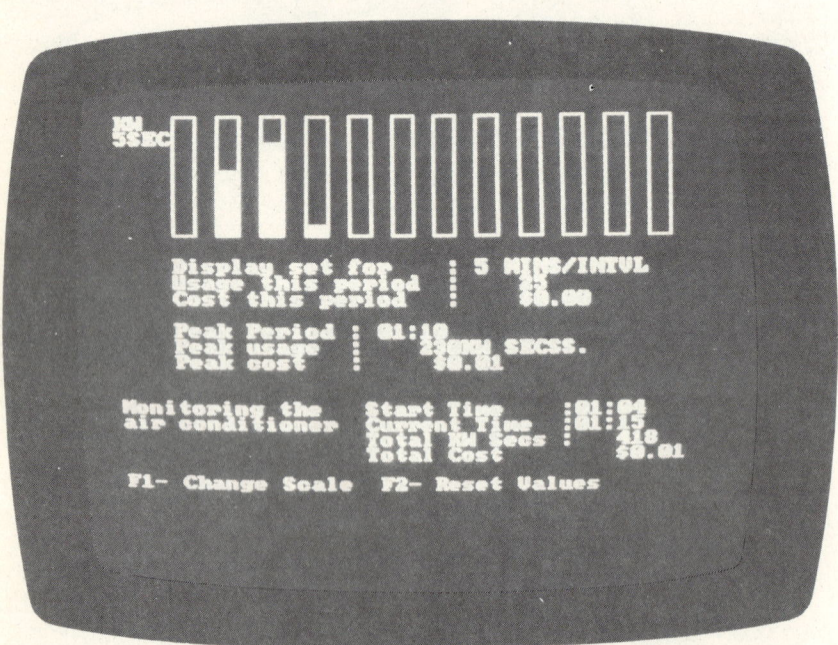

Photo 8: *Monitoring software in operation, demonstrating the power consumed by an air conditioner over a predetermined interval. Time starts from left to right. In this example, each rectangle represents 5 minutes. The current time is indicated by the red block, and increasing consumption is registered in real time for the current period and for total elapsed time. After 15 minutes, $0.01 has been consumed.*

Listing 1: *The source code for the power-usage display system.*

```
10 CLS
15 KEY OFF
20 WIDTH(80)
30 PRINT "Power Usage Display System"
40 PRINT "For use with the Circuit Cellar Power Monitor and the
    IBM P.C."
50 PRINT
60 PRINT "system time is now :";TIME$
70 PRINT
80 PRINT "If you have NOT set the system time, go back and do
    it now."
90 PRINT
100 PRINT
110 PRINT "Enter the name of the device being monitored."
120 INPUT P$
130 IF LEN(P$) > 16 THEN P$=MID$(P$,1,16)
140 PRINT
150 PRINT "Enter your cost of electricity per Kilowatt Hour."
160 INPUT C
170 C=C/3600 :REM scale to KW seconds.
180 PRINT
190 PRINT "shall I display in  :"
200 PRINT
210 PRINT "  1 — 1 HOUR INTERVALS"
220 PRINT "  2 — 5 MINUTE INTERVALS "
230 PRINT
240 INPUT S
250 IF S=1 OR S=2 THEN 270
260 GOTO 180
270 REM set up the screen
280 CLS
290 SCREEN 1
300 COLOR 0,0
350 LOCATE 1,1
360 PRINT "KW";
370 LOCATE 2,1
380 IF S=1 THEN PRINT "MIN ";
390 IF S=2 THEN PRINT "5SEC";
400 LOCATE 10,5
410 IF S=1 THEN PRINT "Display set for     : HOURS";
420 IF S=2 THEN PRINT "Display set for     : 5 MINS/INTVL";
430 LOCATE 11,5
440 PRINT "Usage this period    :";
450 LOCATE 12,5
460 PRINT "Cost this period     :";
470 LOCATE 14,5
480 PRINT "Peak Period   :";
490 LOCATE 15,5
500 PRINT "Peak usage    :";
510 LOCATE 16,5
520 PRINT "Peak cost     :";
530 LOCATE 19,18
540 PRINT "Start Time        :";
550 LOCATE 19,1
560 PRINT "Monitoring the";
570 LOCATE 20,18
580 PRINT "Current Time  :";
590 LOCATE 21,18
600 PRINT "Total KW Secs:";
610 LOCATE 20,1
620 PRINT P$;
630 LOCATE 22,18
640 PRINT "Total Cost    :";
650 LOCATE 24,1
660 PRINT "F1- Change Scale  F2- Reset Values";
700 DIM T(2,12),PK$(2),PU(2)
710 M$="$$####.##":N$="######"
720 GOSUB 10000
725 ON KEY(2) GOSUB 10000
727 KEY(2) ON
730 ON KEY(1) GOSUB 9000
740 KEY(1) ON
1000 IF (INP(&H3BD) AND 16) < >0 THEN GOSUB 2000
1010 IF T(S,PE)>PG THEN GOSUB 3000
1020 T$=MID$(TIME$,4,2)
1030 IF T$< >MO$ THEN GOSUB 4000
1040 GOTO 1000
2000 OUT &H3BC,1:OUT &H3BC,0
```

(continued)

Listing 1 continued

```
2010 A=A+1:PU=PU+1
2020 IF A/60 = INT(A/60)THEN T(1,PH)=T(1,PH)+1
2030 IF A/5  = INT(A/5) THEN T(2,PM)=T(2,PM)+1
2040 LOCATE 11,25
2050 PRINT USING N$;PU;
2060 LOCATE 12,25
2070 PRINT USING M$;PU*C
2080 LOCATE 21,33
2090 PRINT USING N$;A;
2100 LOCATE 22,32
2110 PRINT USING M$;A*C;
2120 RETURN
3000 PG=T(S,PE)
3010 DRAW "bm"+STR$(32+PE*24)+",60"
3020 FOR X=1 TO PG
3030 DRAW "c2 u1 r12 l12"
3040 NEXT X
3050 RETURN
4000 MO$=T$
4010 T=VAL(T$)
4020 IF T/5 = INT(T/5) THEN GOSUB 5000
4030 LOCATE 20,33
4040 PRINT MID$(TIME$,1,5)
4050 RETURN
5000 KEY (1) OFF
5003 KEY (2) OFF
5005 IF T(2,PM)*5<=PU(2) THEN 5030
5010 PU(2)=T(2,PM)*5
5020 PK$(2)=PS$
5030 MT=PM
5040 PM=PM+1
5050 IF PM>=12 THEN PM=0
5060 IF S=2 THEN PS$=MID$(TIME$,1,5)
5070 T(2,PM)=0
5080 IF MID$(TIME$,1,2)=HO$ THEN 5190
5090 HO$=MID$(TIME$,1,2)
5100 IF T(1,PH)*60<=PU(1) THEN 5130
5110 PU(1)=T(1,PH)*60
5120 PK$(2)=PS$
5130 PS$=MID$(TIME$,1,5)
5140 HT=PH
5150 PH=PH+1
5160 IF PH>=12 THEN PH=0
5170 T(1,PH)=0
5180 GOTO 5200
5190 IF S=1 THEN GOTO 5410
5200 PU=0
5210 IF S=1 THEN PE=PH:OP=HT:GOTO 5230
5220 PE=PM:OP=MT
5230 DRAW "bm"+STR$(32+OP*24)+",60"
5240 DRAW "c1 u60 r12 d60 l12"
5250 FOR X=1 TO T(S,OP)
5260 DRAW "u1 r12 l12"
5270 NEXT X
5280 DRAW "bm"+STR$(32+PE*24)+",60"
5290 FOR X=1 TO 12
5300 DRAW "c0 u60 r1 d60"
5310 NEXT X
5320 DRAW "l12 c2 u60 r12 d60 l12"
5325 PG=0
5330 LOCATE 14,19
5340 PRINT PK$(S);
5350 LOCATE 15,19
5360 PRINT USING N$;PU(S);
5370 LOCATE 15,25
5380 PRINT "KW SECS";
5390 LOCATE 16,19
5400 PRINT USING M$;PU(S)*C
5410 KEY (1) ON
5415 KEY (2) ON
5420 RETURN
9000 IF S=1 THEN S=2:GOTO 9020
9010 S=1
9020 LOCATE 10,25
9030 IF S=1 THEN PRINT "HOURS      ";:GOTO 9050
9040 PRINT " 5 MINS/INTVL";
9050 LOCATE 2,1
9060 IF S=1 THEN PRINT "MIN  ";:GOTO 9080
9070 PRINT "5SEC";
9080 FOR X=0 TO 11
```

```
9090 DRAW "bm"+STR$(32+(X*24))+",60"
9095 FOR Y=1 TO 12
9100 DRAW "c0 u60 r1 d60 "
9105 NEXT Y
9110 DRAW "l12 c1 u60 r12 d60 l12"
9120 FOR Y=1 TO T(S,X)
9130 DRAW "u1 r12 l12"
9140 NEXT Y
9150 NEXT X
9151 IF S=1 THEN PE=PH:GOTO 9153
9152 PE=PM
9153 DRAW "bm"+STR$(32+PE*24)+",60"
9154 DRAW "c2 u60 r12 d60 l12"
9155 PG=T(S,PE)
9156 FOR Y=1 TO PG
9157 DRAW "c2 u1 r12 l12"
9158 NEXT Y
9160 IF S=1 THEN PU=T(S,PE)*60+(A-INT(A/60)*60):GOTO 9180
9170 PU=T(S,PE)*5+(A-INT(A/5)*5)
9180 LOCATE 11,25
9190 PRINT USING N$;PU;
9200 LOCATE 12,25
9210 PRINT USING M$;C*PU
9220 LOCATE 14,19
9230 PRINT PK$(S);
9240 LOCATE 15,19
9250 PRINT USING N$;PU(S);
9260 LOCATE 15,25
9270 PRINT " KW SECS";
9280 LOCATE 16,19
9290 PRINT USING M$;PU(S)*C
9300 RETURN
10000 REM reset system routine
10005 OUT &H3BC,0
10010 FOR X=1 TO 2
10020 PK$(X)="START":PU(X)=0
10030 FOR Y=0 TO 11
10040 T(X,Y)=0
10050 NEXT Y
10060 NEXT X
10070 PG=0:PU=0:A=0
10100 FOR X=32 TO 300 STEP 24
10110 DRAW "bm"+STR$(X)+",60"
10120 FOR Y=1 TO 12
10122 DRAW "c0 u60 r1 d60"
10124 NEXT Y
10130 DRAW "l12 c1 u60 r12 d60 l12"
10140 NEXT X
10150 T$=TIME$
10160 HO$=MID$(T$,1,2):MO$=MID$(T$,4,2):PS$=MID$(T$,1,5)
10170 T=VAL(HO$)
10180 IF T>=12 THEN T=T-12
10190 PH=T
10200 T=VAL(MO$)
10210 PM=INT(T/5)
10220 IF S=1 THEN PE=PH:GOTO 10240
10230 PE=PM
10240 DRAW "bm"+STR$(32+(PE*24))+",60"
10250 DRAW "c2 u60 r12 d60 l12"
10260 LOCATE 11,25
10270 PRINT USING N$;PU;
10280 LOCATE 12,25
10290 PRINT USING M$;C*PU;
10300 LOCATE 14,19
10310 PRINT PK$(S)
10312 LOCATE 15,25
10314 PRINT " KW SECS."
10320 LOCATE 15,19
10330 PRINT USING N$;PU(S);
10340 LOCATE 16,19
10350 PRINT USING M$;PU(S)*C;
10360 LOCATE 19,33
10370 PRINT MID$(T$,1,5)
10380 LOCATE 20,33
10390 PRINT MID$(T$,1,5)
10400 LOCATE 21,33
10410 PRINT USING N$;A;
10420 LOCATE 22,32
10430 PRINT USING M$;C*A
10440 RETURN
```

Figure 3a is the usual connecting circuit between the power monitor and the computer; figure 3b is the specific circuit necessary for the IBM PC and the BASIC-only program described. Figure 4 is the flowchart of how the program works; listing 1 is the actual source code. Photo 7 shows the initialization of the software, and photo 8 shows the software in operation.

IN CONCLUSION

I think the Circuit Cellar power monitor is a step forward in energy management. We've discussed various methods for computerized control. Until now, however, we haven't had a cost-effective means to monitor activity on the power line or specifically determine power consumption.

While I haven't had my power monitor long, I've run a few tests that may be of interest to you. The monitor appears very linear and very accurate at higher loads. The range between 250 and 2000 W contains less than 1 percent of point error. At 100 W, the error is about 5 percent; about 10 percent at 50 W.

While I know there will be a great demand, no kit for the power monitor is available at the present time. The primary reason for this is packaging. Simply providing a PC board could lead to some potentially hazardous user installations. A safe, cost-effective enclosure must be provided before I feel that I can advocate direct powerline connection. Eventually, the power monitor will be available.

Special thanks to Bill Curlew for his software expertise.

AN ULTRASONIC RANGING SYSTEM

Build the SonarTape

Those of you who have followed the Circuit Cellar projects over the years will recognize that I have discussed sonar and ultrasonic ranging before. November 1980's "Home In on the Range! An Ultrasonic Ranging System" used the Polaroid Ultrasonic Ranging System Designer's Kit. This $150 kit, based on a modified, custom-manufactured sonar-ranging circuit board from Texas Instruments for Polaroid's SX-70 camera, greatly simplified the circuitry normally associated with ultrasonics. The November 1980 project described circuit additions that facilitated connection of the ranging kit to a computer's I/O (input/output) port.

To my knowledge, these kits are still available from Polaroid if you are interested in producing the stepper-motor-driven scanning system described in the original article. However, if you are looking for a cost-effective distance sensor for your computer, you'll be happy to know that LSI (large-scale integration) technology did not stand still in the interim. Texas Instruments introduced a new sonar-ranging module that is both cost-effective and simple to integrate into computer-based systems.

I'll first describe this new module in detail and then demonstrate how easily you can attach it to a computer or use it independently with an LCD (liquid-crystal display) to create an electronic tape measure.

THE SONAR-RANGING MODULE

The latest Texas Instruments sonar-ranging module, the SN28827, is actually an updated and higher-functioning version of the original SX-70 module, both of which are shown in photo 1. The newer unit has similar performance characteristics but requires far less support circuitry and interfacing hardware. It is designed to drive a 50-kHz 300-volt (V) electrostatic transducer, which Polaroid is still manufacturing. The module and the transducer are the only components necessary to measure distances from 1.33 to 35 feet with an accuracy of ±2 percent.

In operation, a pulse is transmitted toward a target and the resulting echo detected. The elapsed time between the transmission and echo detection is a function of the distance to the target. Basically, this distance in feet is simply the elapsed time in seconds (actually milliseconds) multiplied by the speed of sound in feet per second

Specifications	
Minimum Transmitting Sensitivity at 50 kHz 300 V AC peak to peak, 150 V DC bias (dB re 20 µPa at 1 meter)	110 dB
Minimum Receiving Sensitivity at 50 kHz 150 V DC bias (dB re 1 V/Pa)	−42 dB
Suggested DC Bias Voltage	150 V
Suggested AC Driving Voltage (peak)	150 V
Maximum Combined Voltage	400 V
Capacitance at 1 kHz (typical) 150 V DC bias	400–500 pF

Operating Conditions
  Temperature
  Relative Humidity
	
	32°–140° F
	5%–95%

Standard Finish
  Foil
  Housing
	
	Gold
	Flat Black

Figure 1: *Specifications for the Polaroid electrostatic transducer.*

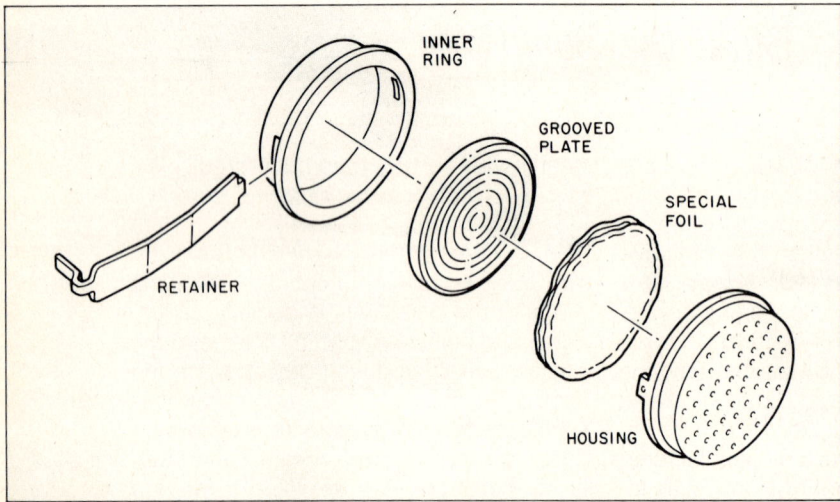

Figure 2: *The parts of the transducer.*

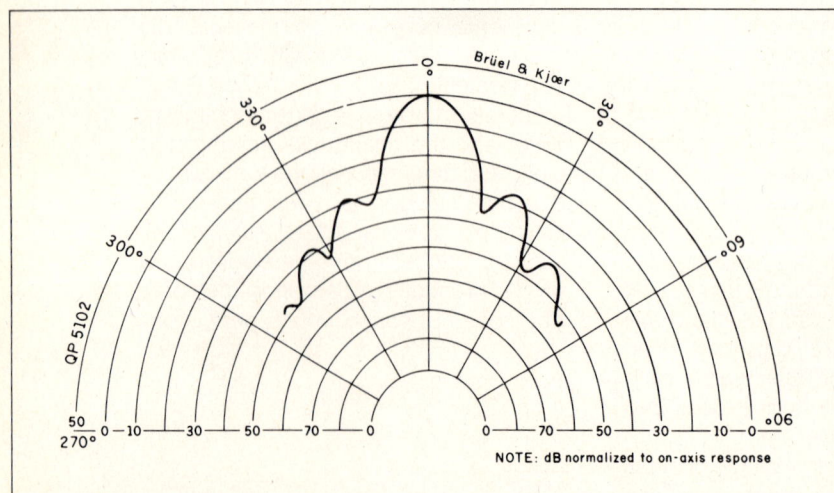

Figure 3: *The transducer's beam pattern at 50 kHz*

(approximately 1100 ft/sec). More specifically, the rate is 1.78 milliseconds (ms) per round-trip foot. It takes 3.55 ms for a pulse to leave the transducer, strike a target 2 feet away, and return to the transducer.

The transducer most frequently used with this module is the instrument-grade Polaroid electrostatic transducer (see figure 1 for its specifications), which acts as a speaker in the transmit mode and a microphone in the receive mode. The transducer (shown in photo 2 and disassembled into its component parts in figure 2) is 1.5 inches in diameter and consists of a 3-millimeter gold-plated foil stretched over a concentrically grooved aluminum disc.

The foil, electrically insulated yet bonded closely to the metallic backplate, forms a capacitor. The foil is the moving element in the transducer that converts electrical energy into sound and the returning echo into electrical energy.

The Polaroid sensor is larger and, in my opinion, considerably more expensive than other 50-kHz transducers. However, it is designed this way for a specific purpose. (I've tried 50-kHz transducers from two other manufacturers with no success.) The diameter of the transducer determines its directional sensitivity. The Polaroid unit is extremely directional, as indicated in the graph of acoustical-signal strength shown in figure 3.

OPERATING THE MODULE

The sonar-ranging module (pictured in photo 3 and diagramed in figure 4) is a two-chip, 2- by 2-inch module. It is a 12-component, custom-manufactured module built around the Texas Instruments TL851 and TL852 sonar-ranging controller/receiver chip set. In small quantities, the assembled module is far less expensive than the individual components, and it eliminates the ever-present aggravation of finding the correct coils and chokes. Electrical connection is made through an eight-conductor flat ribbon cable or direct solder attachment to the connector base, which contains three output lines, three input lines, power, and ground.

The two basic modes of operation for the module are single echo and multiple echo. The single-echo mode

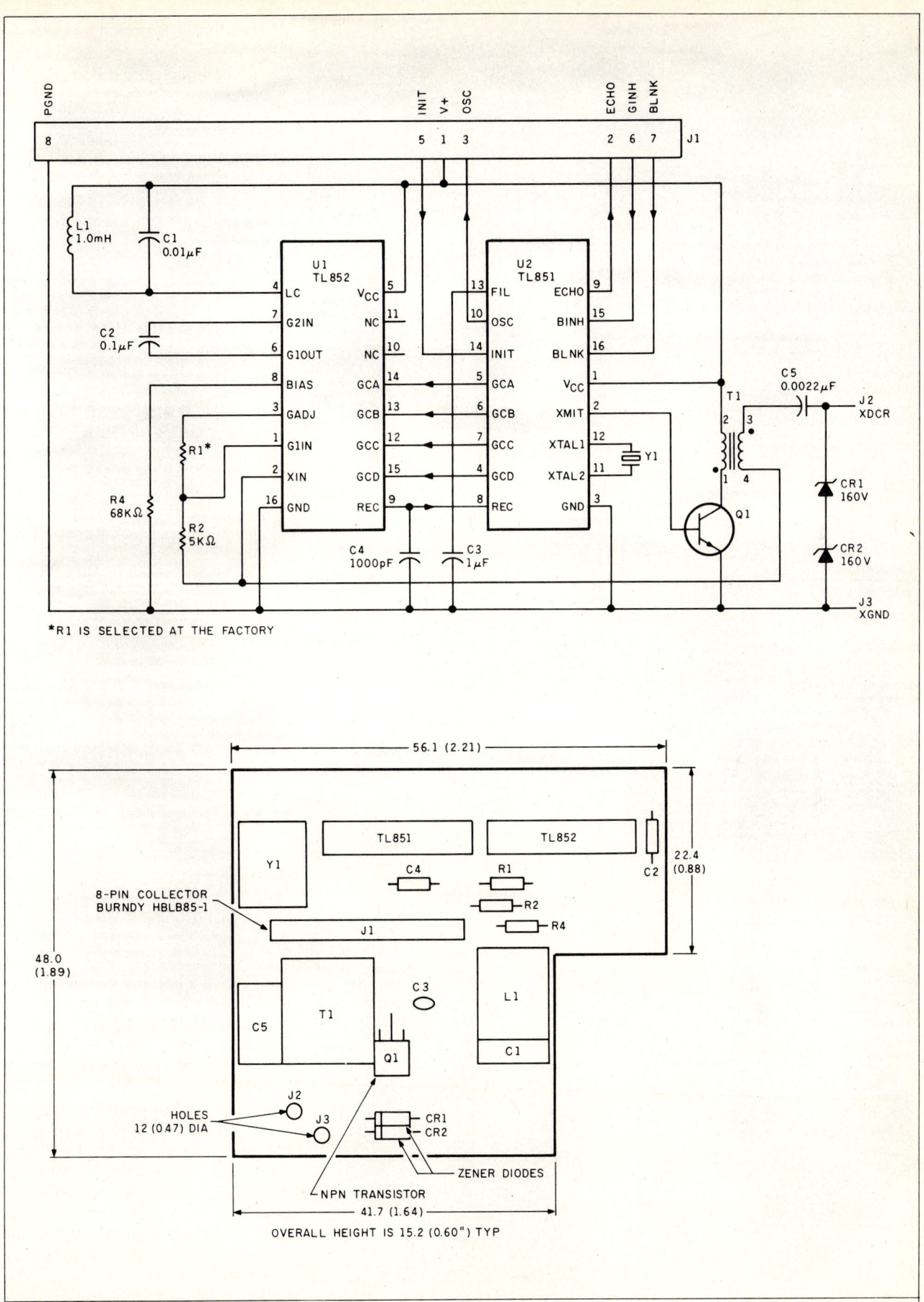

Figure 4: *The schematic and layout of the Texas Instruments sonar-ranging module.*

Photo 1: *The original Texas Instruments ultrasonic ranging module (left) provided in the experimenter's kit from Polaroid compared to the new TI ranging module (right) that is the basis of this article.*

Photo 2: *A close-up of the Polaroid ultrasonic transducer.*

Photo 3: *A close-up of the TI ranging module.*

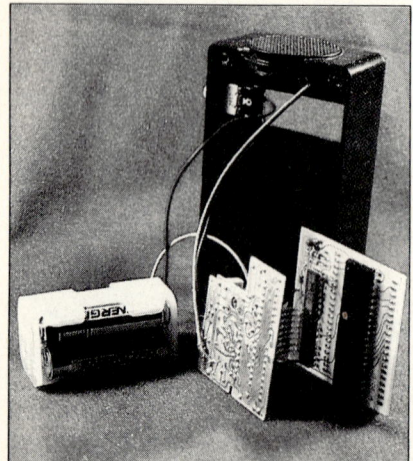

Photo 4: *A rear view of the disassembled tape measure.*

Photo 5: *A front view of the disassembled tape measure.*

implies that only one target exists and that a single ranging value is desired. In the multiple-echo mode, the echo monitoring time is extended to "hear" the echo from objects farther away than the closest target. Differentiating among these echoes can be a data-reduction nightmare, so I will limit my discussion and the use of the ranging module to the single-echo mode.

Figure 5 is the basic timing diagram of the ranging module. Within 5 ms of applying power (at 4.5 V to 6.8 V), the internal circuitry has reset and stabilized and the module is ready. The distance-measuring sequence is activated by raising the INIT input line to a logic 1 state. This enables the 420-kHz on-board ceramic resonator, and 16 cycles of a 300-V 49.4-kHz (420 kHz divided by 8.5) signal are generated. The 16 cycles are passed through a driver/transformer combination that boosts the signal's magnitude to 300 V at the transducer. At the end of the 16 cycles, a DC bias of 150 V remains on the transducer for optimum operation, and the oscillator output steps up to 93 kHz (420 kHz divided by 4.5), where it remains as long as INIT is high.

In order to prevent any ringing in the transducer from being detected as a return signal, the receive input of the ranging control IC (integrated circuit) is inhibited by an internal blanking signal for 2.38 ms. This blanking interval corresponds to 1.33 feet, which is the minimum distance that the ranging module can sense without external control intervention. To detect objects closer than 1.33 feet, the blanking inhibit line, BINH, can be taken high prior to 2.38 ms to enable the receiver.

In the single-echo mode, all that must be done is to wait for the echo from the target. After INIT, the transmitted pulse travels at a rate of 0.9 ms per foot. When the ranging module hears the echo, the ECHO output line goes high. The difference in time between INIT and ECHO both going high is a measure of the distance to the target. This elapsed time can be measured through a timing loop in a computer or gated with the oscillator output to increment a counter and drive a display. I will explore both of these techniques.

One final note on the ranging-module electronics. If you have experimented with ultrasonics, you've probably found, as I have, that fixed-distance reception is far easier to accomplish than variable-distance reception. Sound intensity decreases geometrically proportional to increases in distance. If you have designed a transmitter-and-receiver system that works well at 6 feet, it may not function at 12 feet without substantially increasing the receiver sensitivity (gain) to account for the reduced echo amplitude. Leaving the sensitivity at a high level and reducing the distance again invites other interference and false-echo-detection problems. To adequately compensate for changing distances, the sensitivity setting must also be adjustable with distance.

The TI ranging module dynamically tailors the receive amplifier's sensitivity. Lower amplification is needed for close objects, higher amplification for distant echoes. Twelve gain steps within the range of 0 to 35 feet are automatically incremented as the time between INIT and ECHO lengthens. If the distance is 6 feet, the receiver will be at its second gain setting. At 20 feet it will be at its sixth level. The twelfth level is sufficient for unaided reception of echoes from 35 feet. The addition of a direction cone to the transducer, which will further improve sensitivity, will facilitate measuring distances beyond 35 feet.

Computer Connection

Figure 6 is the schematic for connecting the ranging module to the parallel port of a computer. The entire single-echo-mode interface requires only 1 input bit and 1 output bit. The output bit connects to the INIT line; the input bit connects to the ECHO line. To measure distance, the computer merely raises the INIT line and measures the time until ECHO goes high. The repetition rate depends on the distance. The cycle is repeated when INIT is lowered and raised again. This can occur at any point after ECHO. For short distances, a 100-Hz repetition rate is possible. To allow a full 35-foot range, however, the repetition rate should be limited to 10 Hz.

Unlike the Polaroid Designer's Kit, which is specified to run on 6 V, the TI ranging module can function between 4.5 and 6.8 V. If you use a 5-V supply, the ranging module I/O is TTL

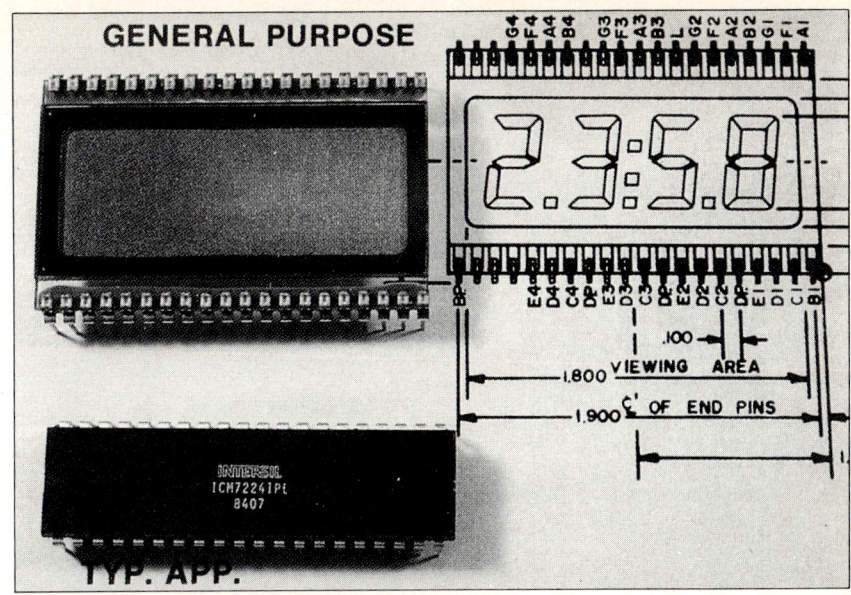

Photo 6: *With the addition of the 4-digit LCD, such as the UXD unit shown, and a 4½-digit 7224 LCD counter/decoder/driver chip from Intersil, a hand-held tape measure can be easily constructed.*

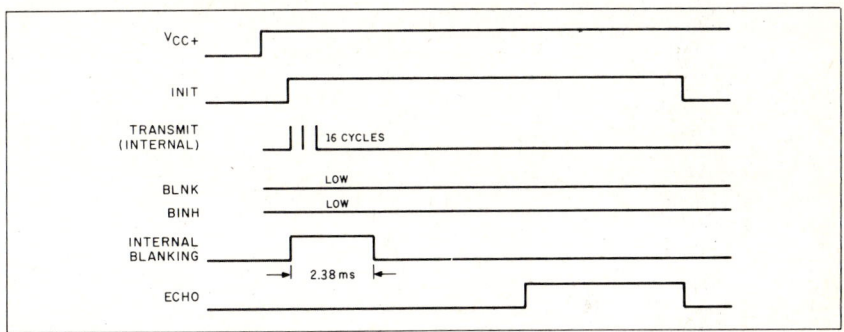

Figure 5: *A timing diagram of the sonar-ranging module.*

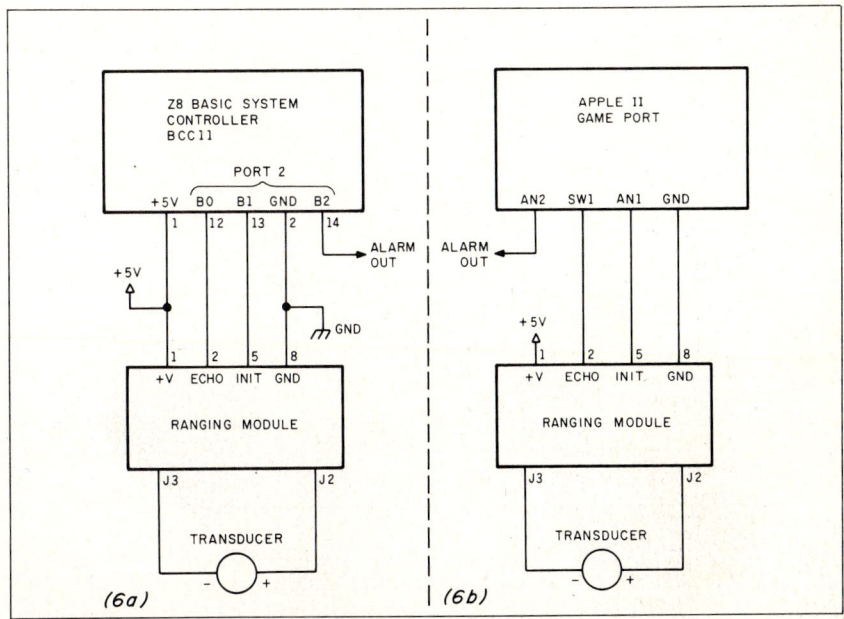

Figure 6: *The sonar-ranging module's connection to the Z8 BASIC system controller (6a) and to the Apple II system controller (6b).*

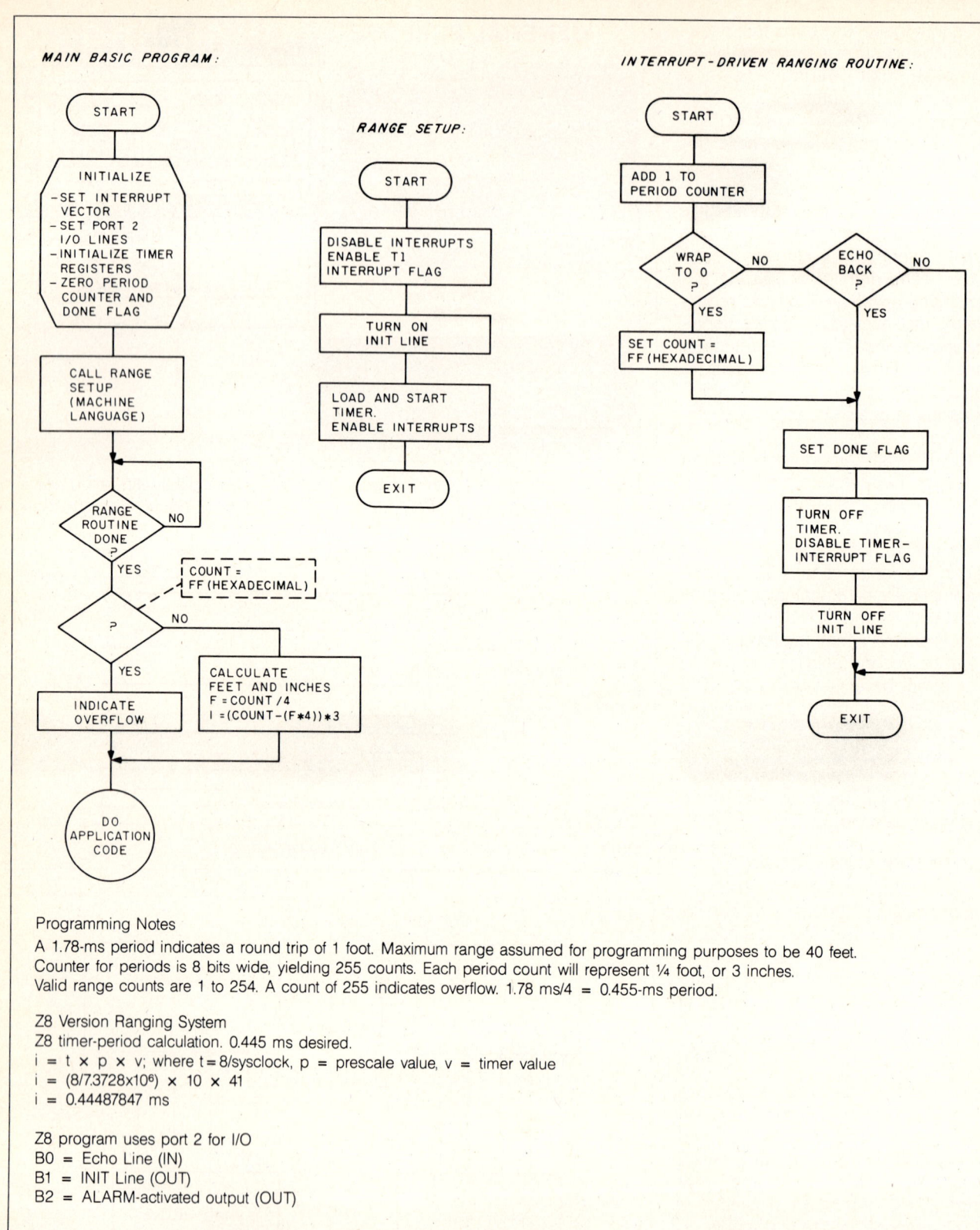

Figure 7: A flowchart of the range-finder program for the Z8 controller.

(transistor-transistor logic) compatible and can be connected directly to most computers. The module normally requires about 100 milliamperes (mA) except during the 16-cycle transmission period, when it can reach 2 amperes (A). Any power source intended for use with the module should have an intermittent power rating this high. For portable operation, you can use either standard AA or larger alkaline batteries or Polaroid's Polapulse high-current batteries.

For the purposes of this article, I chose to demonstrate connection of the ranging module to an Apple II and the Z8 system-controller board, which is an improved descendant of the original Z8-based computer controller presented in the June and July 1980 Circuit Cellar articles. The description of connecting the ranging module to the Z8 system controller through a parallel I/O port, measuring the elapsed time in a callable machine-language routine, and manipulating the results within the tiny BASIC interpreter are transferable in principle to any computer. (I have since had a Z8 FORTH chip produced that is interchangeable with the Z8671 tiny BASIC chip. The FORTH chip could also have been used.) By using the Z8 board, however, I can produce a dedicated measuring system that can sense and average a number of readings, make decisions, activate specific control outputs, and communicate readings to larger host systems serially.

Figure 7 is a block diagram of the necessary steps to initialize the module and measure distance with the Z8. Figure 8 is a block diagram of the Apple II version of the same code. Figure 9 is the block diagram of a typical proximity-detector application with alarm outputs. Listings 1, 2, and 3 show the code to perform these tasks.

ELECTRONIC TAPE MEASURE

While describing the electrical characteristics and computer connection of the ranging module might ordinarily suffice as a Circuit Cellar project, I was intrigued by the simplicity of using the ranging module and decided to build the electronic tape measure shown in photos 4 and 5. The circuit shown in figure 10 required only three additional CMOS (complementary metal-oxide semiconductor) chips and can be constructed as a hand-held device.

The Circuit Cellar Sonar Tape Measure, hereafter called SonarTape, consists of the TI ranging module, a CD4049 inverter, a CD4029 counter, and an Intersil ICM7224 4½-digit LCD counter/decoder/driver chip (shown in photo 6). When pointed at an object or a wall and activated, the SonarTape transmits once and holds its reading (so that you can turn the unit and see the display if it wasn't already in view). The LCD indicates the distance in feet and tenths of feet.

The SonarTape is activated by turning on the power (6 V provided by

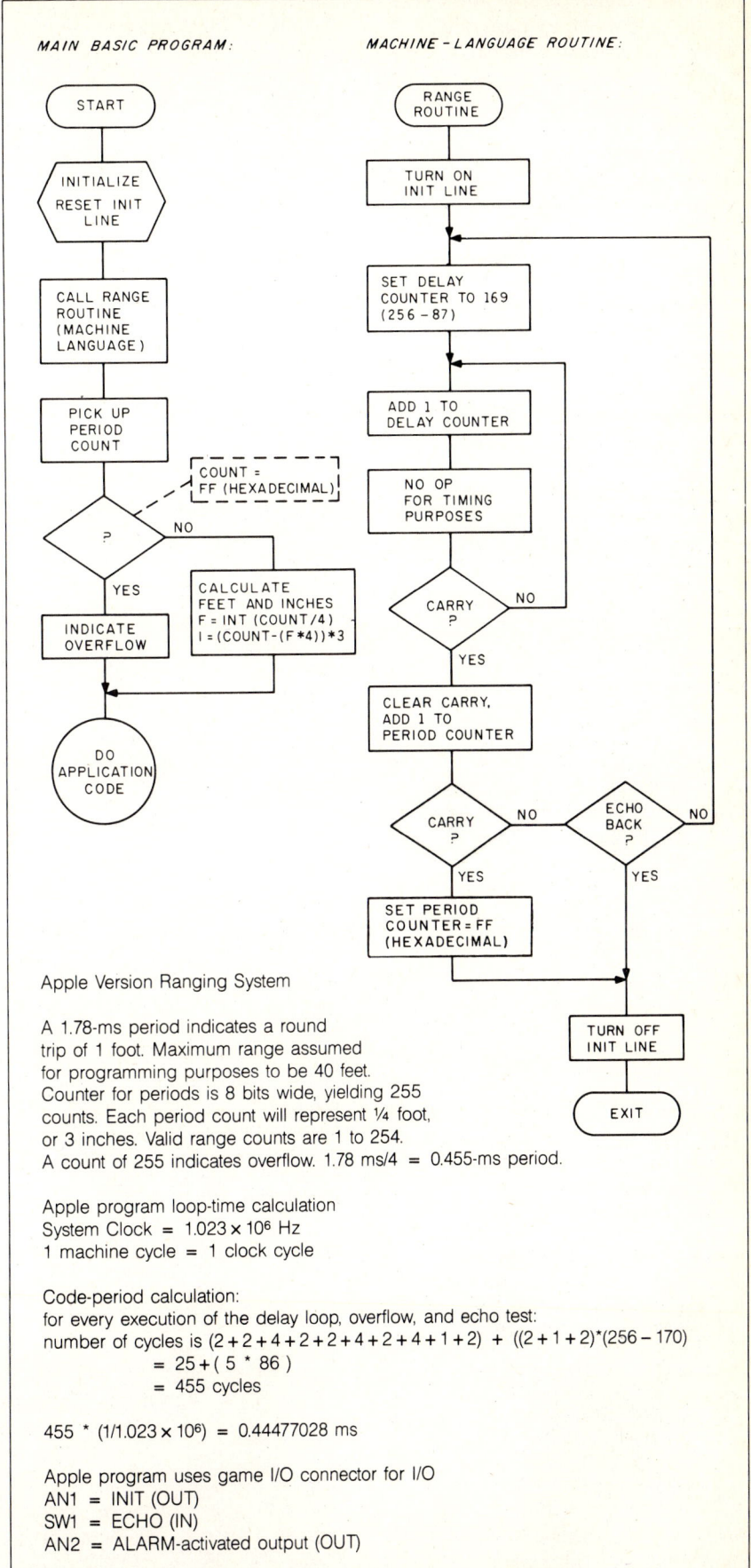

Apple Version Ranging System

A 1.78-ms period indicates a round trip of 1 foot. Maximum range assumed for programming purposes to be 40 feet. Counter for periods is 8 bits wide, yielding 255 counts. Each period count will represent ¼ foot, or 3 inches. Valid range counts are 1 to 254. A count of 255 indicates overflow. 1.78 ms/4 = 0.455-ms period.

Apple program loop-time calculation
System Clock = 1.023×10^6 Hz
1 machine cycle = 1 clock cycle

Code-period calculation:
for every execution of the delay loop, overflow, and echo test:
number of cycles is $(2+2+4+2+2+4+2+4+1+2) + ((2+1+2)*(256-170)$
$= 25 + (5 * 86)$
$= 455$ cycles

$455 * (1/1.023 \times 10^6) = 0.44477028$ ms

Apple program uses game I/O connector for I/O
AN1 = INIT (OUT)
SW1 = ECHO (IN)
AN2 = ALARM-activated output (OUT)

Figure 8: *A flowchart of the range-finder program for the Apple II.*

four AA alkaline cells) to the circuit through a momentary push button. A resistor/capacitor timing circuit connected to V_{cc} and attached to pin 3 of IC1 resets IC3, presets IC2, and keeps the INIT input line to the ranging module low until the 5-ms power-on stabilization time has passed. When the capacitor charges up to a logic 1 level, the INIT line goes to a logic 1, allowing the ranging module to transmit.

Once the INIT line goes high, the clock output, CLOCK, from the ranging module is enabled, and a 93.333-kHz clock (after the first 16 cycles at 49.4 kHz) is presented to the clock input of IC2. Configured as a divide-by-16 counter, the carry output, pin 7, will be at 5.83333 kHz. This frequency is connected to the count input of the counter/decoder/driver chip, IC3, which continues to count input pulses (started when INIT went high) until the ECHO line, connected through an in-

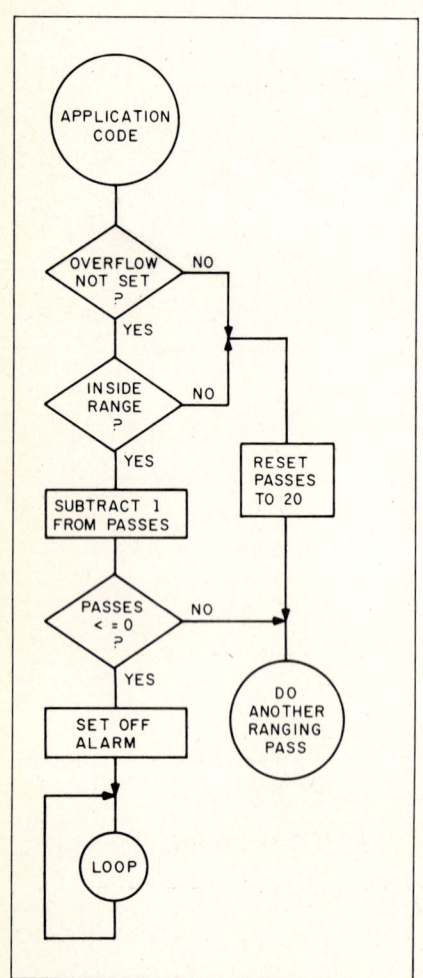

Figure 9: *A flowchart of the alarm-monitor program.*

Listing 1: *The Z8 programs to initialize the ranging module, measure the distance, and print the results. The BASIC program (1a) initializes the module and calls the machine-language routine, represented here in assembly language (1b). The machine-language program oversees the distance measurement. Finally, the BASIC program prints the results (1c). Remember to delete the REM statements.*

(1a)

```
1 REM Z8 BASIC PROGRAM FOR RANGE FINDER
10 @%100F=%8D:@%1010=%15:@%1011=%0C    :REM SET UP INTERRUPT
                                        VECTOR
20 @246=%F9:@2=0                        :REM SET UP PORT 2 I/O.
                                        RESET INIT LINE
30 @243=%2B:@242=41                     :REM SET UP TIMER
                                        REGISTERS
40 @%30=0:@%31=0                        :REM ZERO COUNTER
                                        AND DONE FLAG
50 GO @%1500                            :REM CALL RANGE-SETUP
                                        ROUTINE
60 REM LOOP UNTIL DONE FLAG SET
70 IF @%31=0THEN60                      :REM TEST DONE FLAG
                                        AND LOOP
80 C=@%30                               :REM PICK UP COUNT
                                        VALUE
90 F=0:I=0:O=0                          :REM CLEAR FEET,
                                        INCHES, OVERFLOW
100 IF C=%FF THEN O=1:GOTO 130          :REM SET OVERFLOW IF
                                        COUNT=%FF
110 F=C/4                               :REM CALCULATE FEET
120 I=(C−(F*4))*3                       :REM AND INCHES
130 REM APPLICATION CODE STARTS HERE
140 IF O=1 THEN PRINT "OVERFLOW":GOTO 40
150 PRINT F;" FEET ";I;" INCHES"
160 PRINT C;" .445-MS PERIODS COUNTED"
170 GOTO 40
:
:
:NOTE: DO NOT ENTER THE :REM STATEMENTS, OR THE PROGRAM WILL NOT
RUN.
```

(1b)

```
0010 ;
0020 ;  Z8 VERSION RANGE-FINDER ROUTINES
0030 ;
0040 ;  DRIVE INIT LINE AND SET UP CLOCKS
0050 ;
0060         ORG     1500H
0070 INIT    EQU     $
0080         DI                      ; SHUT DOWN INTERRUPTS
0090         OR      IMR,#20H        ; ENABLE IRQ 5 (TIMER 1)
0100         LD      02H,#02         ; TURN ON INIT LINE
0110         OR      TMR,#0CH        ; LOAD AND START TIMER 1
0120         EI                      ; ENABLE INTERRUPTS
0130         RET                     ; RETURN TO BASIC
0140 ;
0150 ;
0160 ;  THIS IS THE INTERRUPT-DRIVEN RANGING ROUTINE
0170 ;
0180 ;
0190 RANGE   EQU     $
0200         INC     30H             ; ADD 1 TO PERIOD COUNTER
0210         JR      NZ,ECHO         ; IF NO WRAP, CHECK ECHO
0220         LD      30H,#0FFH       ; SET COUNT TO %FF
0230         JR      DONE            ; AND SHUT DOWN RANGE ROUTINE
```

(continued)

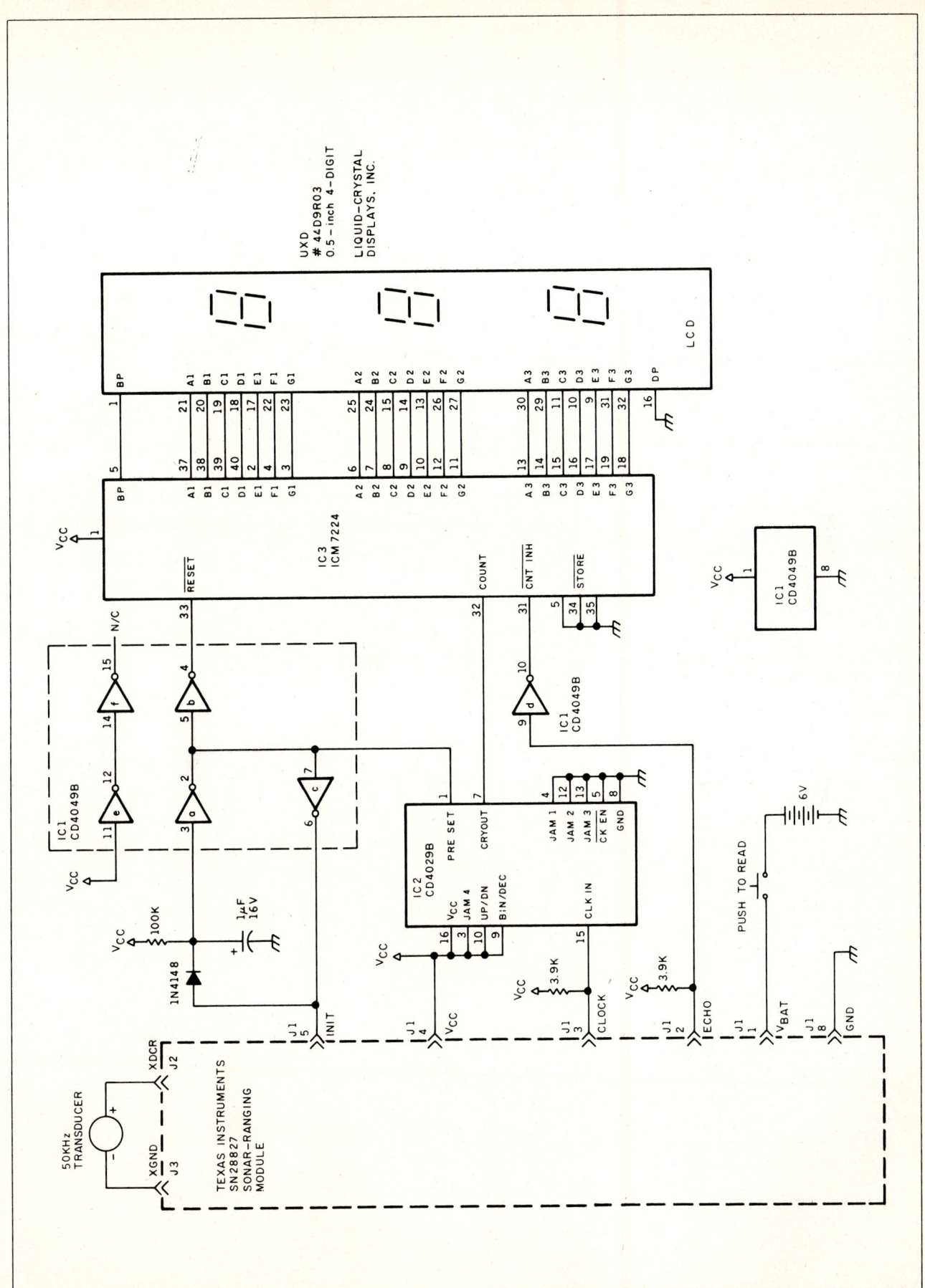

Figure 10: *The schematic of the SonarTape.*

A sonar sensor would let a sightless person "hear" a picture of the environment.

verter to the count-inhibit line on IC3, goes high, indicating reception of the echo. Clock pulses will therefore have been counted during the elapsed time period between INIT and ECHO.

If the elapsed time were 6 ms, about 34 clock cycles would have been counted by IC3. With the decimal point permanently set between the least and next most significant digit, the indication on the LCD would have been 3.4 feet. Similarly, 31 ms of elapsed time would allow about 175 cycles to occur, which would indicate 17.5 feet.

In Conclusion

I did think about commercial applications for SonarTape. Just because I'd like one, however, doesn't mean it has any potential. The high cost of the additional circuitry, packaging, and LCD make it either a high-priced novelty or cost-effective only in mass production. I'll stick with my prototype and wait for the next development iteration so that I can cover this subject in another four years.

One additional application that I started prototyping but discontinued as not completely germane to this article was a blind sonar sensor. Such a device (for the blind or any sightless application, such as firemen entering a smoke-filled room) would conceivably allow a sightless person to "hear" a picture of the environment by sweeping a hand-held ranging unit in front as he or she walked. Rather than a visual display, the distance measurement would be a tone whose frequency was a function of distance. My initial design consisted of continuously triggering the INIT line with a 5-Hz oscillator and applying the INIT and ECHO signals to an AND gate to produce a single pulse output whose pulse width was the elapsed time between INIT and ECHO. Next, using an integrator circuit, or even a simple RC (resistor-capacitor) combination, the pulse width is converted to a DC voltage level. The longer the pulse width, the higher the DC level.

```
0240 ECHO    EQU   $
0250         TM    02H,#01        ; IS ECHO BIT ON?
0260         JR    Z,EXIT         ; NO, LET ANOTHER PASS OCCUR
0270 DONE    EQU   $
0280         LD    31H,#0FFH      ; TURN ON DONE FLAG
0290         AND   TMR,#0F3H      ; TURN OFF T1 TIMER
0300         AND   IMR,#1FH       ; DISABLE IRQ 5 INTERRUPT
0310         LD    02H,#0         ; SHUT OFF INIT LINE
0320 EXIT    EQU   $
0330         IRET                 ; RETURN FROM INTERRUPT
0340 ;
0350 ; END OF MACHINE-LANGUAGE ROUTINES
0360 ;
0370 ZZZZ    EQU   $
```

(1c)

```
RUN
32 FEET 3 INCHES
129 0.445-MS PERIODS

32 FEET 3 INCHES
129 0.445-MS PERIODS
```

Listing 2: *The Apple II versions of the programs in listing 1. The output is the same as for the Z8.*

(2a)

```
1 REM APPLE II VERSION RANGING SYSTEM
10 POKE – 19293,0: REM SHUT OFF INIT LINE
30 PRINT CHR$ (4);"BLOAD RANGE.0": REM LOAD OBJECT PROGRAM
40 POKE 12328,0: REM CLEAR COUNT
50 CALL 12288: REM CALL RANGING SUBROUTINE
60 C = PEEK (12328): REM GET COUNTER VALUE
70 F = 0:I = 0:O = 0: REM CLEAR FEET, INCHES, OVERFLOW
80 IF C = 255 THEN O = 1: GOTO 120: REM SET OVERFLOW IF
   COUNT = $FF
90 F = INT (C / 4): REM CALCULATE FEET
100 I = (C – (F * 4)) * 3: REM AND INCHES
120 REM APPLICATION CODE STARTS HERE
130 IF O = 1 THEN PRINT "OVERFLOW": GOTO 40
140 PRINT F;" FEET ";I;" INCHES"
150 PRINT C;" .455-MS PERIODS": PRINT
160 GOTO 40: REM AND MAKE ANOTHER PASS
```

(2b)

```
0010 *----------------------------------------------------------------
0020 * APPLE II   VERSION RANGING SYSTEM
0030 *----------------------------------------------------------------
0040 *
0050         .OR   $3000
0060 RANGE.EQ *
0070         LDA   $C05B          * TURN ON INIT LINE
0080 DELAY.EQ *
0090         LDA   #184           * SET DELAY COUNT TO 184
0100 LOOP .EQ *
0110         ADC   #1             * ADD 1 TO DELAY COUNTER
0120         NOP                  * NO OPERATION FOR TIMING
0130         BCC   LOOP           * LOOP UNTIL CARRY
0140         CLC                  * CLEAR CARRY FLAG
0150         LDA   COUNT          * PICK UP OLD COUNT
0160         ADC   #1             * ADD 1 TO IT
```

(continued)

```
0170        STA  COUNT      * AND SAVE IT
0180        BCC  ECHO       * IF NO WRAP, CHECK ECHO
0190 * WRAP OF PERIOD HAS OCCURRED. SET OVERFLOW.
0200        LDA  #$FF       * GET A $FF
0210        STA  COUNT      * WRITE TO COUNTER
0220        JMP  EXIT       * AND LEAVE THIS ROUTINE
0230 ECHO   .EQ  *
0240        LDA  $C061      * PICK UP ECHO BYTE
0250        AND  #$80       * TEST ECHO BIT
0260        BEQ  DELAY      * IF NO ECHO, DO ANOTHER LOOP
0270 EXIT   .EQ  *
0280        LDA  $C05A      * SHUT OFF INIT LINE
0290        RTS             * RETURN TO BASIC
0300 *-----------------------------------------------------
0310 COUNT       .BS $1     * COUNTER DATA BYTE
0320 *-----------------------------------------------------
0330 *-----------------------------------------------------
0340 * END OF RANGE SUBROUTINE
0350 *-----------------------------------------------------
0360 ZZZZ   .EQ  *
```

Listing 3: *To modify the Z8 program in listing (1a) for the alarm application, add the BASIC statements in (3a), except for the REM statements. To modify the Apple II program in listing (2a) for the alarm application, add the BASIC statements in (3b).*

(3a)

```
5   R = 120:P = 20                    :REM SET RANGE AND
                                       NUMBER OF PASSES
140 IF O = 1 THEN P = 20:GOTO 40      :REM RESET PASSES IF
                                       OVERFLOW
150 IF ((F*12)+I) > R THEN P = 20:GOTO 40 :REM RESET PASSES IF
                                           OUTSIDE RANGE
160 P = P - 1                         :REM PASSES = PASSES - 1
170 IF P > 0 THEN 40                  :REM IF PASSES NOT DONE,
                                       TRY AGAIN
180 REM OBJECT WITHIN RANGE FOR NUMBER OF PASSES. ALARM
190 @2 = 4                            :REM SET ALARM BIT
200 REM LOOP
210 GOTO 200                          :REM LOOP FOREVER
:
:
:
:AS BEFORE, DO NOT ENTER THE :REM STATEMENTS
```

(3b)

```
2 REM TO MODIFY THE APPLE PROGRAM FOR THE APPLICATION, ENTER
  THE FOLLOWING.
3 R = 120:P = 20: REM SET RANGE TO 10 FEET, PASSES TO 20
20 POKE -16291,0: REM SHUT OFF ALARM
80 IF O = 1 THEN P = 20: GOTO 40: REM RESET PASSES IF OVERFLOW
90 IF ((F * 12) + 1) > R THEN P = 20: GOTO 40:
   REM RESET PASSES IF OUTSIDE RANGE
100 P = P - 1: REM PASSES = PASSES - 1
110 IF P > 0 THEN GOTO 40: REM IF NOT THROUGH ALL PASSES, DO
    ANOTHER
120 POKE 16292,1: REM TURN ON ALARM
130 REM LOOP HERE
140 GOTO 130
```

When ECHO goes high, the integrator output would be sampled and held (until the next ranging sample) with a sample-and-hold circuit. The output of the sample-and-hold circuit is connected to a voltage-controlled oscillator such as the XR4151 or XR2206. Short distances produce low tones; long distances result in high tones. A little bit of experience recognizing tone patterns should allow any of us to walk without seeing.

I chose not to pursue the design at this time. If any of you build a working unit, however, I'd like to know about it. Among the hundreds of letters I receive each month are some from readers who might benefit by the information.

Finally, I'm not ending this article by discussing all the possible applications for the TI ranging module. The number is so great that they are impossible to list. Now that you know the unit exists and how to attach it to a computer, perhaps you'll demonstrate some novel uses. For me, it's on to the next project.

The following is available from

The Micromint, Inc.
25 Terrace Drive
Vernon, CT 06266
For orders: (800) 635-3355
For information: (203) 871-6170

One Texas Instruments SN28827 sonar-ranging module and one Polaroid electrostatic transducer. Data sheets are included. $57

Please include $3 for shipping in the continental United States, $10 elsewhere. New York residents please include 8 percent sales tax.

THE LIS'NER 1000

Build a low-cost, high-performance speech-recognition system

The concept of a computer understanding speech is not new. For years we have watched Capt. Kirk and Mr. Spock on the bridge of the *Enterprise* talking with the ship's computer or have remarked at the diabolical mind of HAL in *2001: A Space Odyssey*. These computers represent the ultimate in automatic speech recognition (ASR). Unfortunately, most of their capabilities are still science fiction.

The ultimate goal of all speech-recognition techniques is to characterize the spoken word into a recognizable pattern. Specifically, ASR is the ability that would let a computer recognize the spoken word. Exactly how the words are spoken, however, determines the hardware cost and analysis techniques employed.

Speech-Recognition Units

The first type of unit is the *speaker-dependent* recognition system, which creates its recognition vocabulary by "listening" to the voice of a single speaker. It then concerns itself only with recognizing the same word as spoken by that speaker.

First, the user speaks into a microphone all the words the machine is to recognize. The acoustical characteristics of each word are analyzed and stored as templates, which are digital patterns used by a recognition algorithm to identify words. The procedure of creating templates is referred to as *training*. Depending upon the available memory and the recognition-algorithm speed, the total vocabulary can be from 4 to 100 words. Generally speaking, the more words in the vocabulary, the longer it takes to recognize a specific word and the more sophisticated the algorithm must be.

The second type of unit is the *speaker-independent* recognition system. Other than HAL or the *Enterprise* computer, few functioning speaker-independent systems exist that have more than a 10-word vocabulary. This system requires no template training by a single speaker. Its speech templates are preprogrammed, and the matching algorithm is supposed to be adaptable to the voices of a variety of speakers and accents.

The third type of unit is *unconnected speech*. Also called discrete-utterance recognition, unconnected speech is simply single words preceded and followed by pauses. This is

Photo 1: *The General Instrument SP1000 voice-recognition chip.*

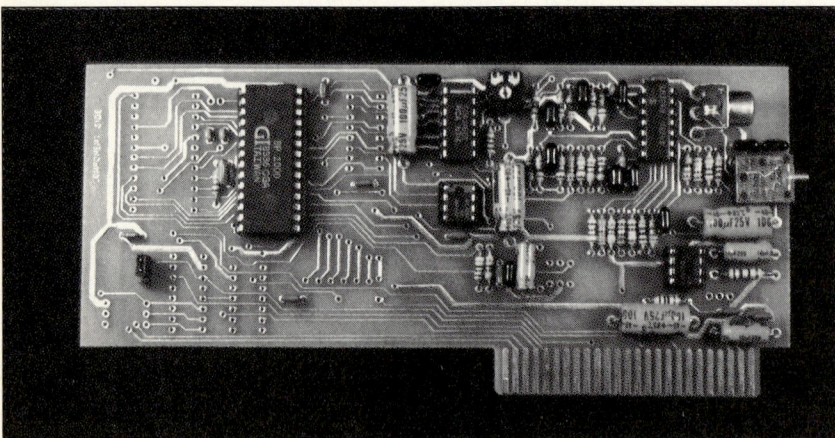

Photo 2: *A prototype printed-circuit board of the Apple II recognition-only Lis'ner 1000 circuit. The two connectors on the top right are for an external speaker (RCA) and the microphone (miniphono). An IBM PC version is in the works.*

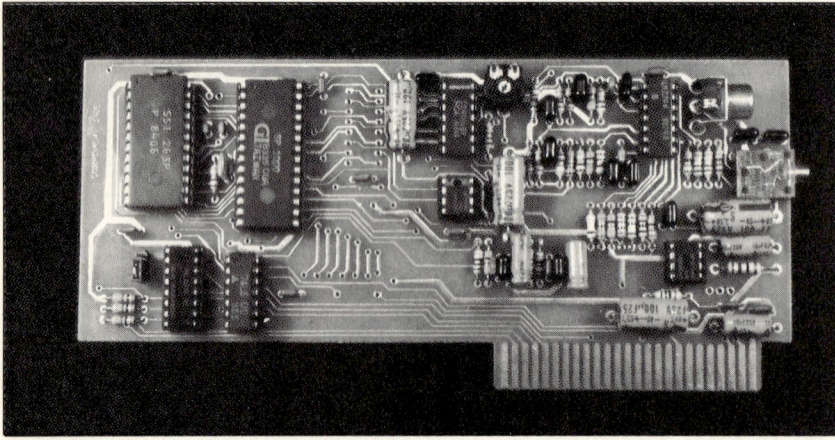

Photo 3: *A prototype printed-circuit board containing Lis'ner 1000 circuitry, which performs recognition, and speech-synthesis circuitry for the Apple II. It contains both the SP1000 and an SSI-263 speech synthesizer with a text-to-speech algorithm. Together they facilitate a functional hands-off computer with complete speech I/O.*

the easiest recognition approach and generally the technique used in most inexpensive systems. Each template is for a single utterance that must be spoken as a discrete word rather than as part of a longer word or phrase.

The final type of unit is *connected speech*. Also called continuous speech, this is the way we normally talk. Unfortunately, much of our understanding is dependent on recognizing the words in context. One major problem for the computer is coarticulation, where words are blended so that there are no distinct word boundaries for direct template matching. The result is costly computing overhead because every template must be aligned with every possible interval of the utterance. While significant advances have been made in this area, connected-speech ASR systems are expensive and generally require some monitoring of context as well.

Today, most speech-recognition systems are discrete-utterance speaker-dependent units. They may be in the form of expansion boards for existing computers or stand-alone black boxes. Ultimately, however, their purpose is singular. When the user speaks, the computer analyzes the acoustical signal, compares it to the stored templates, and decides which most closely resembles the spoken word. Once a candidate is chosen, the computer can itself respond to the user's utterance or output a control signal to another device.

Each stage of the analysis and pattern-matching procedure can be carried out by a variety of techniques. The earliest techniques used a simple zero-crossing detector to produce a pattern somewhat related to frequency. It soon became evident that speech, which is a complex combination of frequencies, could not be so easily represented. The next refinement was to break the voice frequencies out through a series of filters and separately record energy levels. While economically attractive, since it used readily available components, the massive quantities of data gathered proved ponderous and slow to compute. Many systems on the market still use this technique.

One significant advance in estimating the amplitude spectrum of speech is linear predictive coding (LPC). Also known as autoregressive analysis, this method predicts the amplitude of a speech waveform at any instant by combining the amplitudes at a number of earlier instants. The LPC coefficients that best approximate the speech waveform can be mathematically converted to approximate the amplitude spectrum. In speech applications, the LPC analyzer is basically a lattice of filters that approximate a series of resonant cavities, thus simulating the vocal tract.

A Circuit Cellar System

Up until now, it hasn't seemed worthwhile to present a speech-recognition system that merely imitated others. My article in the March 1982 BYTE ("Use Voiceprints to Analyze Speech," page 50) demonstrated the separated-filter and energy-level recording technique in the hopes that I could learn enough to quickly present an ASR system based on that principle. While feasible in theory, I ultimately scrapped the idea as having too many components, even if they were readily available. Since then, I have been watching for any new components that might improve the situation.

Fortunately, the wait has not been in vain. The new SP1000 voice-recognition chip from General Instrument allows me to demonstrate the construction of a low-cost, high-performance voice-recognition system (see photo 1). To my knowledge, this project is one of the first recognition devices using the SP1000.

The Circuit Cellar speech-recognition system, which I've called the Lis'ner 1000, is both a voice-recognition and voice-synthesizer board using the SP1000. The schematics I present are specifically for the Apple II, but they are applicable to other 6502-based systems such as the Commodore 64. The Apple II version plugs into any of the computer's expansion slots, but slot 4 is preferred. The Commodore 64 version (shown in the opening photo) plugs into the rear expansion connector. The Commodore board is configured for recognition only; the Apple II board, shown in photos 2 and 3, supports the LPC speech output from the SP1000 and has optional provision for an SSI-263 phonetic speech synthesizer with a text-to-speech algorithm.

The Lis'ner 1000 hardware forms merely the front end of a recognition system by performing feature extraction of the incoming audio signal. The host microcomputer compares these features with those of the templates stored in memory and makes the recognition decisions. Such a separation of system tasks leaves control of system performance to the system designer. You can use the Lis'ner 1000 board in speaker-dependent or speaker-independent systems with connected or unconnected speech. The designer is not locked into a specific recognition algorithm that may not be suitable for a particular application. Instead, the recognition algorithm is contained in software resident in the host microcomputer and can be easily upgraded to take advantage of advances in recognition techniques without requiring hardware redesign.

In an effort to more fully support

This project is one of the first recognition devices using the SP1000.

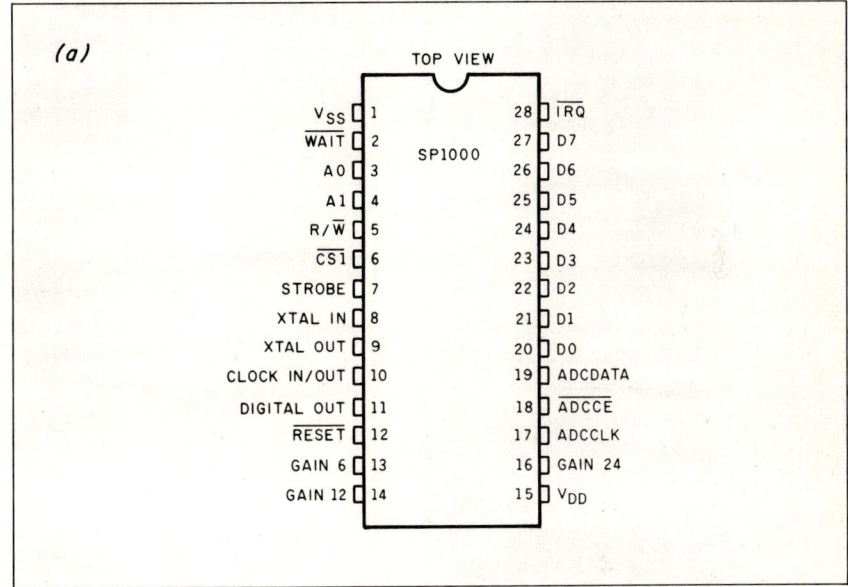

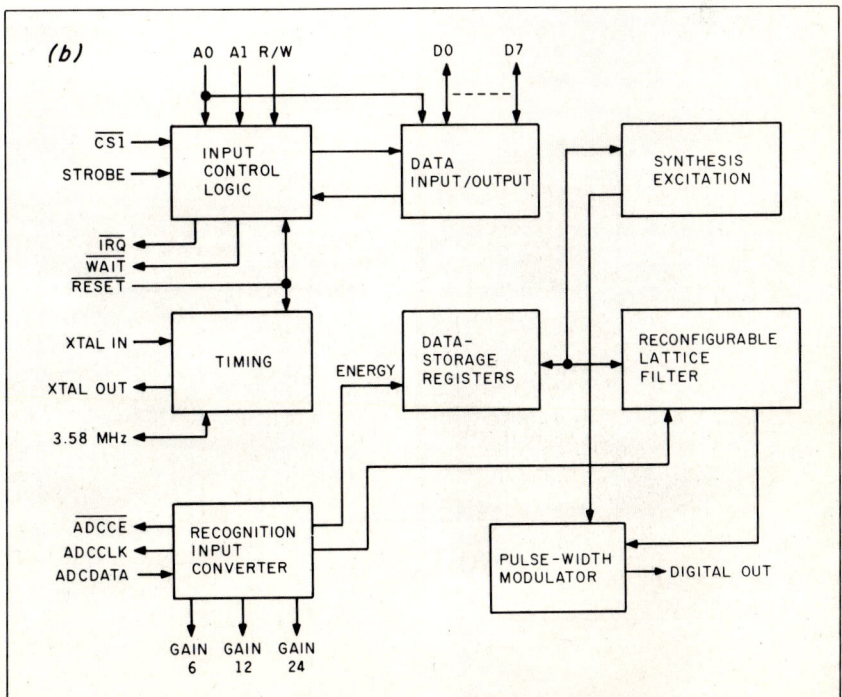

Figure 1: *The SP1000 pin configuration (a) and block diagram (b).*

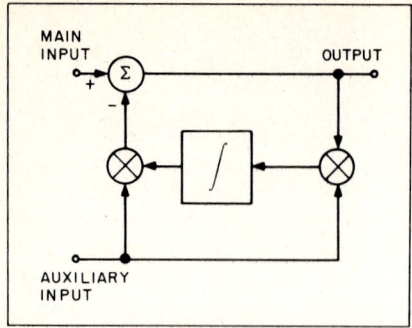

Figure 2: A CCL loop.

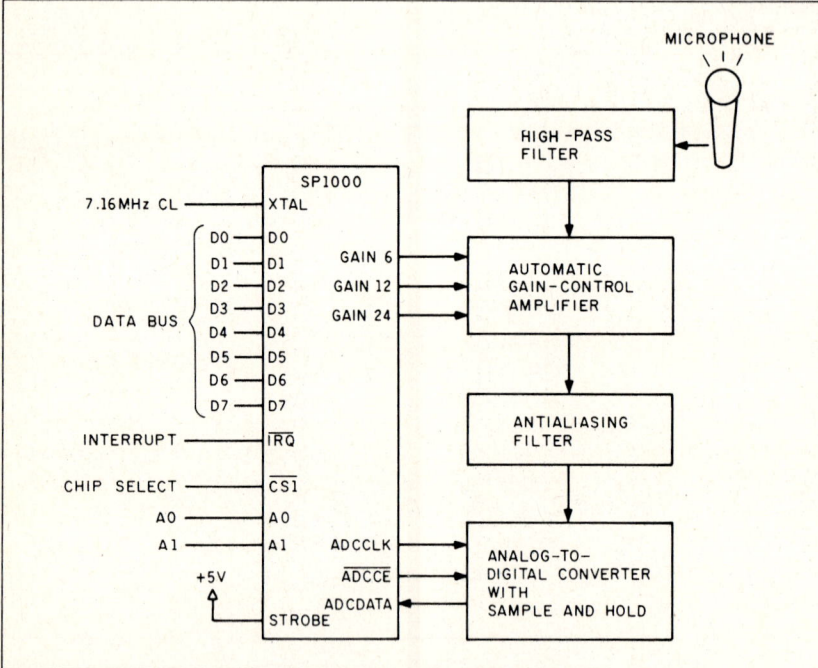

Figure 3: A block diagram of the voice-recognition hardware.

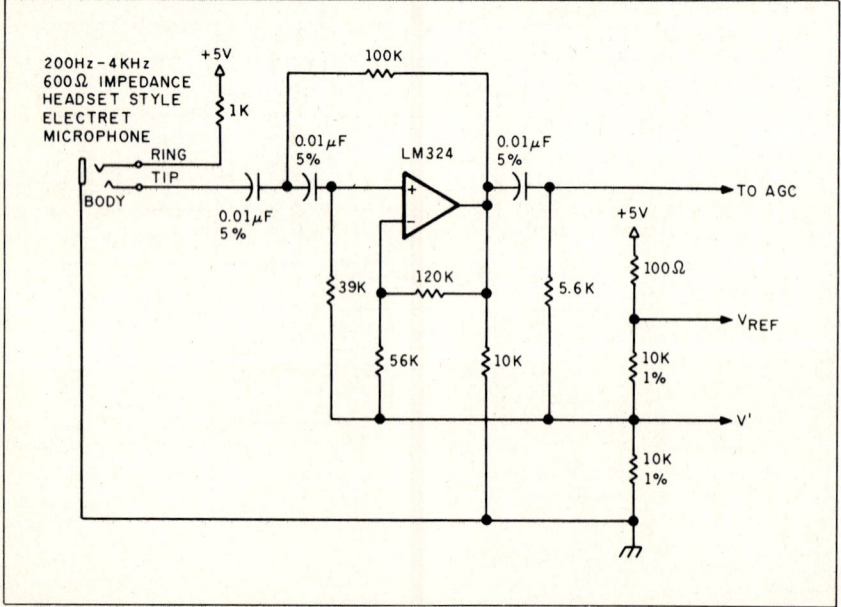

Figure 4: The high-pass filter for the microphone.

the Lis'ner 1000 and make it immediately usable, I've had developed a package of software routines that allows the board to function as a voice-operated keyboard on the Apple II and Commodore 64. Not to lose touch with true experimenters, however, the source code necessary to make the basic system function will be available to those who build the project and want to modify the software (see Experimenter Support on page 205).

The software I provide makes the Lis'ner 1000 a speaker-dependent, discrete-utterance recognition system. The present software supports 64 words in two groups of 32. You'll need a disk drive for either unit to function with the present software.

Before I get too far ahead, however, let me describe the SP1000 chip itself and what's necessary to build the Lis'ner 1000.

General Instrument SP1000

The SP1000, block-diagramed in figure 1, is a 5-volt (V) 28-pin NMOS (negative-channel metal-oxide semiconductor) microprocessor peripheral chip that can be used for both speech recognition and LPC speech synthesis. Using a bidirectional data bus and control lines, the SP1000 interfaces to most 8-bit processors as a memory-mapped peripheral device.

The unique aspect of the SP1000 is its ability to do LPC analysis in real time. LPC analysis solves for the coefficients Ai in an equation of the form:

$$X'_K = A_1 X_{K-1} + A_2 X_{K-2} + \ldots + A_N X_{K-N}$$

where X'_k is an approximation of X_K.

Typical techniques used to solve for the coefficients involve matrix calculations and manipulations. Such techniques, which require vast amounts of memory and extensive calculations, preclude their use on an inexpensive device at this time.

The SP1000 uses a modified form of the correlation cancellation loop (CCL) shown in figure 2 in a reconfigurable lattice structure. The CCL approach can be used to operate directly on the incoming data stream without extensive buffering of data or exorbitant processing power. The predictor coefficients (Ai) are taken from the integrator output of each stage. The stages can be cascaded for higher-order analysis and multiplexed in low-bandwidth applications.

In simpler terms, by modifying the feedback-control scheme within the filter itself, the ultimate number of computations is reduced. The CCL approach requires 300 bits of working storage versus 3 kilobits for a standard covariance or autocorrelation analysis. It also has the interesting property of being able to run backward with a minimum of reconfiguration. This property allows the SP1000 to be used as a speech synthesizer as well as an analyzer for recognition.

The SP1000 can perform useful speech analysis with a relatively inexpensive 8-bit A/D (analog-to-digital) converter. The major reason for this is the use of an on-board automatic-gain-control (AGC) algorithm. The three gain outputs from the SP1000 are used to control a variable gain amplifier. The SP1000 tests the 2 most significant bits of each incoming sample and lowers the gain if they are too high. The net effect is to keep the amplitude of the analog signal within the dynamic range of the A/D converter, preventing distortion and stabilizing the signal level entering the lattice filter.

When used as a synthesizer, the filter is presented with LPC coefficients of the speech frame to be synthesized. Typically, these coefficients are computed on a minicomputer and stored as files to be loaded into the microprocessor's memory. Eventually, General Instrument intends to supply an allophone set that will let the user synthesize any word using a text-to-speech algorithm or dictionary table. The functional use of the Lis'ner 1000 in recognition applications is not dependent on this software, which can be added when it is available.

The desire for user-programmable voice-output capability immediately did not go unnoticed, however. I anticipated the interest in a functional recognition/synthesizer board and purposely designed the Lis'ner 1000 to perform as one. While the project described is for an SP1000-only device, the Apple II printed-circuit board for this project is also etched to accommodate an SSI-263 phonetic speech-synthesizer chip (see "Build a Third-Generation Phonetic Speech Synthesizer," page 107). Adding the SSI-263 and the text-to-speech algorithm facilitates true voice I/O (input/output) and supports both phonetic-generated and allophone-generated (LPC) speech.

Building the Lis'ner 1000

Figure 3 is a block diagram of the recognition portion of the Lis'ner 1000, which interfaces to the Apple II and Commodore 64 through an 3-bit bidirectional bus and a few control lines. The SP1000 occupies four address locations and is written to or read from as any other peripheral device at that address. Data is transferred through the data lines whenever the chip-select line is active. The read/write line determines the direction of the transfer, and the two address lines specify the particular register within the chip. Of the four registers, three are read/write and one is write only.

A typical system consists of the SP1000 and an assortment of analog components. The analog interface consists of filters, amplifiers, switches, and an A/D converter. The purpose of the circuitry is to convert the utterances spoken by the user into a form that the chip can understand. The entire circuit is designed to run on +5 V and, except for the SP1000 connection to the host computer, is virtually the same for all applications.

The first section (see figure 4) contains the microphone input and high-pass filter. For best performance, you should use a 600-ohm-impedance, condenser-type electret microphone.

> The source code that is necessary to make the basic system function will be available.

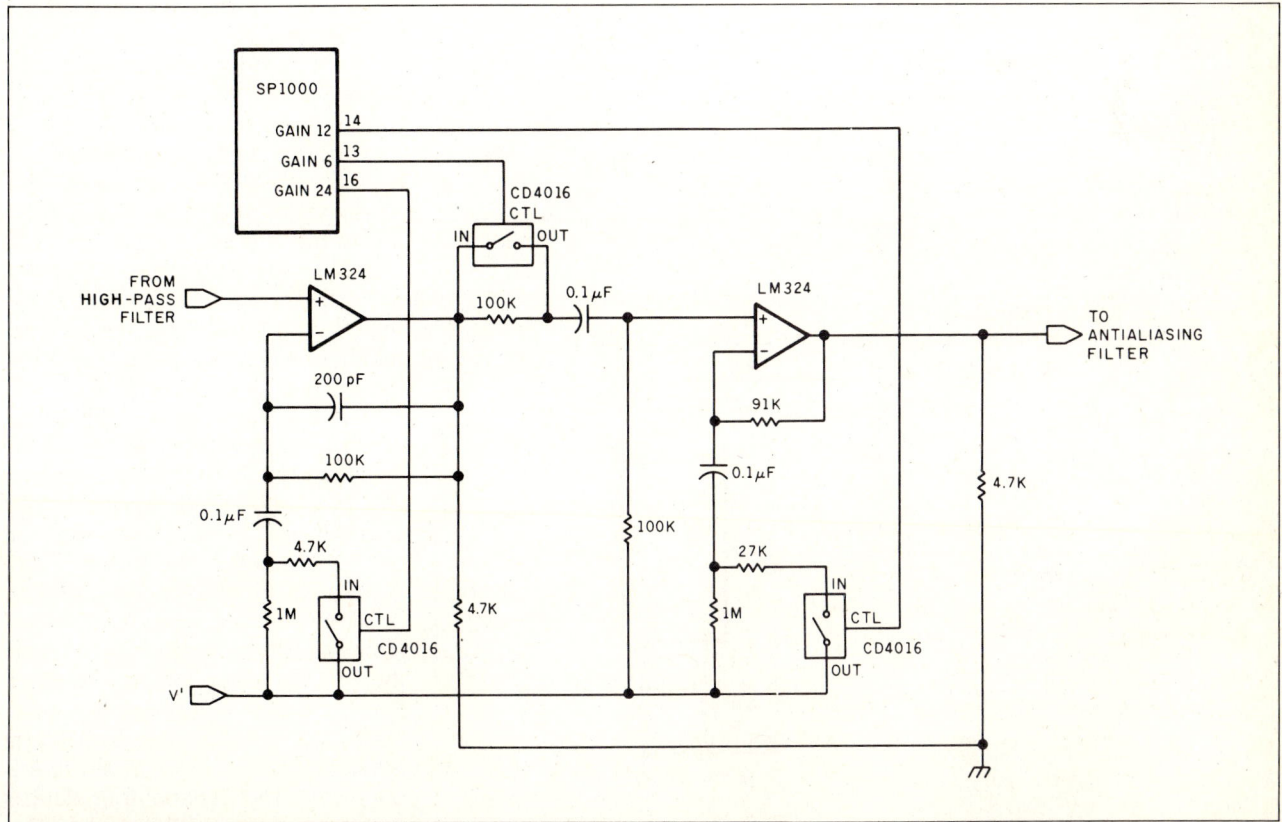

Figure 5: *A programmable automatic-gain-controlled amplifier.*

Table 1: *Signal amplifications possible with different combinations of the SP1000 gain pins, which are shown in figure 5.*

Gain Pins			Gain
24	12	6	
0	0	0	0 dB
0	0	1	6 dB
0	1	0	12 dB
0	1	1	18 dB
1	0	0	24 dB
1	0	1	30 dB
1	1	0	36 dB
1	1	1	42 dB

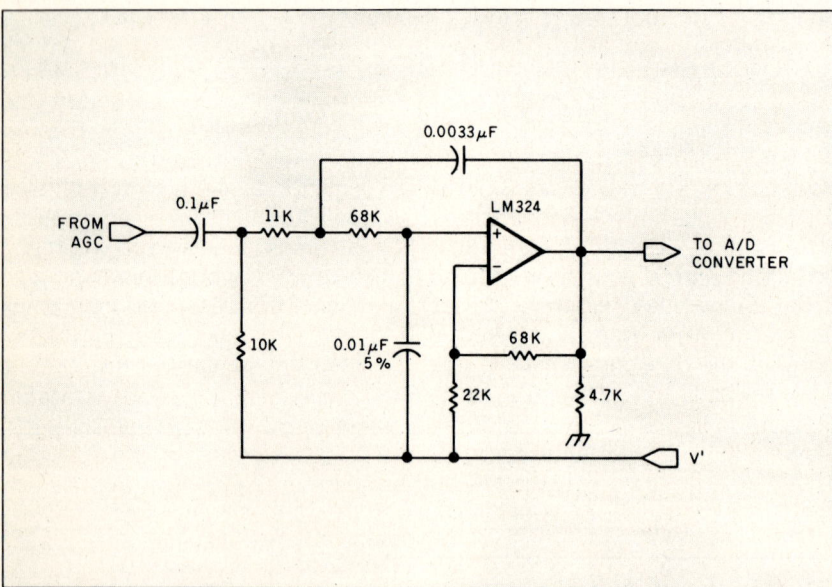

Figure 6: *An antialiasing filter.*

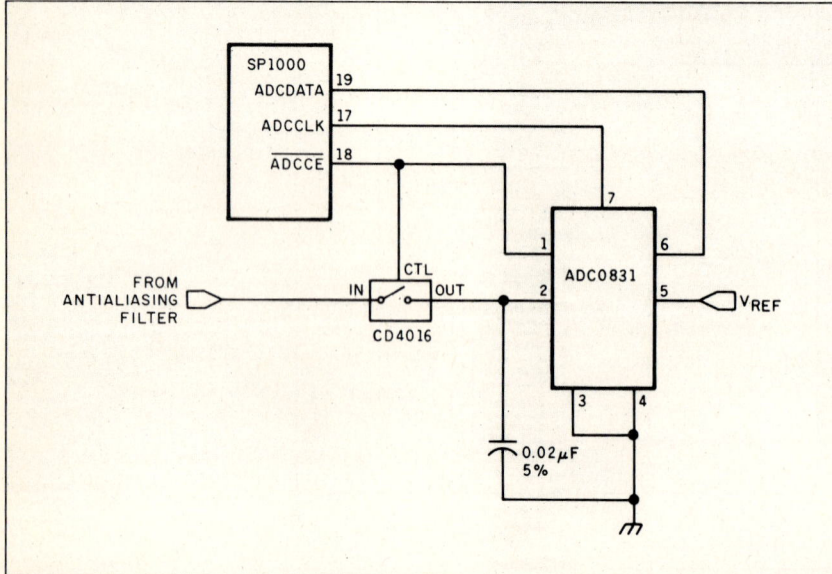

Figure 7: *The sample-and-hold 8-bit A/D converter.*

To avoid background noise pickup, I suggest the microphone headset combination shown in the opening photo. This keeps the microphone close to the mouth and limits interference.

The high-pass filter removes all sounds below 250 Hz.

The output from the high-pass filter is connected to an automatic-gain-controlled amplifier (see figure 5). The SP1000 provides three output lines that control switches to vary the resistor values within a circuit consisting of two noninverting operational amplifiers connected in series. These signals are GAIN 6, GAIN 12, and GAIN 24, corresponding to 6-, 12-, and 24-decibel (dB) signal levels (this is a voltage gain of 2, 4, and 15.8, if you are interested). See table 1 for the gain produced by combining these pins.

The SP1000 updates these signals at a predetermined interval, depending upon the value of the digital output from the A/D converter. The three lines create eight combinations of signal amplification from 0 dB to 42 dB in 6-dB steps. The purpose of the AGC is to monitor and modify the incoming signal amplitude so that it always stays within the range of the A/D converter.

Switching these resistors in and out in the AGC produces high-frequency transients known as *aliases*. These unwanted frequencies are removed by a two-pole, 3200-Hz, low-pass, antialiasing filter (see figure 6) before going to the A/D converter.

Once a conditioned signal with all the extraneous noise removed is obtained, it is directed to the sample-and-hold A/D converter to be read by the SP1000 (see figure 7).

The SP1000 provides two signals for controlling the A/D converter and the sample-and-hold circuit: ADCCLK and ADCCE. The ADCCE signal provides an active-low chip select that turns off the sample switch (the switch, which is normally closed, opens when the A/D converter reads the voltage level stored on the capacitor) and enables the A/D converter. I used a National Semiconductor ADC0831 8-bit serial-output converter clocked at 150 kHz provided through ADCCLK as the A/D converter. The serial-output data is read through the ADCDATA line. You can program the SP1000 to read the input data at 5k to 16k samples per second. As configured in this project, the sample rate is 6.25k samples per second.

A complete schematic showing the recognition part and LPC-synthesis portion of the Apple II Lis'ner 1000 is shown in figure 8. Figure 9 shows the circuit changes necessary to add the SSI-263 specifically for the Apple II. Figure 10 is the Commodore 64 version.

SP1000 Software

Figure 11 is a flowchart of the basic software control of the Lis'ner 1000. The routines described assume that the SP1000 is implemented as a discrete-utterance speaker-dependent unit. The software can be segmented into two major functions: the creation of training templates and actual recognition of utterances relative to the training templates previously created.

Training

The purpose of training is to create a set of patterns, each of which represents a specific utterance. (Note that an utterance may be a single word or a phrase.) When recognition is performed, these patterns are compared to a pattern created from the word to be recognized. The pattern or template from the training utterances that is closest to the word to be recognized is the one the system chooses as the recognized word.

A well-designed training process will create templates that capture the unique features of an utterance in a form simple enough to facilitate the matching process and the efficient use of a system's memory. With this in mind, let's examine the training process implemented here.

The first step is initialization of the hardware and software. The SP1000 is an extremely flexible device that allows the user to specify several parameters that govern its analysis calculations. The parameters include the sample rate (6.25 kHz), the analysis-frame duration (20 milliseconds [ms]), and the gain-update period (10 ms). Once these parameters have been specified and the software has enabled the interrupts, the SP1000 will provide the processor with a fresh analysis frame at the end of each frame period.

The software initialization consists of setting the counters for the number of templates and the number of training passes for each template. The number of templates (utterances) is variable. The system uses two training passes for each template.

After initialization, the program enters the endpoint-detection process. Since this is a discrete-utterance recognizer, it must identify the start and finish of each utterance it "hears." This applies to both training and recognition. The endpoint-detection algorithm is designed around a finite-state machine with four states: silence, rising, plateau, and falling.

The SP1000 continually analyzes the audio input and sends its analysis data to the host processor. Whenever no speech is reaching the microphone, the SP1000 will be analyzing the ambient room noise. This represents the silence state. While in this state, the processor is constantly calculating a noise level based on the average energy of the last 16 frames of silence. If an incoming frame has an energy 6 dB or more above the noise level, the machine enters the rising state. Similar energy measurements control the state transitions throughout the duration of the utterance until the machine exits back to silence, indicating that the end of the utterance has been reached.

Once the machine enters the rising state, it saves all the analysis frames generated by the SP1000 until the end of the utterance has been found. At that point, the data collected is tested with criteria pertaining to minimum duration and dynamic range to confirm its legitimacy as speech input and pinpoint the endpoints more closely. A normalization process is also performed on the energy coefficients to equalize weighting.

At the end of this process, we have captured a parametric representation of the utterance. The next step is to include that representation in a training template.

It is worth noting that the data collected for an utterance with one second of duration is calculated as follows: (8 bits/coefficient) × (9 coefficients/frame) × (50 frames/second) × (1 second) = 3600 bits of data (450 bytes). Utterances of 3 seconds in duration would generate 1350 bytes. If left in this form, a few dozen utterances would take a sizable quantity of memory just for storage.

Fortunately, the system need save only the unique characteristics of an

> *The purpose of training is to create a set of patterns, each representing a specific utterance.*

utterance in order to perform good recognition. The unique sounds that constitute a particular utterance will usually be several frames in duration. Thus, the algorithm tests the utterance data one more time, essentially to perform a type of averaging in which adjacent frames with similar coefficient values are combined to form one new frame that replaces the two old ones. This process reduces the total number of frames in an utterance to 12. Theoretically, these 12 frames are representative of the unique speech sounds that occurred in the utterance. This process also provides a time normalization for all utterances. Since all utterances are reduced to 12 frames, they all have an identical length for comparison purposes.

The resulting 12 frames constitute a template. Since different repetitions of an utterance are never exactly alike, even when spoken by the same person, the software averages two templates created from two repetitions of the utterance in order to form a more general template. This is stored as the training template for that utterance. The final size of the training template is 12 frames of 9 coefficients each (or 108 bytes of data).

Recognition

Recognition is performed with the same front-end software as the training. It uses the finite-state machine and post-processing functions to identify the endpoints of the utterance and performs time normalization to create a 12-frame unknown utterance template. It then tries to find the best match among the training templates previously stored. The two key elements of the matching process are the frame-to-frame distance measure and nonlinear time alignment.

Distance Measure

I have mentioned the closeness of templates, which is used to determine the best match. But just how do you determine the closeness of two templates? The answer lies in a frame-to-frame distance measure, which is used to build a template-to-template distance. The smaller the template-to-template distance, the closer the two templates are to one another.

A Chebyshev distance measure is employed as the frame-to-frame distance measure. The equation ABS VAL $(A_i - B_i)$ is summed for all i, where i is the distance from frame A to frame B, A_i is the Ith element of frame A, and B_i is the Ith element of frame B.

Thus, frame-to-frame distance measurement consists of simply summing the magnitudes of the differences of corresponding elements in the frames being compared.

To find a template-to-template distance, the frame-to-frame distance measure is applied within the context

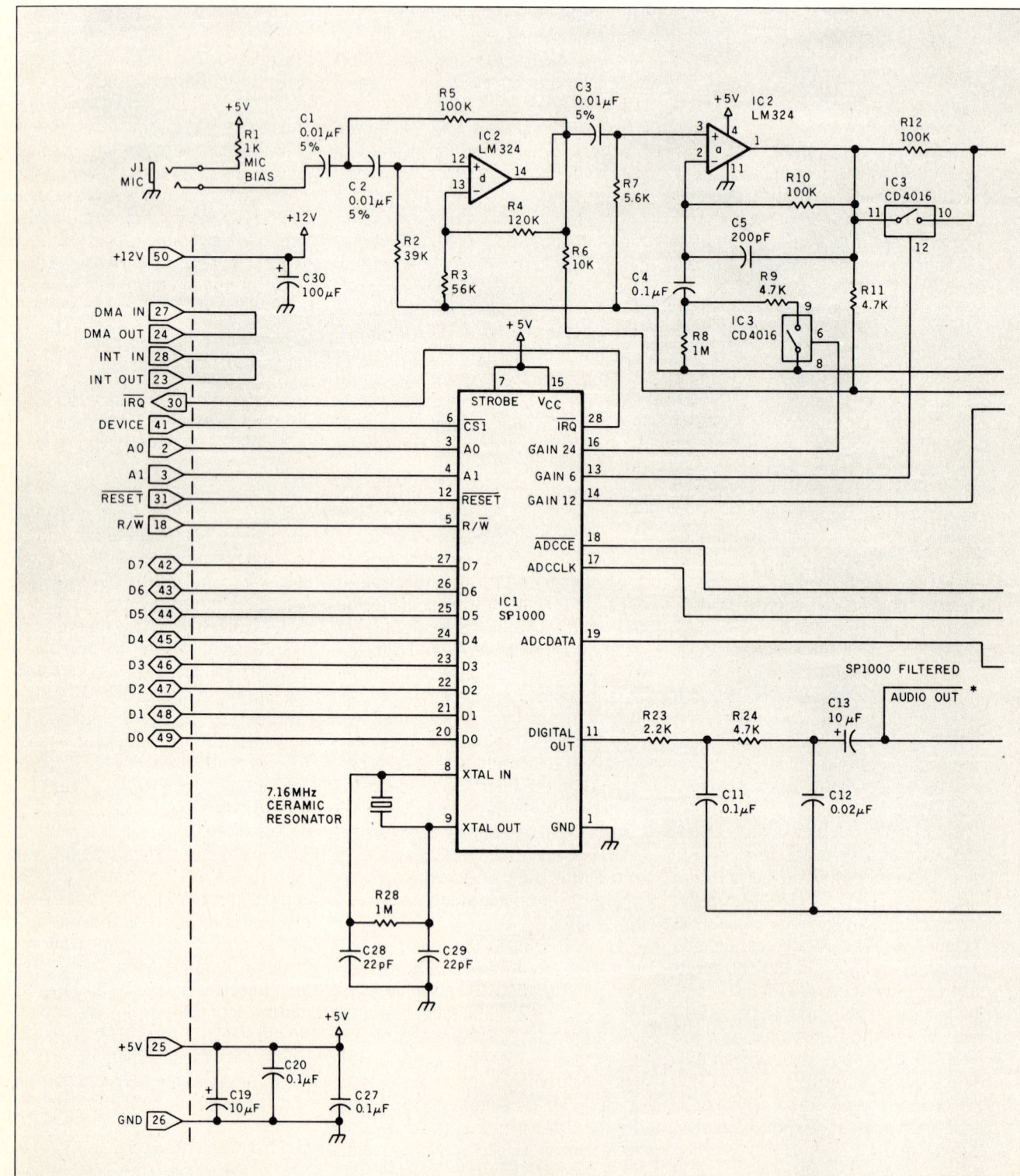

Figure 8: *The Lis'ner 1000 schematic without the SSI-263 for the Apple II.*

of the nonlinear time alignment of the frames.

NONLINEAR TIME ALIGNMENT

The need for nonlinear time alignment arises because human beings do not speak the same words exactly the same way each time. Volume and duration of words obviously vary, but a more subtle variation is very significant to a speech recognizer. The individual speech sounds comprising a word vary in duration relative to one another in different repetitions of the same word. This time distortion is nonlinear because simply stretching or compressing one entire repetition will not time-align the boundaries of the speech sounds with those of another repetition of the same utterance.

Consider a word with two syllables, such as table. Two repetitions of this

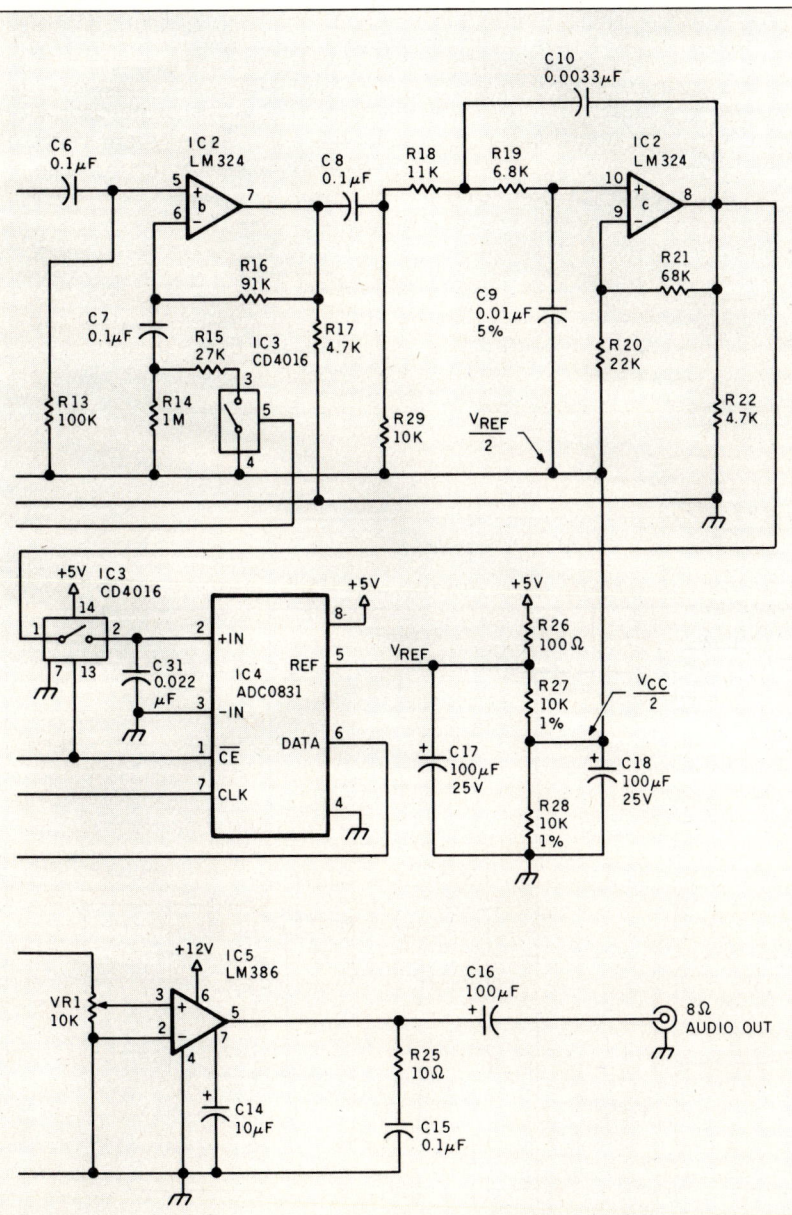

word may have the same total duration, but the first syllable may constitute 50 percent of repetition one and only 30 percent of repetition two. If we created templates for the two repetitions and compared them on a frame-by-frame basis, we would not get the best match because at some point we would be comparing parts of syllable one with parts of syllable two.

On different occasions, the timing of these patterns may vary considerably, but they must all be present in the described order if the utterance is to count as a reasonable rendition of the word table. The misalignment can be corrected by stretching the template in some places and compressing it in others, so that a mathematically optimum match is found. This procedure is called dynamic time warping (DTW).

(See "Speech Recognition: An Idea Whose Time Is Coming," BYTE, January 1984, page 213.)

REJECTION THRESHOLD

Once we have a best match, we have to determine if it is usable or not. One method is to qualify the match by setting a rejection threshold. This allows the recognizer to request that an input be repeated because it is not confident of a good match. With no rejection, the recognizer is forced to make a choice. The rejection threshold itself is the degree of confidence necessary to consider a match valid. The use of rejection criteria implies a trade-off between two types of error, the incorrect match versus no match at all. Rejection criteria can enhance the performance of the recognizer if they are adjusted to suit specific applications. The problems caused by the two types of error are application dependent.

As part of the recognition process, a template is made of the word just spoken, and it is compared to the templates made during training. For each comparison, a distance is computed that is used to determine the best fit to the spoken word. In order to reduce the number of false alarms (i.e., extraneous room noises being recognized as words), a method of rejection is used. Three parameters are used during rejection: the lower limit, the upper limit, and the rejection threshold.

The lower limit specifies a distance below which a word is automatically accepted and no more rejection tests are performed. This is useful in reducing recognition times and allows the

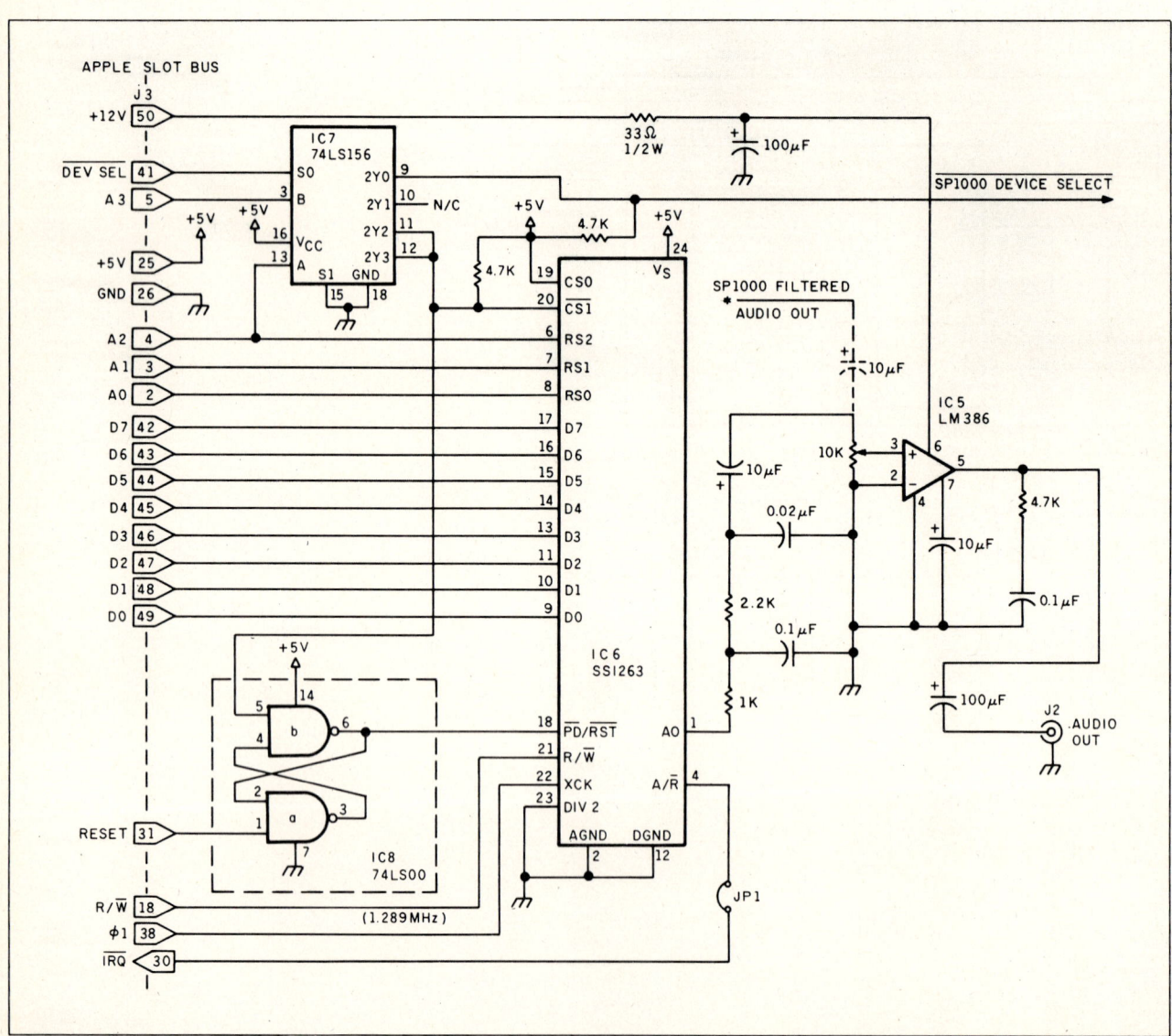

Figure 9: *The Lis'ner 1000 schematic with the SSI-263 for the Apple II.*

obvious correct matches to pass through. However, if it is set too high, many false alarms may occur.

The upper limit specifies just the opposite, the distance above which a word is automatically rejected. This too is helpful in speeding reaction times and discards obvious room noises such as clapping. If this number is set too low, a large incidence of rejecting good words will occur, resulting in a good deal of frustration for the user.

The last parameter is the rejection threshold, which is used to control just how close the spoken word may be to the two next closest reference templates. In short, a small rejection threshold results in a higher degree of rejection; a large rejection threshold is more forgiving and rejects less.

These three parameters are combined to tailor the system to the user's particular needs. If a highly speaker-dependent system is desired, a small lower limit, a small upper limit, and

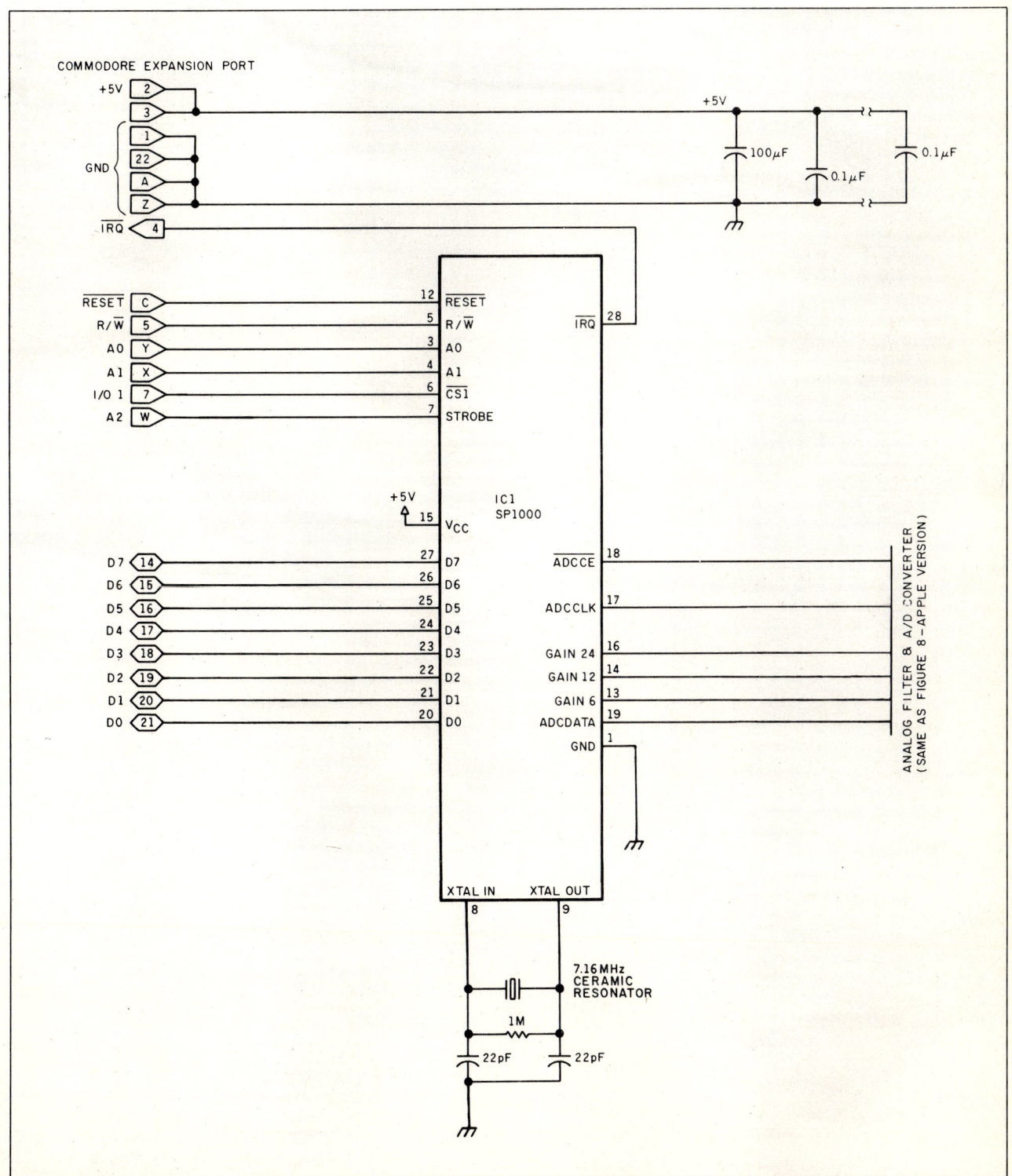

Figure 10: *The Lis'ner 1000 schematic for the Commodore 64.*

a small rejection threshold should be used. The result would be that only the person who trained the system would have good recognition results.

Lis'ner 1000 Software

The Lis'ner software consists of a combination of BASIC and assembly-language routines. The recognition algorithm and template matching are handled in assembly language to speed execution. Training and other infrequently used housekeeping functions are done in BASIC.

The purpose of the software is to function as a parallel voice input for application programs or normal operation of the computer. When first starting the system, for example, you are prompted to train a preselected vocabulary of DOS (disk operating system) and system commands. Rather than typing CATALOG and a carriage return, you merely have to say "catalog" and "return" (you can still type any part of it if you wish). In effect, the Lis'ner 1000 can be programmed to send a sequence of characters to the keyboard input handler as if they had been typed. This function can be turned on and off at will or used at specific points in application programs.

The process of selecting and training a vocabulary is prompted by a menu. You start by entering your own list of words to be recognized. Up to four groups of 8 words, or 32 words, are entered at one time, as shown in photo 4. A total of 64 words may be entered into the system. Next, you are asked for each spoken word followed by its corresponding command sequence, as shown in photo 5. The command sequence is the group of characters that the recognizer routine will respond with when it hears this particular utterance. The command sequence may contain any combination of letters, numbers, punctuation, and control characters.

The recognition software responds with the preset command sequence when it hears a particular word, regardless of whether it is appropriate. For example, you could make one of the speech commands a phrase such as "DIRECTORY, PLEASE." In response, the command sequence would print CATALOG and a carriage return for an Apple. (If you plan to use the device to simulate the direct function of discrete keyboard keys, it is best to use words such as APPLE, BAKER, CHARLIE, etc., rather than the single-syllable letters A, B, C, etc.).

Once all the words have been entered and a series of questions regarding rejection levels has been answered, you are prompted to train the system by saying each of the words two times. When this is done, the computer knows your voice and saves the templates to disk. The following is a list of editor commands, which can be spoken or typed. Their function is to aid in producing a vocabulary that approaches 100 percent recognition accuracy.

TEST—Enter test mode. This option, shown in photo 6, is useful in testing how well each word was trained. After each word is spoken, the letter "A" (accept) or "R" (reject) is displayed next to the word. If an "R" is displayed, the match between the spoken word and the word as the computer knows it is totally unacceptable. It may be due to the fact that the word that was recognized wasn't even the word spoken, or two words that sound alike may keep being confused. If any word or words consistently get low scores, that word or words should be retrained. To get back to command mode, hit any key.

EDIT—Add, delete, replace, and retrain any of the words.

LOAD—Load prestored templates so that editing and training may be performed.

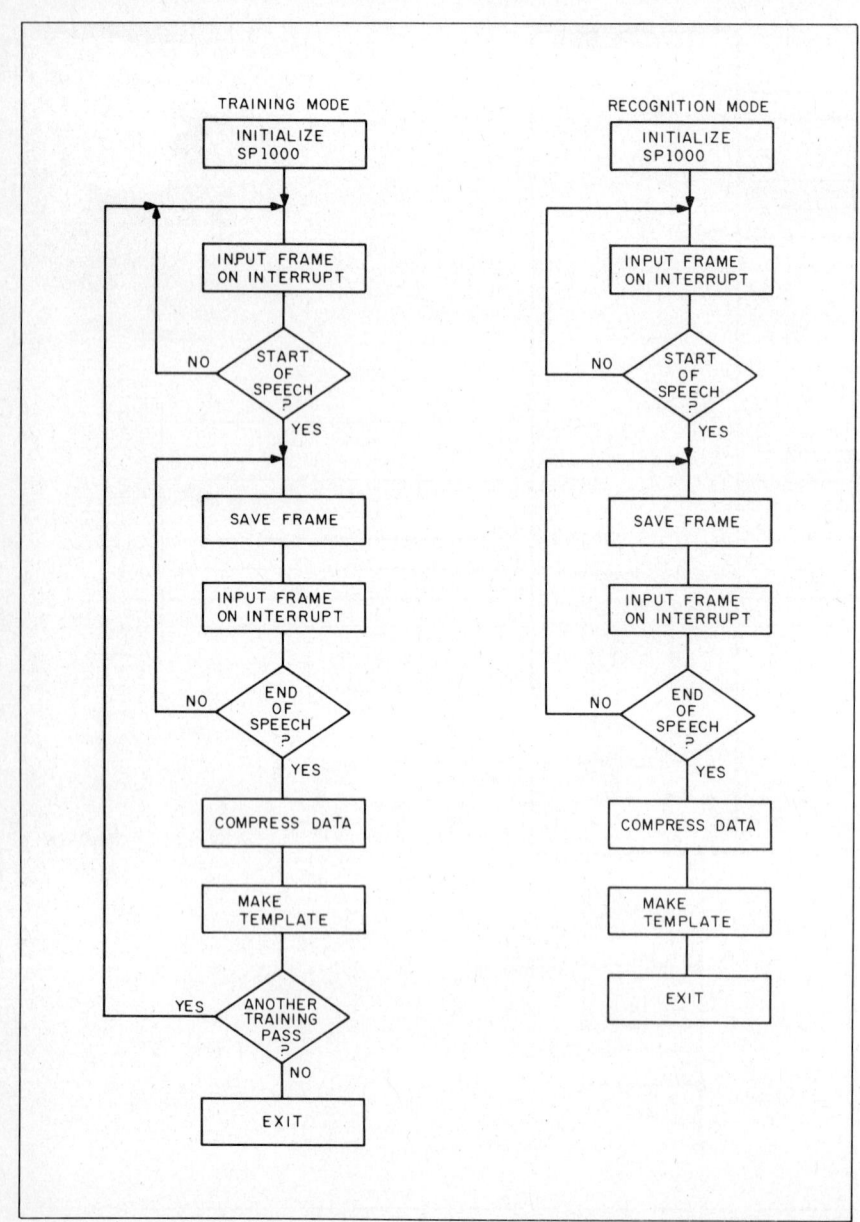

Figure 11: *The SP1000 recognition-software flowchart.*

SAVE—Save the templates being worked on for later use. This is used to save all the work you've done up to now. These templates may then be loaded by one of the Hello programs at some later date for use in your application programs.

QUIT—Leave the editor and return to BASIC. Once all your editing and saving are done, you may enter BASIC with the recognize routine and DOS templates still active.

Software design is of course dynamic. Some aspects of the Lis'ner software I've described here may have been modified by the time you read this.

Experimenter Support

I try to support the individual experimenter as much as I possibly can, and this project is no exception. To aid you in building the Lis'ner 1000 or an SP1000-based system, I have coordinated parts and software suppliers.

The Lis'ner software package consists of a combination of source-code and executable-only code files that are much too lengthy to print for distribution or to be published here. The Lis'ner software is supplied as BASIC source code with assembly-language executable code. Since I expect that many of you won't be happy until you've personally experimented with dynamic time warping and converted the routines to run on a different processor, I am making available demonstration source code for an SP1000 recognition algorithm for the Apple II. This code, which is less complicated than the Lis'ner software, was written by General Instrument. Also included are LPC coefficient files that will demonstrate the SP1000's synthesis capability.

Although this software is well annotated, it is unsupported and distributed for its educational value only. It contains all the necessary structure should you care to roll your own. (If you do convert these routines, I would be very interested in seeing your handiwork.)

Finally, while it isn't a requirement that you include a picture this time when you write to me, I'd like to see your finished product so that I can add your picture to the many hundreds I've received on previous projects.

Conclusion

The toughest part about writing this article was deciding how much to say about recognition techniques. I have

Photo 4: *In this training mode, you select and train up to 32 words at a time. Multiple overlays of these template dictionaries result in potential recognition vocabularies of thousands of words. In practice, 64 concurrent-available words is a reasonable search vocabulary that maintains a high response reaction time.*

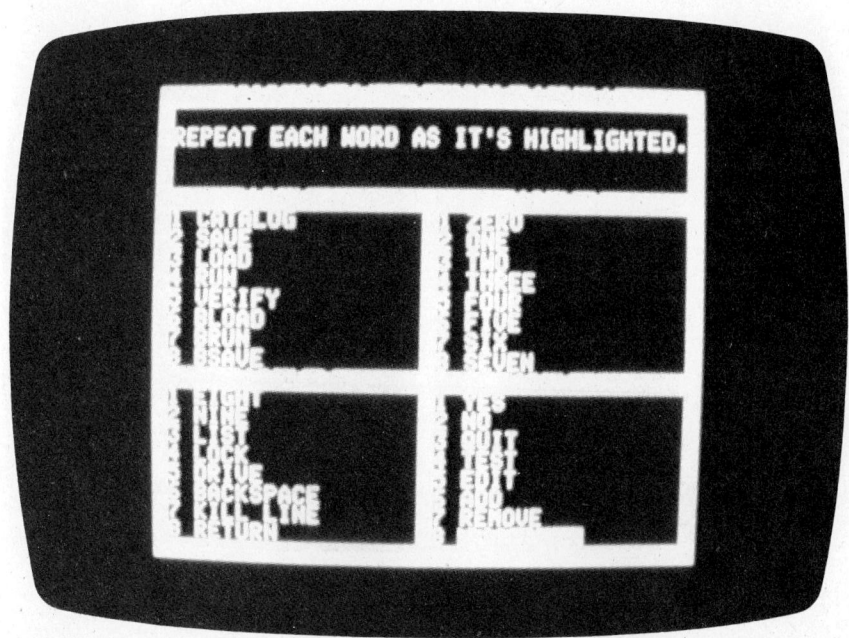

Photo 5: *These standard DOS commands comprise one of the vocabularies that the user is directed to train. Once trained, many keyboard entries can now be verbal.*

barely scratched the surface in my explanation.

It is equally difficult for me to list and describe the multitude of potential applications for computerized voice recognition. Besides the obvious aids for the disabled, the Lis'ner 1000 can be used in order-entry systems, voiceprint-security systems, video games, and telephone communications. Also, many people subscribe to the notion that the world needs a voice-operated typewriter. In my opinion, it will be a long time before voice entry becomes commonplace in an office environment, but there have been inroads.

I intend to apply the board to telephone communications so that I can call and correspond with my computer. Since the Apple II Lis'ner has both recognition and synthesis, it would seem natural that all conversation over the phone with the Apple should be spoken. "Hello, computer, how are you?" "Fine, Steve, your house is still here."

While this is a possibility, the quality of the telephone lines suggests that an alternate means of backup communication also be used. Some time ago I wrote an article about DTMF (dual-tone, multiple-frequency) decoders ("Build a Touch Tone Decoder for Remote Control," December 1981, page 42). In it, I suggested that one way to communicate with your computer was through an auto-answer device with a DTMF decoder. Once the computer answers, simply send your message by pressing the Touch-Tone keys on the telephone.

Your first thought might be to add a DTMF decoder in parallel with the recognition board, but it is quite unnecessary. DTMF tones and spoken words are all sounds as far as Lis'ner is concerned. It is simply a matter of pressing the telephone buttons while in the template training mode to program the Lis'ner to respond to the DTMF tones. Adding a few select words in addition will make it a truly unique answering system. Using just DTMF tones will allow invited subscribers a certain level of access to your system, but combining speaker-dependent voice recognition with DTMF tone recognition will allow you to reserve certain functions only for yourself.

Special thanks to Dennis Intravia for his work on the recognition software.

Diagrams and data specific to the SP1000 are reprinted courtesy of General Instrument.

The following items are available from

The Micromint, Inc.
25 Terrace Drive
Vernon, CT 06266
For orders: (800) 635-3355
For information: (203) 871-6170

1. Apple II Lis'ner 1000 with SP1000 recognition/synthesis components only—includes headset-style microphone and software on disk.
VR01 assembled and tested $189
VR02 complete kit $149
2. Apple II Lis'ner 1000 with SP1000 recognition/synthesis components and SSI-263 phoneme synthesizer chip with text-to-speech algorithm—includes headset-style microphone and software on disk.
VR03 assembled and tested $259
VR04 complete kit $219
3. VR01/VR02 phoneme-synthesis upgrade to VR03. Includes SSI-263, miscellaneous components, and text-to-speech algorithm on disk.
VR05 VR01/VR02 upgrade kit $79
4. Commodore 64 Lis'ner 1000 with SP1000 recognition/synthesis components—includes headset-style microphone and software on disk.
VR10 assembled and tested $149
VR11 complete kit $119
5. Apple II speech experimenter's kit—includes SP1000, 7.16-MHz ceramic resonator, ADC 0831 A/D chip, Lis'ner manual, and Lis'ner software on disk.
VR20 complete kit $60

Please include $4 for shipping and handling in the continental United States, $10 elsewhere. New York residents please include 8 percent sales tax.

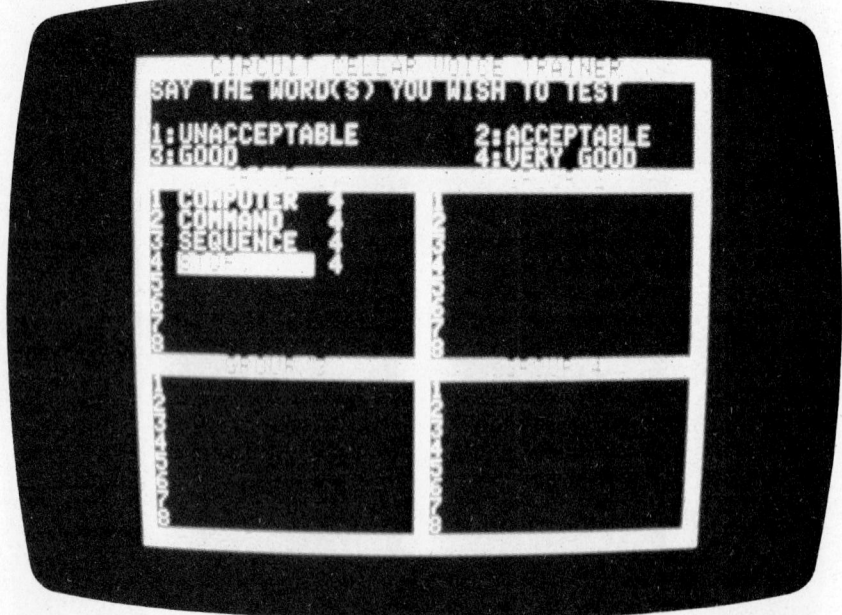

Photo 6: *One of the features of the Lis'ner software is the ability to make and test a recognition vocabulary. In the modified editor program shown here, as the words are spoken, the acceptance level is noted so the user can select words with less interference. Words like "computer" and "sequence" have few differentiation problems. "Nine" and "mine" would present difficulties, as would "next" and "text."*

BUILD THE POWER I/O SYSTEM

Controlling the power to the real world with your computer

If you've been reading the Circuit Cellar for any period of time, you've probably noticed that I have a definite prejudice toward computer control. You'll never read about the merits of various DOS (disk operating system) utilities in a Circuit Cellar project, but it is conceivable that you'll come across a computerized, time-of-day-activated dog feeder.

Seriously, though, over the years I've presented a variety of sensors, monitors, and controllers that could turn your computer from a mild-mannered games machine into the HAL 9000 of the neighborhood. These capabilities, be they menacing or beneficial, are directly the result of making the computer aware of the real world.

I define "real world" as conditions that occur external to the computer. A 100-watt table lamp next to your computer is in the real world. The computer is unaware of the lamp's presence because the computer is not connected to the real world. Unless something happens within the address and memory space of the computer, however, it is unaware that anything else exists.

To remedy this condition of ignorance, we must construct an interface that allows the computer to recognize the occurrence of real-world activities and respond accordingly. This real-world interface is a translational device of sorts: the computer sees it as simply another addressable peripheral device, such as a cassette recorder or printer, yet the information communicated comes from the real world or is directed to the control of real-world events, such as turning on the lamp.

In the case of the lamp, the appropriate control element would be an electronic substitute for the mechanical switch to turn the light on and off. However, is the light really on? Of course we can see it, but, unless we provide additional sensing capability, the computer knows only that it has turned the light on, not that it *is* on.

Additional real-world conditions must be monitored to know this with certainty. One way is to physically read the 115 volts (V) AC applied to the bulb or monitor a thermostatic sensor attached to the bulb.

All this might sound absurd, but that is because we take for granted that it is easy to turn a light on and see that it is in fact on. In critical control situations, more than simple on/off activation is required. Frequently, as in industrial applications, both the actions and the results must be monitored to produce reliable control conditions. (Open-loop control systems such as the BSR X-10 are generally unsuitable for industrial applications.)

One of the prime components of any real-world interface is discrete-bit AC/DC power input and output, which is on/off control and monitoring of 115-V AC or 5- to 48-V DC devices. With an AC/DC power I/O (input/output) interface, we can control and monitor motors, lights, high-voltage AC systems, and process-control and monitoring devices.

This month's project is a discussion of the design and construction of an AC/DC power I/O (power I/O hereafter) interface with particular emphasis on the internal configuration of the solid-state relays (SSRs) and receivers. The emphasis is on an industrial quality-control interface that meets the needs of both the experimenter and the industrialist. Beyond the homebrew relays and receivers that you can build, I'll describe a true closed-loop power I/O control system using the Circuit Cellar Z8-based computer system (which I've described and used numerous times) and commercially available components.

A Discrete-Bit Interface

Generally speaking, most computers are parallel in function. If you are using an Apple II or IBM PC, communication between the processor and its peripheral devices is handled 8 bits at a time presented in parallel. A parallel printer, for example, receives its character data as a 7- or 8-bit parallel word and sends its status and operating conditions back in a similar manner.

Most peripheral devices use all 8 bits at a time because they are most often communicating 8-bit data or ASCII (American Standard Code for Information Interchange) characters. Externally connected devices such as a light and a thermostatically controlled switch are single-bit devices. However, since the internal function of the computer is word rather than bit wide, each bit of the 8-bit word is used to separately receive or control an external event. If it were a 16-bit computer such as the 68000, each word would have 16 discrete bits of I/O.

Within the computer, one or more memory locations (or I/O port locations if it's a Z80 or 8080) are set aside and the addresses decoded as parallel I/O ports. If configured for output, each bit in a port is then connected to a discrete module that converts the TTL (transistor-transistor logic) level presented to it to a high/low, on/off voltage-level output. If the module is for AC control, it will convert a TTL high to 115 V AC and a TTL low to 0 V AC. DC output modules function as simple contact closures with the voltages dependent upon your proposed application.

When the addressed location is an input port, each bit is attached to an input receiver that converts a high-voltage input level to a TTL logic 1 and a low-voltage input level to a logic 0. The exact range and switch point

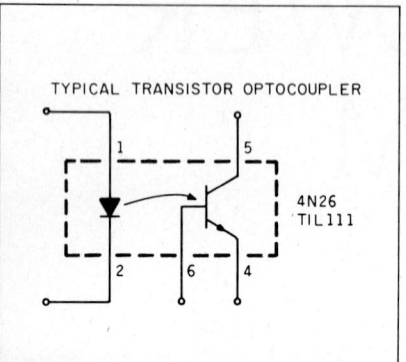

Figure 1: *A typical transistor optocoupler.*

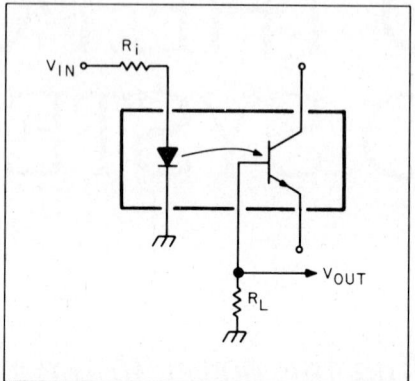

Figure 2: *Figure 1 used as a high-speed diode-diode optocoupler.*

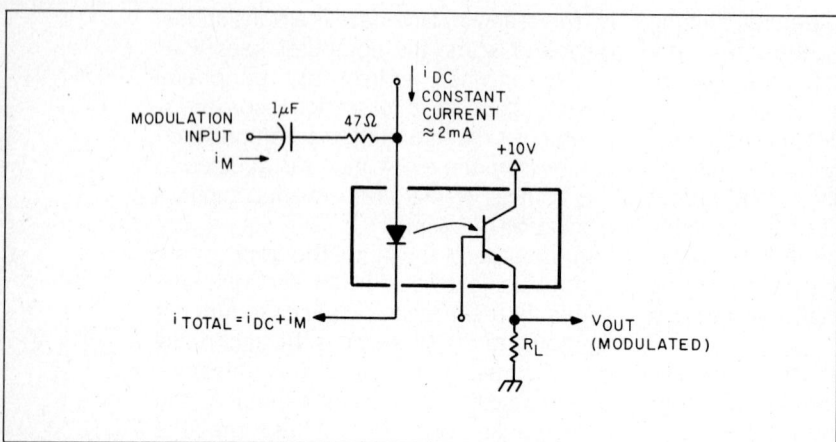

Figure 3: *Figure 1 used as an analog-signal optocoupler.*

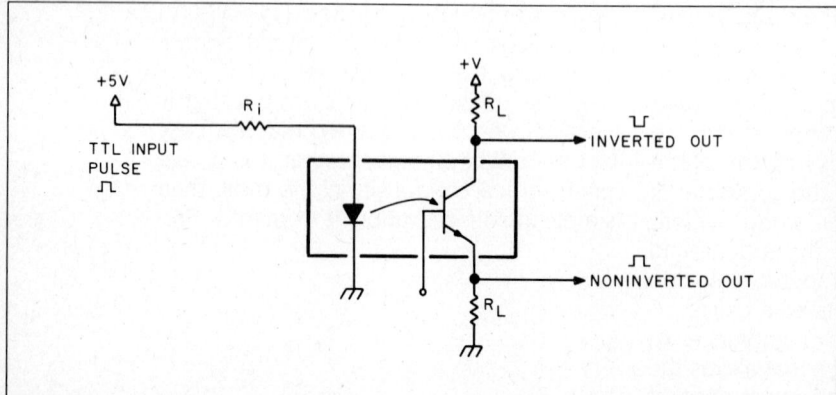

Figure 4: *Figure 1 used as a digital-signal optocoupler.*

of the module have to be selected for the application, and there are differences depending on whether the applied voltage is AC or DC.

With a single parallel I/O location (port), eight separate devices can be controlled and eight discrete events monitored. To properly coordinate the activity, bit rather than word manipulation becomes essential.

The I/O modules provide level conversion and isolation between the computer and the external device. Depending upon the components employed, I/O interfacing need not be a prodigious task.

Isolation Is the Key

The most important factor in I/O interfacing, especially with AC line voltages, is isolation. The computer you are using most likely operates on 5 V. If 115 V AC is applied to an unisolated input port, you are definitely going to produce smoke! High-voltage inputs must be safely converted to 5 V, and output devices must have no way to inadvertently feed 115 V AC back into the computer.

The simplest isolation device is the electromechanical relay. You can easily attach a reed relay to each output bit and the isolated contacts used to switch the AC line. Similarly, you can connect the external voltage to a relay whose contacts are attached to an input bit. When the input level is high enough (determined by series resistors), the contacts close and the computer senses the condition.

There is nothing wrong with using relays. For many years, this was the only method available, and it still works, to a point. However, relays are large, expensive, slow, electrically noisy, and subject to wear. They have been replaced with solid-state optoelectronic components that are small,

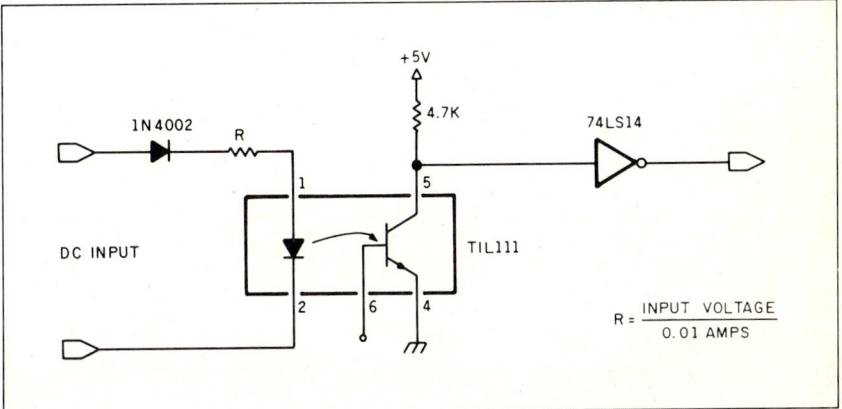

Figure 5: A DC input receiver.

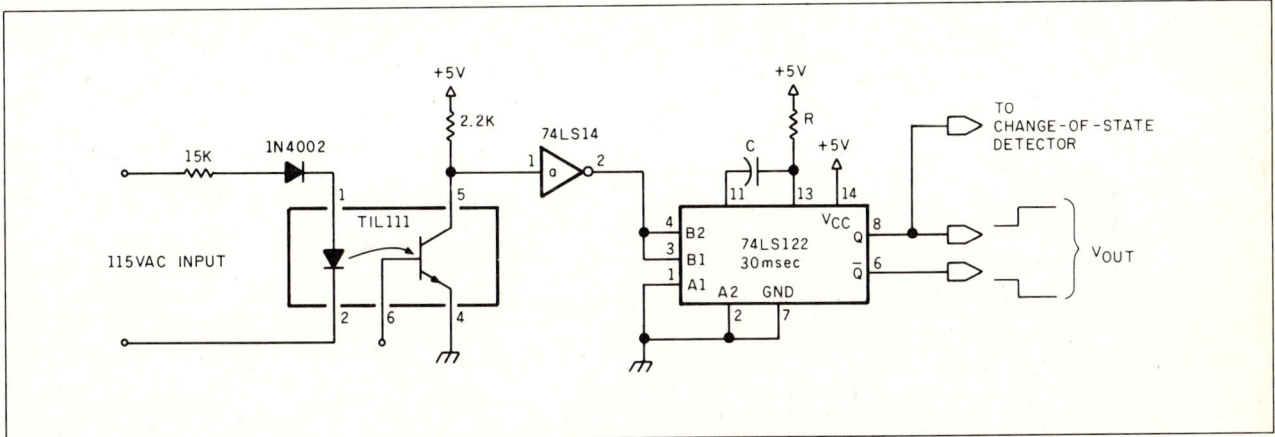

Figure 6: A half-wave AC input receiver with constant-level output.

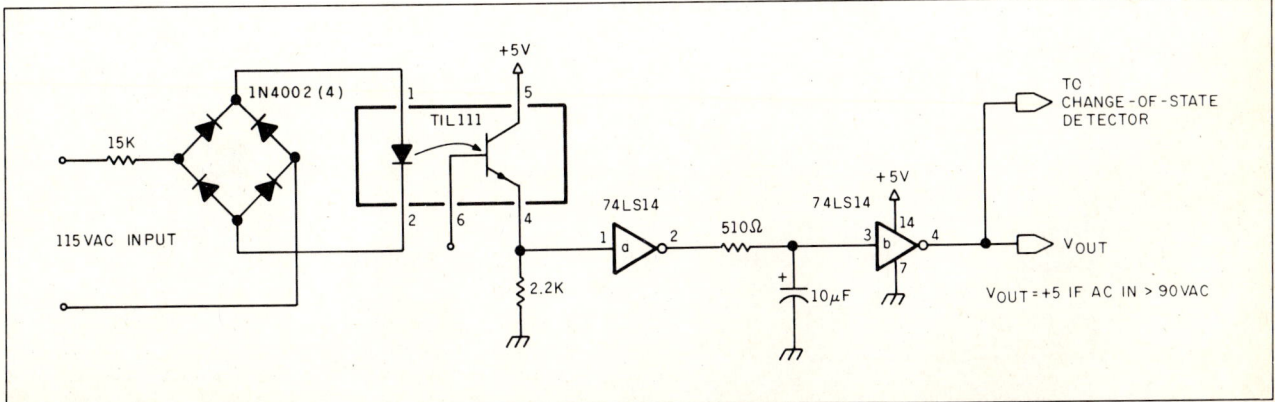

Figure 7: A full-wave AC input receiver with constant-level output.

Generally speaking, optocouplers are not tremendously fast due to the large photosensitive base area and the resulting junction capacitance.

inexpensive, fast, optionally noiseless, and have no wear in proper use.

Inside an Optocoupler

The essential ingredient in this project is the optoisolator and its use as an I/O control device for AC and DC voltages. With it, we can configure high-performance solid-state relays and solid-state input receivers. The exact configuration, as I will explain, depends upon the application and the excitation source.

The simplest and most frequently used optoisolator is the transistor optocoupler. It consists of a GaAs (gallium arsenide) infrared LED (light-emitting diode) and a silicon NPN phototransistor separated by a glass partition. The thickness of the glass determines the isolation level of the component. A typical isolation value is 1500 V. This means that the potential difference between the input and output sides of the optocoupler must be less than 1500 V, or it will break down and expose the computer to hazardous voltages. (While this seems unusually high, remember that these relays often switch inductive loads that produce high-voltage transients. Proper "snubbing" and transient suppression must be employed or these limits can be exceeded.)

Figure 1 is the typical transistor optocoupler. A current is applied to the LED, which induces a base current in the phototransistor proportional to the light radiated by the LED. This in turn allows current to flow between the collector and emitter of the transistor. A typical LED current is 10 to 50 milliamperes (mA). A 10-mA current greatly extends component life, however.

Generally speaking, optocouplers are not tremendously fast due to the large photosensitive base area and resulting junction capacitance. In solid-state-relay applications, however, speed is not an issue, and it will not present any problems for us. As a matter of education, though, various connection methods for the optocoupler are available, depending upon the excitation signal and response required.

If speed is a consideration, the optocoupler should be connected for

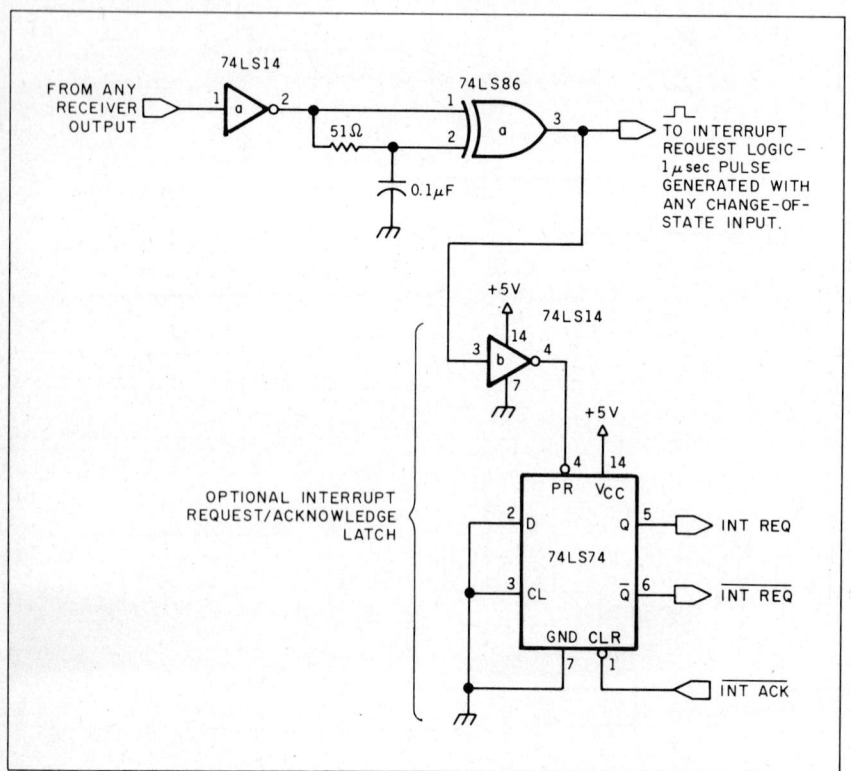

Figure 8: *The change-of-state detector and interrupt-request latch.*

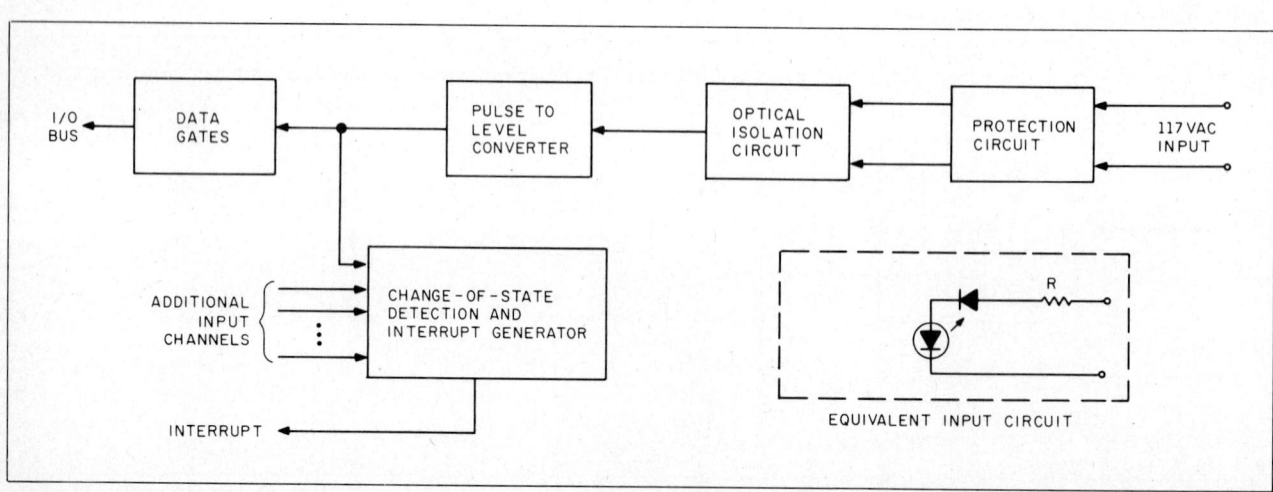

Figure 9: *The block diagram of an optoisolated input circuit.*

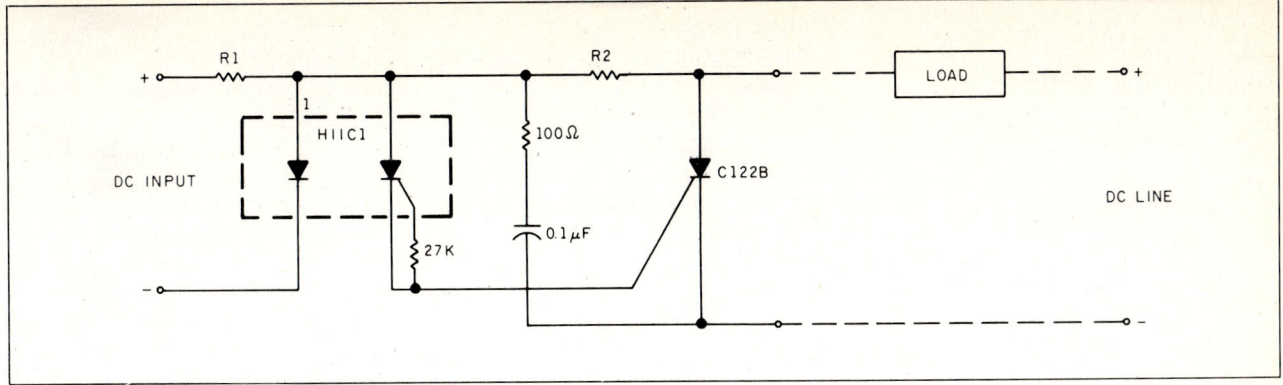

Figure 10: A simple DC output module.

Table 1: Resistor values for figure 10.

DC INPUT VOLTAGE	R1	R2
5	390	200
12	1.1k	200
24	2.4k	470
48	4.7k	1k
120	12k	2.2k

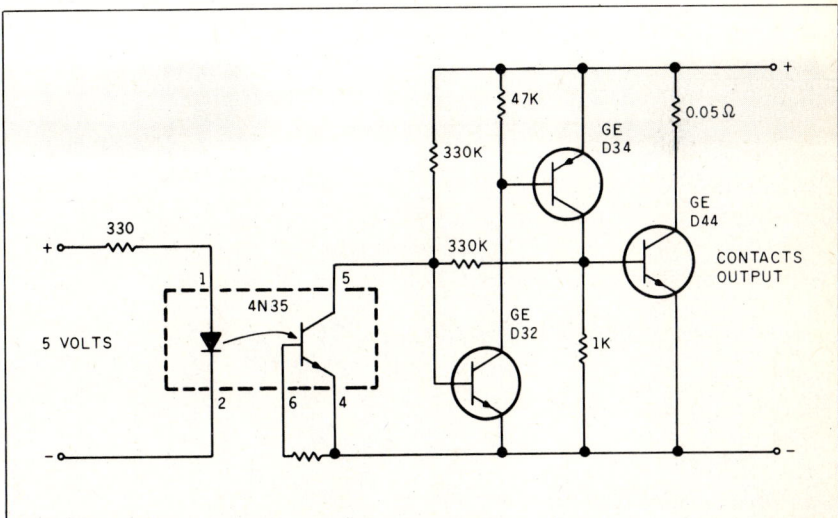

Figure 11: A normally open DC output module.

Figure 12: A normally closed DC output module.

diode-diode operation, as in figure 2. The output signal is directly received at the base connection. Typical response time is 2 to 5 microseconds (µs) as a diode-transistor coupler but only 50 to 100 nanoseconds as a diode-diode coupler. The one disadvantage is the much lower output current, which must be amplified.

While most experimenters think of optocouplers as digital devices, a transistor optoisolator can also be used with analog signals, as shown in figure 3. A constant bias current is applied to the LED to turn the transistor on enough to be within its linear range. Next, an analog input signal (modulation voltage) is also applied to the LED, which varies its light output proportionally to the modulated input. The emitter current in the phototransistor similarly follows this variation.

Most optocouplers are used for digital isolation, as shown in figure 4. A 10-mA LED current simply turns the transistor on or off with the inverted and noninverted responses available at the collector and emitter, respectively.

While the transistor optocoupler is the one I've chosen to describe, optocouplers are available with transistors, silicon-controlled rectifiers (SCRs), and Triacs as output devices. In some applications, the latter devices are more appropriate.

Discrete-Input AC/DC Receivers

The discrete-input DC receiver is by far the easiest module to construct. As demonstrated in figure 5, it is nothing more than an LED and a current-limiting resistor. I have added a series-blocking diode to protect the optocoupler from reverse connection and a 74LS14 Schmitt trigger to provide cleanly switched levels to the computer.

DC input receivers are generally preset as 5-, 12-, 24-, 36-, or 48-V detectors. The series-input resistor

211

Due to the sinusoidal properties of the signal, acquiring AC inputs is somewhat more complicated.

sets the range, which is selected to allow a current of 10 to 20 mA through the LED at the desired DC sense point. While various input-protection methods exist, the resistor should be selected for a single high-level input rather than a range of inputs. When the input is at the desired input voltage, the output will be an LSTTL (low-power Schottky transistor-transistor logic) logic 1; otherwise, it will be a logic 0.

Due to the sinusoidal properties of the signal, acquiring AC inputs is somewhat more complicated. Unlike relays, which are too slow to be seriously affected, optocouplers are fast enough to respond to every cycle of the input producing a pulse rather than a constant-level output. Proper reception by an even faster computer requires that these signals be integrated to an on/off steady-state level, as described in figures 6 and 7. Figure 6 is a half-wave detector that uses a 30-millisecond (ms) retriggerable one-shot; figure 7 is a full-wave detector that employs a simple RC (resistor-capacitor) circuit to integrate the pulse output. In either case, a logic 1 output signifies that a 115-V AC input is present.

CHANGE-OF-STATE DETECTION

One infrequently mentioned but important issue regarding discrete-input receivers is change-of-state detection. "Change of state" means simply that a receiver has changed its input level since the last time you looked at it. This might seem trivial if you have only one input module but is quite essential if you are monitoring 64 devices.

Figure 13: *A discrete-component non-ZVS solid-state relay.*

Photo 1: *The circuit board is a prototype of a typical non-ZVS solid-state switch like the one described in figure 13. A simpler alternative is to use the solid-state relay shown to the right of the board.*

The change-of-state condition can be determined and indicated either through hardware logic or software programming. Figure 8 is a hardware change-of-state detector. Whenever the input module's level goes from 0 to 1, or 1 to 0, a 1-µs pulse is generated at the output of the 74LS86 exclusive OR gate. This signal can be used to directly interrupt the processor or set a change-of-state flip-flop, as shown. The flip-flop retains its set condition until reset by the interrupt program. If eight input modules are used, there would be eight sets of this hardware with the outputs combined to generate a single "someone changed" interrupt. Figure 9 diagrams this approach. The advantages of the hardware change-of-state detection are that it is transparent to the user and requires little processor overhead.

An alternative approach is to scan the inputs in software periodically and compare the old and new readings to find changed states. In sophisticated control systems, a background interrupt routine periodically scans the input channels. Any changed states are represented as a byte in a table available to the application program. More on this technique later in our real application.

DC Power Output Control Devices

As previously mentioned, mechanical relays have been and can still be used in power-control applications. In new designs, however, the cost-effective approach is to use SSRs.

Solid-state relays come in a variety of flavors, depending upon the application. Unlike mechanical relays, which are nonpolarized, SSRs can be either polarized or nonpolarized. DC SSRs are normally polarized; AC SSRs are not.

Figures 10, 11, and 12 illustrate three kinds of DC output control modules. While they technically are SSRs, polarized switches such as DC output control modules are quite different in component configuration and are generally referred to as DC output units rather than SSRs. Figure 10 is a very simple DC module using an H11C1 photo SCR, which in turn triggers a higher-current SCR. Because it uses an SCR, this type of circuit simulates a latched-output relay. A voltage level applied to the LED input turns on the SCR and allows current to flow through it. The amount of voltage that will turn on the SCR is determined by R1 and R2, as shown in table 1. Both the LED input signal and the external load current must be removed to turn off the SCR.

This seems to be an absurd situation if the purpose of the DC output module is in fact to control the external current. Because of this, SCR relay devices are not normally used for DC resistive loads and are reserved in-

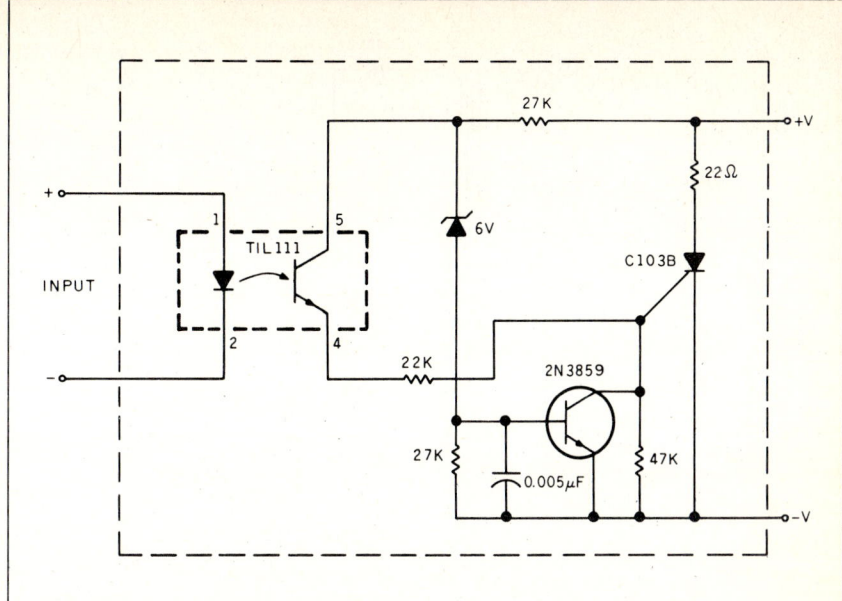

Figure 14: *A normally open ZVS circuit.*

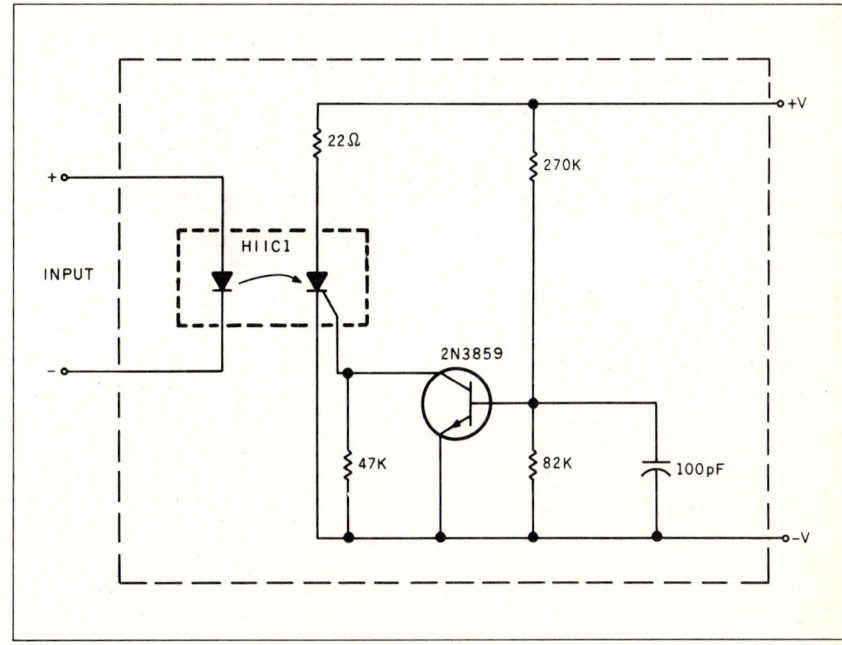

Figure 15: *Another normally open ZVS circuit.*

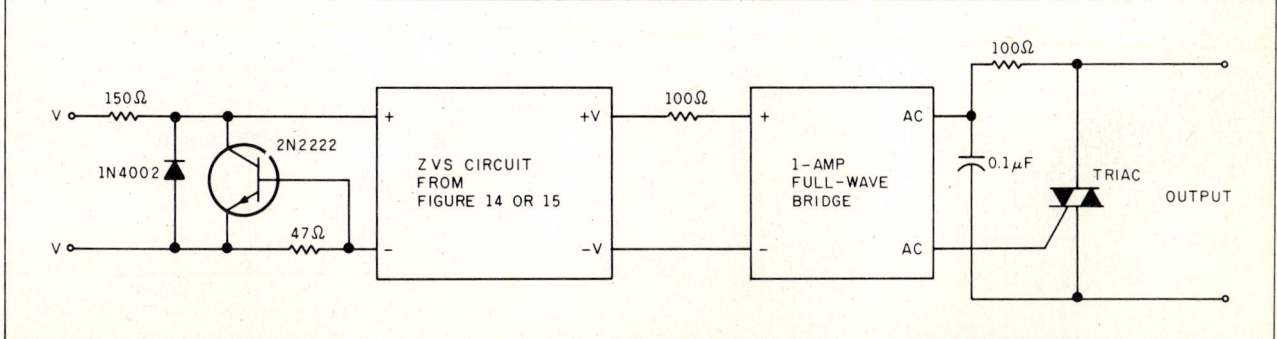

Figure 16: *A ZVS stand-alone SSR.*

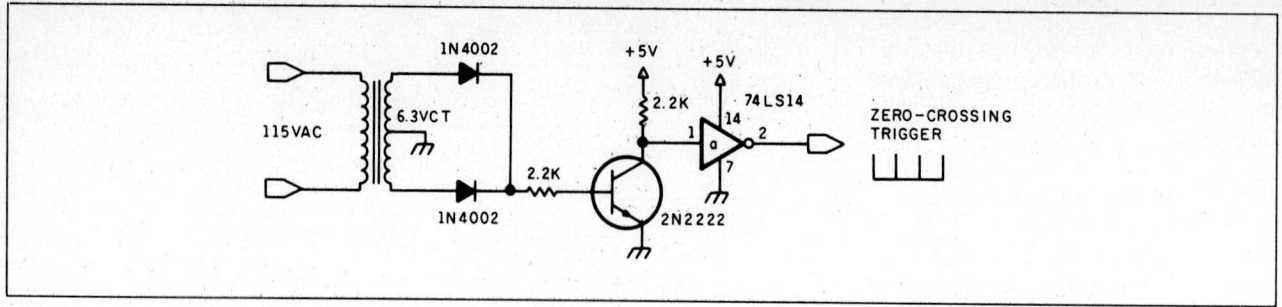

Figure 17: *An isolated zero-voltage sensor.*

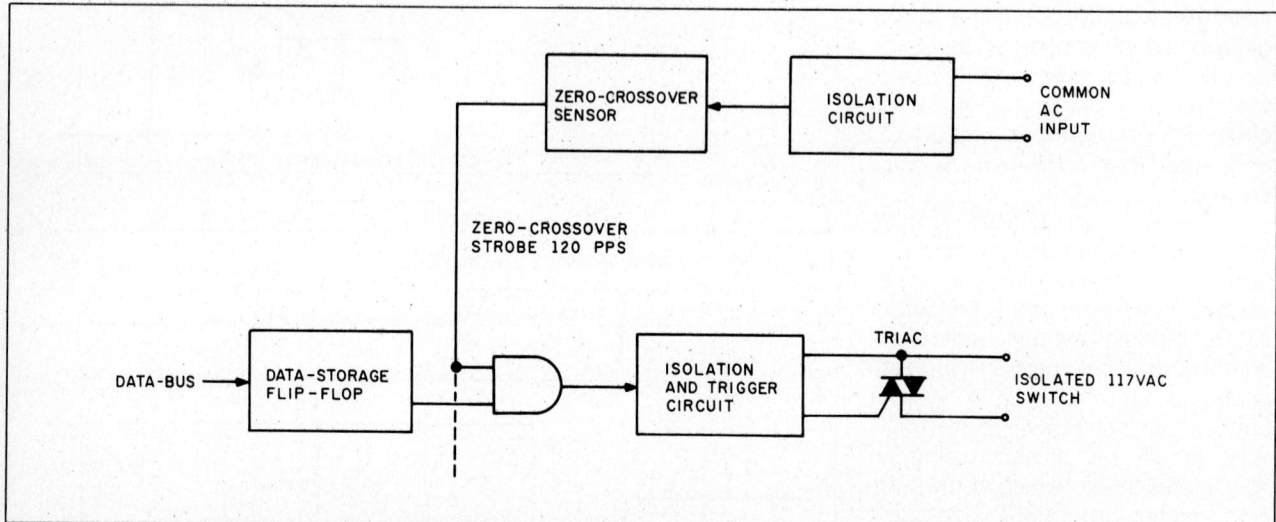

Figure 18: *The block diagram of a typical externally controlled zero-crossover AC output circuit.*

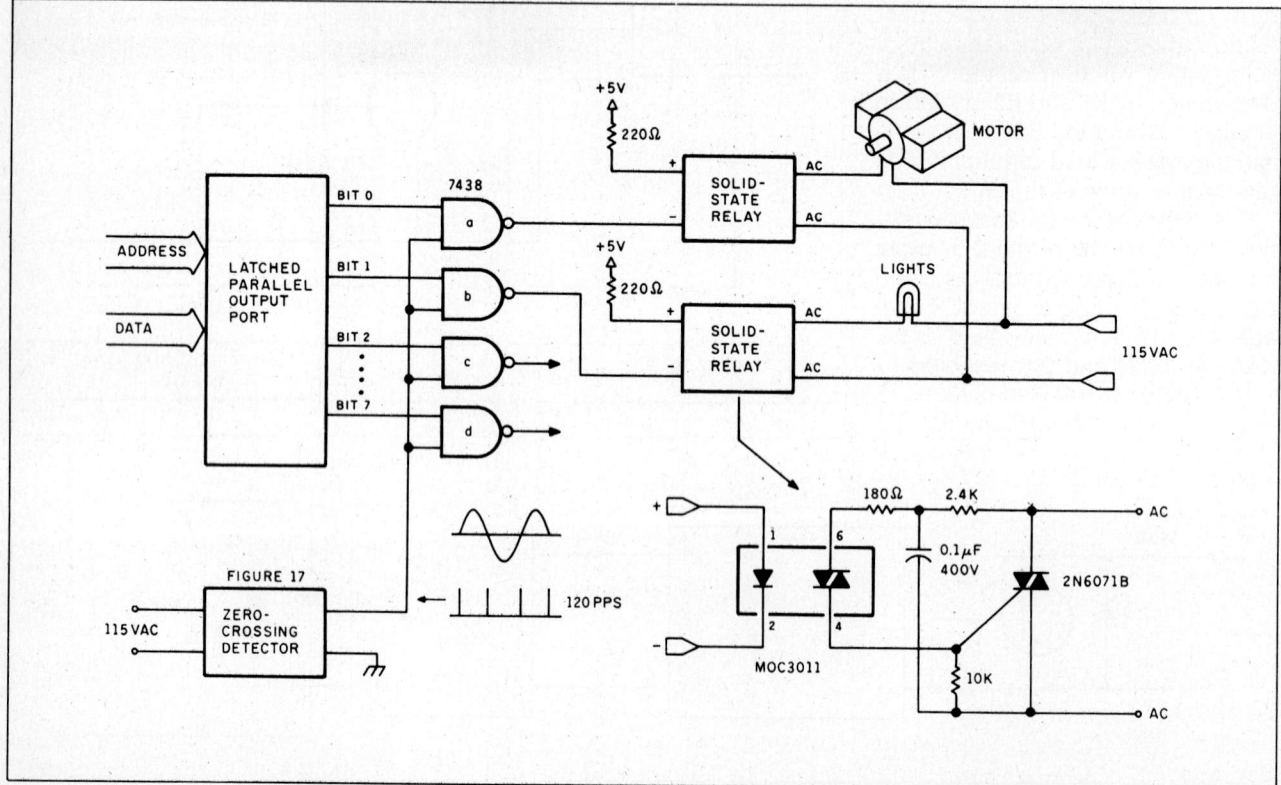

Figure 19: *The typical computer-controlled SSR output with ZVS.*

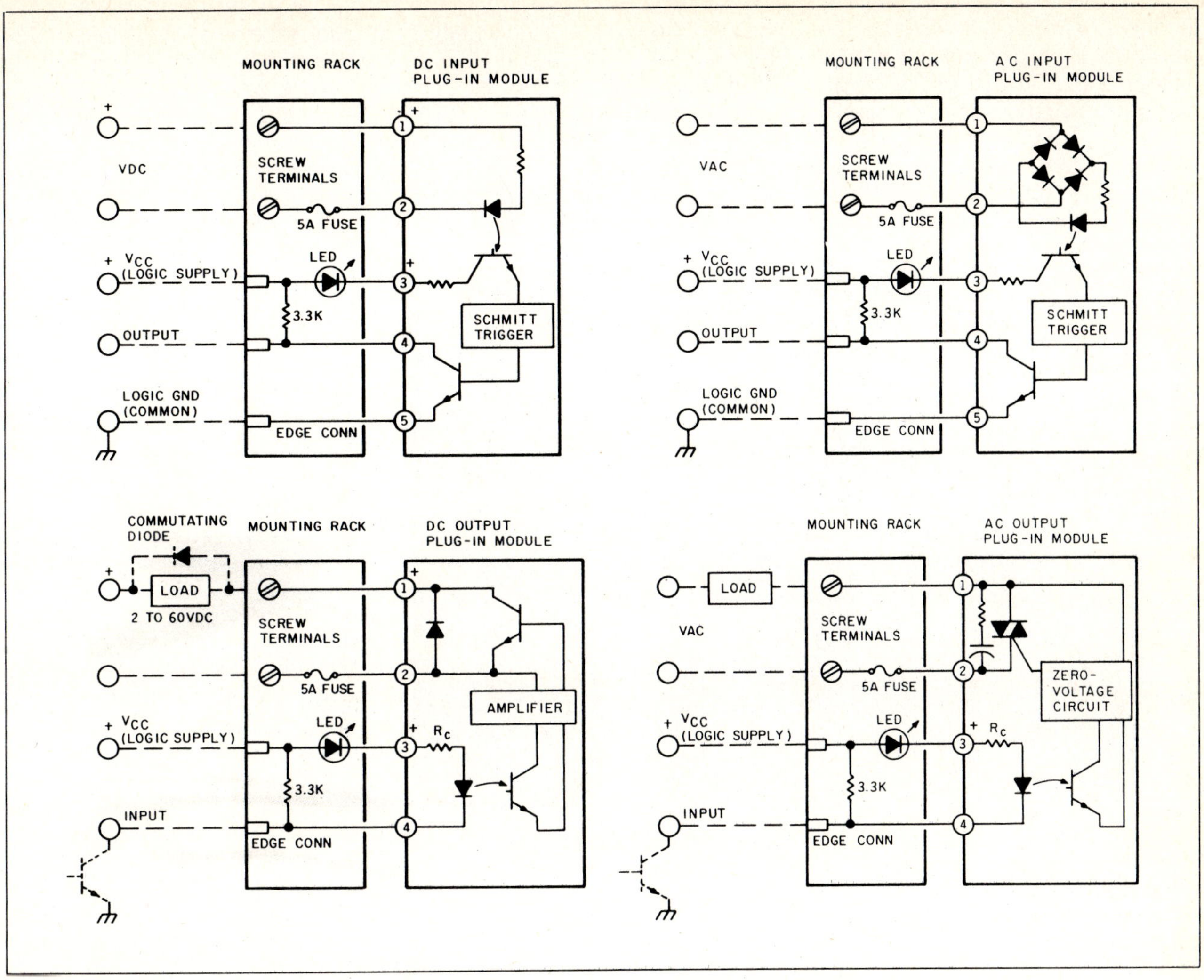

Figure 20: Mounting-connection diagrams of four Gordos Arkansas power I/O interface modules.

stead for commutating loads such as motors. With a DC commutating motor, the output current is interrupted many times a second as the motor shaft turns, allowing the SCR to turn off when the LED is extinguished.

All other DC control applications rely on transistor control elements, which exhibit fewer peculiarities but involve more components. Figures 11 and 12 demonstrate two typical 25-V DC output-module designs. Figure 11 is configured to have a normally open output; figure 12 has a normally closed output. These units are nonlatching and can be turned on or off in direct response to the logic levels from a parallel output port.

AC Power Output Control Devices

When we use the term "solid-state relay," we are generally talking about AC power output devices. These SSRs are nonpolarized and intended for use only with AC loads.

Figure 13 and photo 1 show the circuit of a general-purpose off-the-shelf component-configured SSR module suitable for control of lights and light-load appliances. The circuit employs an MOC3011 photoisolated Triac that in turn controls a power-output Triac. Input-protection circuitry has been added to the LED side of the module so that it can be used within a 3- to 25-V input range. (This input circuit can be added to any of the opto-

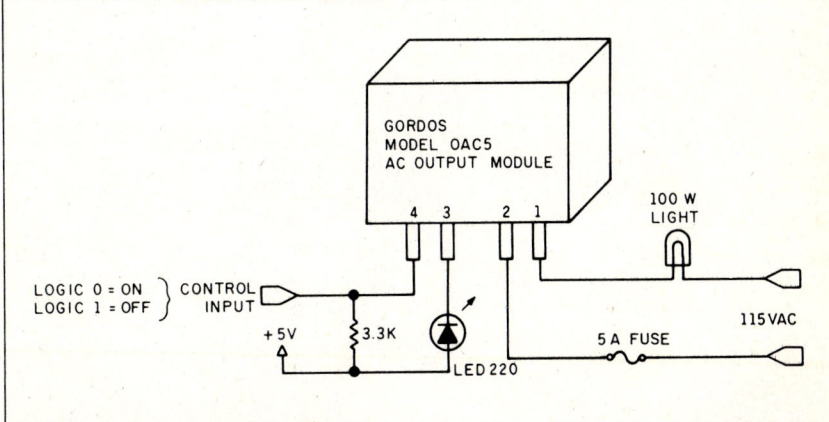

Figure 21: The connections for a Gordos AC output module.

isolators. However, if it is going to be attached only to a computer's parallel output port, you can dispense with this extra circuitry and use the simple resistor-input configuration of figure 9.) The additional resistors and capacitors form a "snubber" network that dissipates the transients produced when the Triac is connected to inductive loads.

As with any relay function, line transients are produced when a voltage is suddenly applied to an inductive load. While snubber networks, varistors, and transient suppressors offer some relief, the solution is to minimize the cause. If the Triac or SCR can be turned on only when the load voltage is at or near zero, no transients will be produced. The circuit that senses this condition is a zero-voltage switch (ZVS) or zero-crossover trigger network.

The ZVS may be built into the individual SSR (increasing its cost considerably if done with discrete components) or applied as a synchronous trigger to noninternal ZVS solid-state relays. Figures 14 and 15 are opto-isolated discrete-component zero-voltage switches that, when individually combined with the circuit in figure 16, produce a stand-alone ZVS output relay. Generally speaking, most commercial relays contain internal ZVS switching to improve performance. To keep costs down, these SSRs are constructed using hybrid technology rather than discrete components.

If you are building an AC I/O interface with commercial modules that include ZVS, it need be of no further concern. If you are building this interface from scratch, however, you can either build each relay module with internal ZVS or provide an external ZVS signal that is routed to an AND gate along with the control signal from the computer and applied to a non-ZVS switch, such as the one in figures 14 and 15. A circuit that detects zero crossing is demonstrated in figure 17, and a block diagram of this synchronous switching concept is presented in figure 18.

Figure 19 is the circuit for a computer-controlled AC output interface using the devices I've described thus far. To turn on the individual output channels, you merely set a logic 1 output at that bit position. This is most easily accomplished with an OUT command in BASIC. Once the bit is set, the SSR will turn on at the next zero crossing of the AC line.

Not Through Yet

Ordinarily, the project would end here. I've demonstrated how to build the I/O modules. A two-line BASIC program with INP and OUT instructions is all that it takes to control them. Unfortunately, knowing how to build a solid-state relay is different from assembling a practical control system.

Rather than leaving these practical matters as an exercise for you, I'd like to describe the 64-channel power I/O

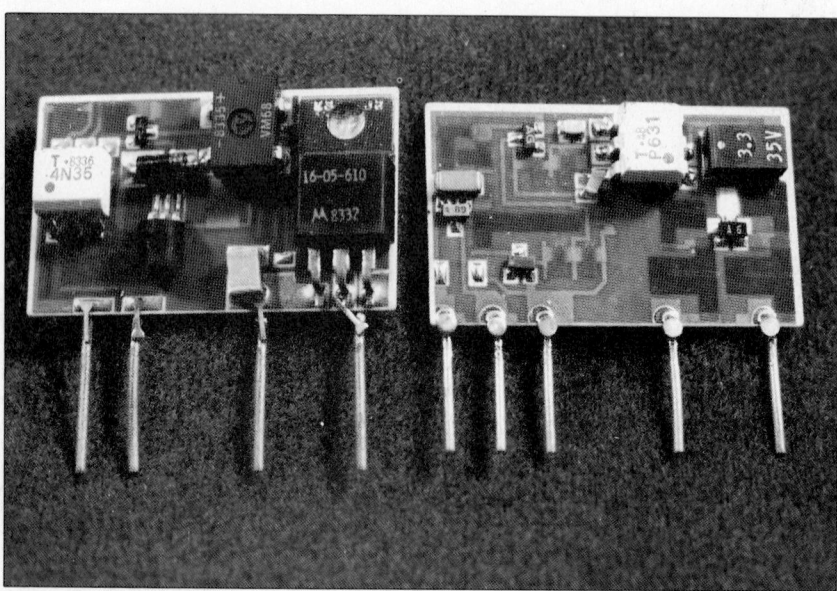

Photo 2: *The commercial I/O modules used in this project are hybrid circuits containing laser-trimmed resistors as well as discrete and integrated components. Two Gordos Arkansas Inc. modules are shown unpotted to demonstrate their complexity. The module on the left is an AC output (type OAC 5) and the unit on the right is an AC input (type IAC 5).*

Photo 3: *Eight of these I/O modules can be mounted together on a single board. Individual boards can be separately configured as AC-DC output or AC-DC input. The Circuit Cellar Z8 Power I/O prototype board in the photo has four AC outputs (black modules) and four DC outputs (red modules).*

system I ultimately configured. The description, though somewhat complex in detail, is intended to provide a basic understanding of the system software necessary to implement a reliable high-performance industrial-grade closed-loop control system. Utilizing the basic concept but substituting commercially available power I/O modules and a dedicated computer, a rather sophisticated programmable power I/O control system can be configured.

The commercial modules I chose are made by Gordos Arkansas Inc. Shown in photo 2, these potted modules are designed using thick-film hybrid technology for high-density packaging. Figure 20 is a diagram of the contents and connections of four typical Gordos power I/O interface modules. Figure 21 demonstrates the physical connections for an AC output module.

The computer I chose is the Z8 system/controller, which I've used in many Circuit Cellar projects. Based on a project presented in July and August 1981, the Z8 system/controller is a 4- by 4½-inch single-board computer with on-board tiny BASIC or FORTH, 6K bytes of RAM (random-access read/write memory) or EPROM (erasable programmable read-only memory), two parallel ports, and one serial port. To the Z8 computer, I've added the Micromint BCC33 memory and parallel I/O expansion board, which adds 8K bytes of memory, three parallel ports, and a cassette-storage interface, to interface to the power I/O modules.

(If you have either built the original Z8 computer/controller and wish to update it to the system/controller configuration or would like to build the memory and I/O expansion board for your existing system, just send me a preaddressed 9- by 12-inch envelope with $2.30 postage [overseas, just send name and address and $4 in international mail coupons] and I will send you the schematics and manuals for the two boards.)

The three parallel ports on the expansion board are configured as an I/O bus with input, output, and control capability. The power I/O modules are separated by function (input or output) and arranged eight modules per I/O card. Both AC and DC modules may be on one card, but only if they are all the same function. Up to 16 boards (64 input and 64 output modules, addressed as input boards 0 through 7 and output boards 0 through 7) can be accommodated with a single BCC33 expansion board. (Eight expansion boards can be put in the system if you are trying to control a small city.)

The computer communicates with the I/O cards through the expansion-board parallel ports. Port A functions as an 8-bit input bus, port B as an 8-bit output bus, and port C as the control lines for the individual I/O cards. Each power I/O card has a set of eight two-position jumpers, a 74LS374 output latch, and a 74LS244 input buffer (see photo 3 and the schematic in figure 22). Photo 4 shows some of the cards mounted in a card cage.

A single jumper selects board address and function. The eight output lines of port C are attached to the center position of the eight jumpers (boards 0 through 7). Only one of these lines is active low at a time; all others are at logic 1. The line that is low enables the power I/O card jumpered to it. Within that enabled card, a jumper installed to the center and left side (O) will enable the LS373. If installed between the center and the right side (I), it selects the LS244. A second jumper Tri-states the LS373 when the board is configured for input. If the jumper were in the #3I position, this would be addressed by bit 3 on port C, and it would be an input-only card.

Figure 23 shows a detailed block diagram of the power I/O system. It can be all AC input, DC input, AC output, DC output, or a mixture (in groups of eight similar functions). When you want to set the eight output modules on board 2, you merely set the bit pattern on port B and then strobe bit 1 on port C (board 2 enable line) to latch that data into the LS373. Conversely, to read the eight input channels on board 3, you would set bit 3 of port C low, read and store the data input to port A, and set bit 3 of port C high again.

The process of interfacing with the power I/O cards is relatively simple and can be accomplished directly in

> *The process of interfacing with the power I/O cards is relatively simple and can be accomplished directly in BASIC if speed is not a critical factor.*

Photo 4: *Up to 16 boards (8 input and 8 output) can be supported from each I/O expansion board in a Z8 system. Up to 4 expansion boards can be mounted in a card cage. The photo shows 3 expansion boards installed. The green connector protruding out of the cage is for external wiring connections.*

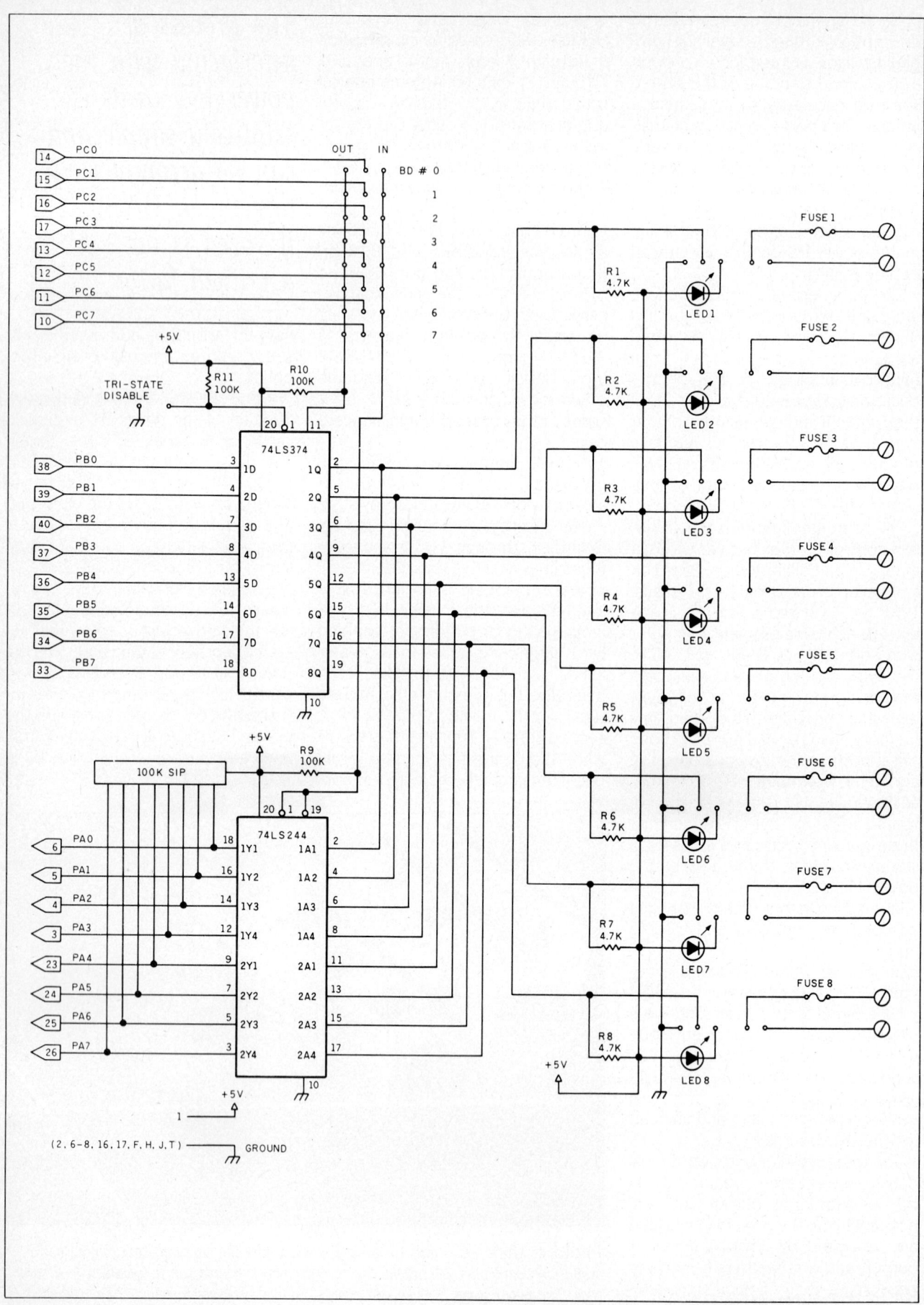

Figure 22: *The schematic of the Z8 power I/O card, as shown in photo 3.*

BASIC if speed is not critical. With 8 or 10 modules, it is not a problem to scan and record change of state, read the real-time clock, and still meet the requirements of the application.

While the simplified hardware for the power I/O system is important, it takes more to produce an industrial-grade control system. It is counterproductive to run time-consuming, repetitive tasks in BASIC that can be done more quickly in assembly language. For that reason, I've added a set of interrupt-driven utilities that greatly simplifies the interaction between user and power I/O system and allows the use of BASIC (unless you prefer assembly language), even with 64 active I/O channels.

These Z8 assembly-language routines, flowcharted in figures 24 and 25, operate as background tasks to any user application programs and are completely transparent. In addition to real-time clock functions, they allow the user to interact with the I/O system through a table of 64 input and output values rather than setting and reading expansion ports. To turn output channel 16 on, we simply load a value greater than 0 into table location 16. To turn output channel 1 off, we load 0 into table location 1.

Conversely, all inputs are continuously scanned and the present values loaded into a similar channel table for examination. In addition to the present value, a separate indication of change of state by board and channel number is also produced. The change-of-state indication is maintained until the user reads the affected channels. The result is a simple BASIC single-byte read-and-compare to find any input channels that have changed and a single-byte write to make a corresponding control output.

The user can drive seven subroutine calls in dealing with the power I/O system. They are

1. System initialization.
2. Read an input channel's change-of-state flag (1 bit).
3. Read an input channel's data bit and reset change-of-state flag.
4. Set an output-channel data bit (1 on or 0 off).
5. Read an input board's change-of-state flags (8 bits).
6. Read an input board's data bits and reset change-of-state flags.
7. Set an output board's data word (8 bits, 1 on or 0 off).

In addition to these subroutines called by the user, other routines

> *By the time you read this, the Circuit Cellar could be rewired. But I might experiment with my new home-control system.*

under interrupt control update the clock/calendar values and the power I/O boards and data/change tables.

The completed software fits on a 2716 EPROM on the I/O expansion board. Unfortunately, I don't have room enough here for a complete program listing, but I will send you one if you write to me.

In Conclusion

What started out as a very simple solid-state-relay project got a little carried away. I use all the devices I design, and this is no exception. By the time you read this, the Circuit Cellar could be completely rewired. On the other hand, I might wait a month and experiment a little more with the new Circuit Cellar Home Control System that's in the works. Com-

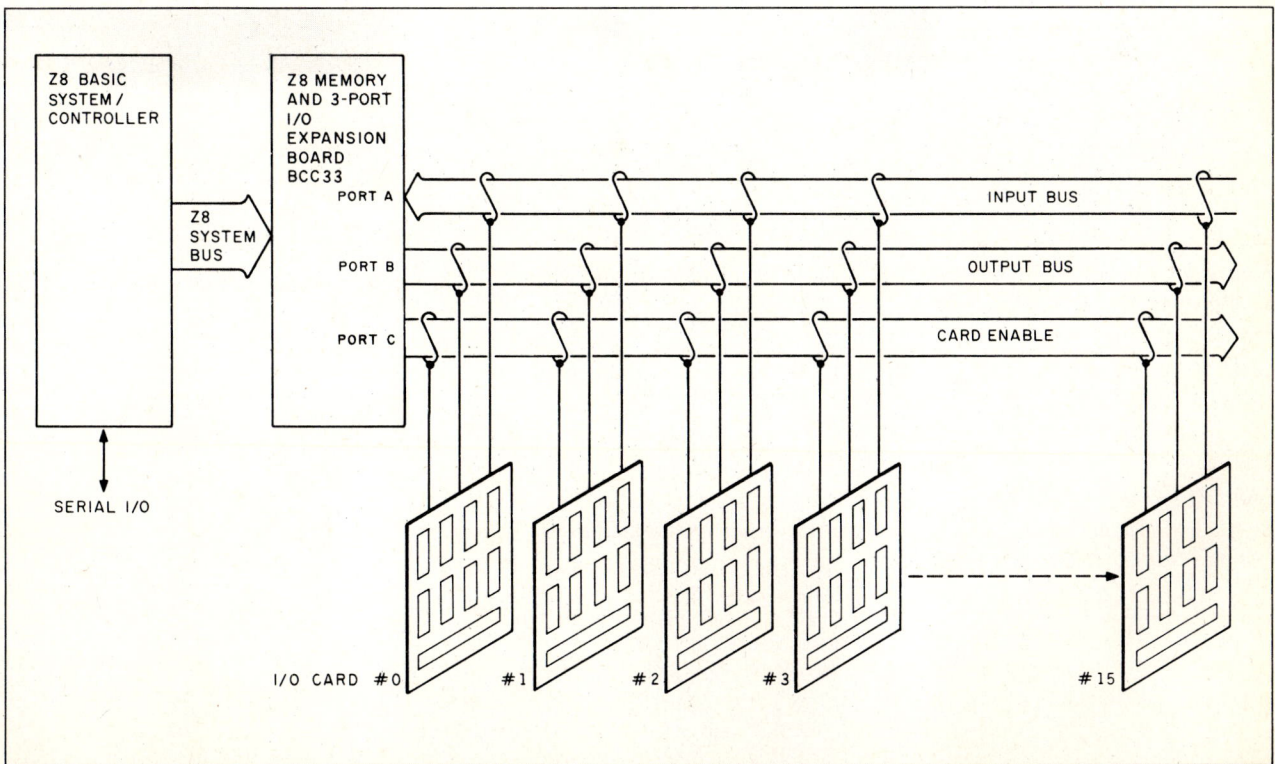

Figure 23: *The block diagram of the Z8 power I/O system.*

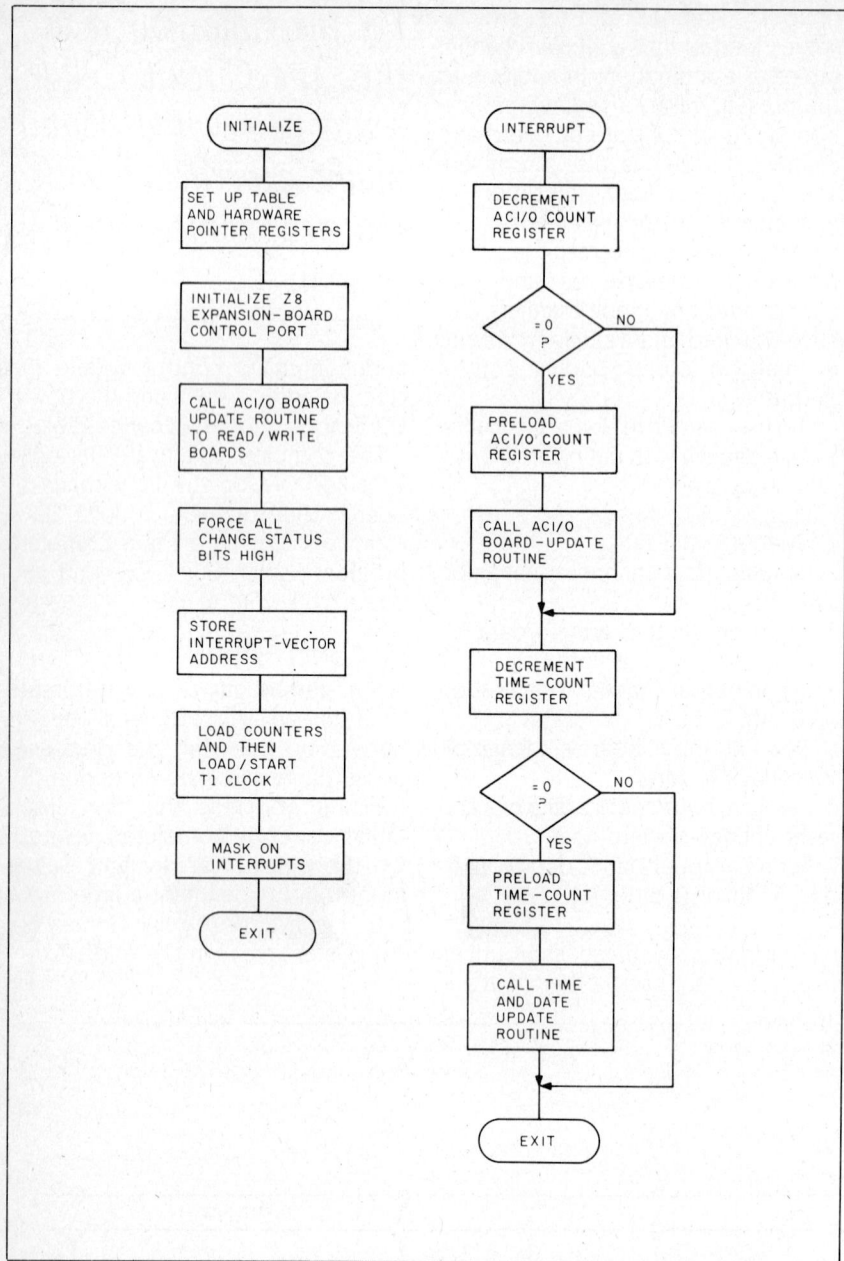

Figure 24: A flowchart of the Z8 AC I/O system initialization and interrupt-handler assembly-language routines.

Figure 25: A flowchart of the Z8 AC I/O board update routine.

puter control fanatics, hold on to your hats, I've just begun.

Special thanks to Bill Curlew for his software expertise. Diagrams of Gordos modules reprinted courtesy of Gordos Arkansas Inc.

The following are available from

The Micromint, Inc.
25 Terrace Drive
Vernon, CT 06266
For orders: (800) 635-3355
For information: (203) 871-6170

1. Assembled and tested Z8 power I/O board without power I/O modules $159
2. AC input or output modules, types IAC5, IDC5, OAC5, and ODC5 . . . . $15 each
3. Eight-slot motherboard $85
4. Z8 system controller $149
5. CC01 10" card cage $59
6. CC02 19" card cage $79

Please include $4 for shipping in the continental United States, $10 elsewhere. New York residents please include 8 percent sales tax.

Index

AC power lines:
  AC power monitor for, 169–179
  AC/DC power I/O system for, 208–220
  carrier-current modem, 12–18
  protection for, 71–79
Adaptive differential pulse-code modulation (ADPCM), 108
Address-descramble circuit in IS32 Optic RAM, 26
Address registers in Micro D-Cam solid-state video camera, 25–26
Addressing by Zilog Z8000 microprocessor, 141
Aliases (high-frequency transients), 198
Alphanumeric LED display, 121–137
Analog signals, transistor optoisolators for, 211
Analog-to-digital (A/D) conversion:
  of human voice, 108
  in measurement of power consumption, 171
  in SP1000 voice-recognition chip, 197
  of video images, 20
Apple II computers:
  Lis'ner 1000 speech-recognition system for, 195, 197
  Micro D-Cam solid-state video camera used with, 32, 35, 50, 53
  phonetic speech synthesizer connected to, 108
  SonarTape ultrasonic ranging system connected to, 186–187
  SS1263 speech-synthesis chip connected to, 109
Applications:
  real-world interfaces for, 207–208
  of RTC-4 real-time controller, 9
  of SonarTape ultrasonic ranging system, 190–191
  of synthesized speech, 111
  of Term-Mite ST Smart Terminal, 101–105
  throughput of, 139–140
  of Trump Card, 150
  voice inputs to, 204
  of voice recognition, 206
Arithmetic/logic units (ALUs) on Intel 8748 microprocessor, 60
ASCII (American Standard Code for Information Interchange) character set, 208
  on H-Com handicapped communicator, 57
  read by Term-Mite ST Smart Terminal, 93
Assembly language, Y multilevel language as, 158–159

Attribute latches, 87
Audio tape, 10–11
Automatic control (*see* Control applications)
Automatic gain-control (AGC) algorithms, 197
Automatic speech recognition (ASR), 193
Automatic telephone dialers, 162
Autoregressive analysis, 194–195
AY-3-1350 tone synthesizer (General Instrument), 167

BASIC (language), 36–49, 138
  interpreters and compilers for, 154–155
  TBASIC compiler for, 153, 155–157
  Trump Card speeding-up of, 140
BASICA (language), 140, 153
  TBASIC compiler and, 155
Batteries (*see* Power supplies)
Bell-103 standard, 13
Bells:
  musical telephone bell, 161–167
  on Term-Mite ST Smart Terminal, 88
Benchmarks:
  for microprocessors, 139
  for TBASIC, 155–156
  for Trump Card, 140
Blackouts, 72
Blinking characters, 97–100
Booting of Trump Card, 150
Brownouts, 72
BSR X-10 Home Control System, 9–10, 12, 208
"Bucket" in Trump Card, 149
Buffers in Term-Mite ST Smart Terminal, 93
Bus ports, 60
Buses in IS32 Optic RAM, 25
Busy signals of telephones, 162

C compiler for Trump Card, 141, 153, 157–158
Capacitors:
  in power-line filters, 76
  in ring-detector circuits, 164
  in SonarTape ultrasonic ranging system, 188
Carrier-current modem, 12–18
CGC (Zilog Z8581 clock generator and controller), 148–149
Change-of-state detection, 212–213
Characters:
  blinking, 97–100

Characters (*Cont.*):
  display of, 86–87
  double-width, double-height, and reverse-video, 100
  processed by Term-Mite ST Smart Terminal, 92–93
  scrolling alphanumeric LED display for, 121–137
Charged-coupled-device (CCD) arrays, 20
Chebyshev distance measurement, 200
Chimes in musical telephone bell, 167
Christensen, Ward, 18
Circuit boards, 9
Clock rates:
  of RTC-4 real-time controller, 7
  in Trump Card, 148–149
Clock signals:
  in Micro D-Cam solid-state video camera, 24
  in RTC-4 real-time controller, 7
  in SonarTape ultrasonic ranging system, 188–190
  for SS1263 speech-synthesis chip, 109
Clocks, RTC-4 real-time controllers as, 7
CMOS (complementary metal-oxide semiconductor) chips:
  in Micro D-Cam solid-state video camera, 24
  ring detectors, 164
  in SonarTape ultrasonic ranging system, 187
  in SS1263 speech-synthesis chip, 109
Color:
  from Micro D-Cam solid-state video camera, 55
  in TBASIC, 155
Command registers, 25
Commands:
  for Lis'ner 1000 speech-recognition system, 204–205
  for Micro D-Cam solid-state video camera, 24–25, 30, 34–35
  in TBASIC, 155, 157
  for Trump Card, 154
  voice input of, 204
Commodore 64 computers, Lis'ner 1000 speech-recognition system for, 195, 197
Communications:
  H-Com handicapped communicator for, 56–70
  musical telephone bell for, 161–167
  power-line carrier-current modem for, 12–18

221

Communications (Cont.):
  between Trump Card and host computer, 149–150
  voice recognition applications for, 206
Compilers:
  C, for Trump Card, 157–158
  interpreters versus, 154–155
  TBASIC, 140, 153
  Y, for Trump Card, 158–159
Computers:
  AC power monitoring by, 176–179
  electric power line protection for, 71–79
  languages for, 138–139
  Micro D-Cam solid-state video camera linked to, 23–24
  real-world interfaces for, 207
  SonarTape ultrasonic ranging system connections to, 185–187
  speech generated by, 108–109
  speech recognition by, 193
  Term-Mite ST Smart Terminal for, 66–77, 92–106
  Trump Card as, 140
  Vidicon cameras and, 20
Configuration:
  of Micro D-Cam solid-state video camera, 32
  of Term-Mite ST Smart Terminal, 95–96
Configuration switches, 96
Connected speech recognition, 194
Continuous speech recognition, 194
Control applications, 207–208
  AC power output control devices, 215–216
  DC power output control devices, 213–215
  RTC-4 real-time controller for, 1–11
Control codes for Term-Mite ST Smart Terminal, 94, 95
Controllers, RTC-4 real-time, 1–11
Corcom power-line filters, 79
Correlation cancellation loops, 196
Counters on Intel 8748 microprocessor, 60
CP/M-80 emulator for Trump Card, 140–141, 153, 159
Crowbar circuits, 77–78
CRT-controller (video-controller) chips, 81
Crystals on Intel 8748 microprocessor, 60
Current, measurement of, 171–174
Cursor, National Semiconductor NS455A Terminal-Management Processor (TMP) control of, 85

Data communications (see Communications)
DC input receivers, 211–212
DC power output control devices, 213–215
Debugger for Trump Card, 141, 153
Demodulators in power-line carrier-current modem, 15–18

Dialing of telephones, 162
Digital-to-analog (D/A) conversion of human voice, 108
Dilks, John, 121
DIP (dual-inline pin) switches on Term-Mite ST Smart Terminal, 95
Discrete-input AC/DC receivers, 211–212
Discrete-utterance recognition, 193–194
Disk emulator for Trump Card, 141, 153
Display drivers in National Semiconductor NS455A Terminal-Management Processor (TMP), 85–86
Displays, 86–87
  on H-Com handicapped communicator, 63
  of image created by IS32 Optic RAM, 30
  from Micro D-Cam solid-state video camera, on Apple II computers, 50, 53
  from RTC-4 real-time controller, 7–9
  scrolling alphanumeric LED, 121–137
  of Term-Mite ST Smart Terminal, 88
Distance:
  SonarTape ultrasonic ranging system to measure, 181–191
  between speech-recognition templates, 199–201
Double-width and double-height characters, 100
Driver-coupler circuits, 14
Dual-tone multifrequency (DTMF; Touch-Tone) telephone dialing, 162
  decoders for, 206
Dynamic memory, electric transients and, 72
Dynamic time warping (DTW), 202

Electric meters, 171
Electric power lines:
  AC power monitor for, 169–179
  AC/DC power I/O system for, 208–220
  protection for, 71–79
Electrical noise, 76
Electromagnetic interference (EMI), power-line filters for, 76
Electronic tape measure (SonarTape ultrasonic ranging system), 187–190
Encoded keyboards, 93
EPROMs (erasable programmable read-only memories):
  in H-Com handicapped communicator, 63
  in Intel 8748 microprocessor, 57
  in Term-Mite ST Smart Terminal, 87–88, 105–106
  for Trump Card, 148, 150
Error-checking systems, 18
Error messages in TBASIC, 155

Errors:
  indicated on RTC-4 real-time controller, 9
  in speech recognition, 202
  in TBASIC, 155
Escape sequences, 93–94
  for Term-Mite ST Smart Terminal, 95–97
Exar Integrated Systems:
  XR-2206 modulator by, 15
  XR-2211 demodulator by, 15–18

Fast Fourier transforms (FFTs), 139
FIFO (first-in, first-out) buffers, 86
Filenames for CP/M-80 emulator, 159
Filters:
  in Lis'ner 1000 speech-recognition system, 198
  power-line, 76–77
  on power strips, 78–79
Fixed-time programs, 5, 8
Floating-point arithmetic in TBASIC, 156
Formant synthesis, 108
Formants, 108
Frequencies, setting, in power-line carrier-current modem, 15–18
Frequency-shift keying (FSK), 13
  by power-line carrier-current modem, 15–16
Full-duplex communications, 13

General Electric Company, metal-oxide varistors (MOVs) by, 78
General Instrument, 108
  AY-3-1350 tone synthesizer by, 167
  software for Lis'ner 1000 speech-recognition system by, 205
  SP1000 voice-recognition chip by, 195–199
Gordos Arkansas Inc., 217
Graphics:
  line, on Term-Mite ST Smart Terminal, 100–101
  in TBASIC, 155
Gray scale from Micro D-Cam solid-state video camera, 39–40, 50–54
GREY16 program, 39–49

Half-duplex communications, 13
Handicapped communicator, 56–70
Hardware:
  change-of-state detection by, 212
  for H-Com handicapped communicator, 63
  for Lis'ner 1000 speech-recognition system, 195–197
  for Micro D-Cam solid-state video camera, 22–30
  for RTC-4 real-time controller, 5–7
  for Term-Mite ST Smart Terminal, 80–91
  for Trump Card, 138–152
H-Com handicapped communicator, 56–70
Helmers, Carl, 121

High accumulator, 85
High-level languages, 138–140
Horizontal synchronization (sync) pulse, 87
Horn on H-Com handicapped communicator, 62, 63

IBM Personal Computers:
  interfacing Micro D-Cam solid-state video camera to, 32–34
  Trump Card for, 140, 150, 153
Image created by IS32 Optic RAM, 29, 30
Infrared light, 55
Inputs:
  AC/DC power I/O system for, 208–220
  to H-Com handicapped communicator, 63
  to scrolling LED display, 127–129
  to text-to-speech algorithm, 115
  visual, 20
  voice for, 204
Instruction sets:
  for Intel 8748 microprocessor, 60, 64
  for National Semiconductor NS455A Terminal-Management Processor (TMP), 82, 85
  for Zilog Z8000 microprocessor, 141
Intel 8008 microprocessor, 80
Intel 8048 microprocessor, 57, 83–85
Intel 8087 numeric processor extension (NPX), 156
Intel 8088 microprocessor, Trump Card and, 149
Intel 8275 CRT controller, 81
Intel 8748 microprocessor, 57–60, 62–70
Interdigit time in telephone dialing, 162
Interfaces:
  AC/DC power I/O system, 208–220
  for Micro D-Cam solid-state video camera, 23–24, 32–34
  for real world, 207
  between Trump Card and host computer, 149–150
Interference, electrical noise as, 76
Interpreters, compilers versus, 154–155
Interrupts:
  in IS32 Optic RAM, 28
  in Micro D-Cam solid-state video camera, 24
  in Term-Mite ST Smart Terminal, 92–93
  to Zilog Z8001 microprocessor, 149–150
Interval timer programs, 5, 8
Intex Micro Systems Corporation, 70
Intonation in simulated speech, 115, 117–119
I/O ports (see Ports)
I/O registers (see Registers)
I/O signal lines (see Signal lines)

IS32 Optic RAM (Micron Technology), 21–29, 31
Isolation in AC/DC power interface, 209–210
  optocouplers for, 210–211

Kernighan, Brian W., 141, 157, 158
Keyboards:
  H-Com handicapped communicator substitute for, 57
  read by Term-Mite ST Smart Terminal, 92
  scanned and encoded, 93
  scanning, 94–95
  for Term-Mite ST Smart Terminal, 88
  voice-operated, Lis'ner 1000 speech-recognition system for, 196
Keypads on RTC-4 real-time controller, 4, 7
Kilowatt-hours, 171

Lancaster, Don, 80
Languages, 138–140
  C, for Trump Card, 157–158
  interpreters versus compilers for, 154–155
  supported by Trump Card, 153
  TBASIC, 155–157
  Y, for Trump Card, 158–159
LDZSYS program for Trump Card, 150, 154
LED displays:
  on AC power monitor, 174
  on H-Com handicapped communicator, 63
  scrolling alphanumeric, 121–137
  in transistor optocouplers, 210, 211
Lenses for Micro D-Cam solid-state video camera, 23
Light-emitting diode displays (see LED displays)
Light pen input to Term-Mite ST Smart Terminal, 91
Lighting, power usage for, 169
Lightning, 71–76
Lightning rods, 73, 74
Line graphics on Term-Mite ST Smart Terminal, 100–101
Line numbers in TBASIC programs, 155
Linear-predictive coding (LPC):
  in SP1000 voice-recognition chip, 196
  for speech recognition, 194–195
  for speech synthesis, 108
Lis'ner 1000 speech-recognition system, 193–206
Loop-damping factor, 18
LSTTL (low-power Schottky transistor-transistor logic), 203

Mark 8 computer, 80
Memory:
  dynamic, electric transients and, 72

Memory (Cont.):
  on H-Com handicapped communicator, 63
  IS32 Optic RAM, 20–29, 31
  in National Semiconductor NS455A Terminal-Management Processor (TMP), 83–85
  throughput and, 139
  on Trump Card, 140, 150, 153
Metal-oxide varistors (MOVs), 78
Micro D-Cam solid-state video camera, 31–32, 36–49
  hardware and RAM for, 20–30
  interfacing, 32–34
  software for, 34–35, 50–55
Micromint, components available from:
  for AC/DC power interface, 220
  for Lis'ner 1000 speech-recognition system, 206
  for Micro D-Cam solid-state video camera, 30, 55
  for RTC-4 real-time controller, 11
  for scrolling LED display, 137
  for SonarTape ultrasonic ranging system, 191
  for Term-Mite ST Smart Terminal, 91, 106
  Whimsi-Bell kit, 167
Micron Technology, Inc., IS32 Optic RAM from, 21–29, 31
Microphones in Lis'ner 1000 speech-recognition system, 197–198
Microprocessors:
  Intel 8008, 80
  Intel 8048, 43, 69–71
  Intel 8748, 57–60, 62–70
  in National Semiconductor NS455A Terminal-Management Processor (TMP), 83–85
  performance of, 139–140
  in SP1000 voice-recognition chip, 196
  in terminals, 80–82
  in TMS1000 series (Texas Instruments), 1
  Zilog Z8, 128
  Zilog Z80, 145
  Zilog Z8000, 141–148
  Zilog Z8001, 126–134, 139
Microvox text-to-speech synthesizer, 107
Modems:
  power-line carrier-current, 12–18
  ring-detector chips in, 166
Modulators in power-line carrier-current modem, 15
Monitors for Term-Mite ST Smart Terminal, 105
MPX-16 computer system, 150–151
Multilevel language compiler for Trump Card, 141, 153, 158–159
Multiplexing in scrolling LED display, 124
Musical telephone bell, 161–167

National Semiconductor:
  DP8350 CRT controller by, 81

National Semiconductor (*Cont.*):
  NS455A Terminal-Management Processor (TMP) by, 82–87, 92–94, 105
Next-address bus, 25
NMOS (negative-channel metal-oxide semiconductor) chips in SP1000 voice-recognition chip, 196
Noise, electrical, 76
Numeric processor extension (NPX; Intel 8087), 156

Oak full-travel membrane (FTM) keyboard, 95
Object code, 154, 155
Optic RAM, 21–29, 31
Optocouplers, 210–211
Optoisolators, 210
Ordered dithering, 53–54
Outputs:
  AC/DC power I/O system for, 208–220
  from H-Com handicapped communicator, 57
  from IS32 Optic RAM, 31
  from RTC-4 real-time controller, 4–8
  from Term-Mite ST Smart Terminal, 88–91
  from video cameras, 20

Parallel interfaces, 208
Parallel ports on H-Com handicapped communicator, 63
Parameters:
  for Lis'ner 1000 speech-recognition system, 199
  for Micro D-Cam solid-state video camera, 34
  for Term-Mite ST Smart Terminal, 95
  of text-to-speech algorithm, 115
Performance:
  of microprocessors, 139–140
  of TBASIC compiler, 155
Phonemes, 108
  programming for synthesis of, 110–111
  rules for, in text-to-speech algorithm, 115–117
Phonetic speech synthesizer, 107–119
Photodiode arrays, 20–21
PIC 1655A microcomputer, 167
Pixels, 86
  in IS32 Optic RAM, 23
  in Micro D-Cam solid-state video camera's display, 53–54
  in resolution of video cameras, 20
Polaroid electrostatic transducer, 182
Polaroid Ultrasonic Ranging System Designer's Kit, 181
Ports:
  for AC/DC power interface, 217
  connecting SonarTape ultrasonic ranging system to, 185

Ports (*Cont.*):
  on H-Com handicapped communicator, 63
  on Intel 8748 microprocessor, 60
  in National Semiconductor NS455A Terminal-Management Processor (TMP), 85
  parallel, 208, 209
  on Z8-BASIC System Controller, 128
Power I/O system, 208–220
Power-line carrier-current modem, 12–18
Power-line filters, 76–77
Power-line irregularities, 72
Power monitor, 169–179
Power output control devices:
  AC, 215–216
  DC, 213–214
Power strips:
  AC power monitor in, 171–172, 175
  filtered, 78–79
Power supplies:
  for AC power monitor, 172
  for H-Com handicapped communicator, 63
  for RTC-4 real-time controller, 7
  for SonarTape ultrasonic ranging system, 186–188
  uninterruptible (UPS), 76
Present-address bus, 25
Prime-number-generator Sieve of Eratosthenes program as benchmark, 140
  for TBASIC, 155–156
Processors:
  in National Semiconductor NS455A Terminal-Management Processor (TMP), 83–85
  performance of, 139–140
  for scrolling LED display, 128–137
  (*See also* Microprocessors)
Programming:
  languages for, 138–139
  for phoneme synthesis, 108, 110–111
  of RTC-4 real-time controller, 7
  of scrolling LED display, 128
  of Term-Mite ST Smart Terminal, 96–101
  (*See also* Languages)
Programs:
  for AC power monitor, 176–179
  in C, 157–158
  in CP/M-80, run on Trump Card, 159
  for H-Com handicapped communicator, 64
  for Micro D-Cam solid-state video camera, 35, 50
  for RTC-4 real-time controller, 4–5, 8
  running, using BASIC compilers, 154
  in TBASIC, 155–156
  for Trump Card, 140, 150
  (*See also* Software)

Protection, electrical, 71–79
Pulse-code modulation (PCM), 108
Pulse telephone dialing, 162

Quasibidirectional signal lines, 60

Radio frequency interference (RFI), power-line filters for, 76
RAM (random-access memory):
  in Intel 8748 microprocessor, 60
  optically sensitive, 21–29, 31
RAM disks, Trump Card for, 140, 141, 150, 153
Ranging system, ultrasonic (SonarTape), 181–191
Real-time controller, RTC-4, 1–11
Real-world interfaces, 207
Refreshing:
  of displays, 86
  of IS32 Optic RAM, 22, 26, 35
  registers for, 25
Registers:
  in Intel 8048 microprocessor, 84
  in Intel 8748 microprocessor, 57–60
  in Lis'ner 1000 speech-recognition system, 197
  in Micro D-Cam solid-state video camera, 25–26, 34
  in National Semiconductor NS455A Terminal-Management Processor (TMP), 87
  in scrolling LED display, 125–126
  in SS1263 speech-synthesis chip, 108, 109, 117
  in Zilog Z8000 microprocessor, 141
Rejection thresholds in speech recognition, 202–204
Relays:
  in AC/DC power interface, 209
  controlled by RTC-4 real-time controller, 7–8
  solid-state (SSRs), 213
Remote control:
  RTC-4 real-time controller for, 1–11
  using AC wiring, 12
  (*See also* Control applications)
Resident data memory in Intel 8748 microprocessor, 57–60
Resistors:
  in AC power monitor, 172–175
  in discrete-input AC/DC receivers, 212
  in power-line carrier-current modem, 15
  in ring-detector circuits, 164, 166
Reverse-video characters, 100
Ring detectors, 163–164
  chips for, 164–167
Ring signals of telephones, 162–163
Ritchie, Dennis M., 141, 157, 158
ROM (read-only memory):
  BASIC interpreters in, 138–139
  erasable programmable (*see* EPROMs)

ROM (read-only memory) (Cont.):
  in National Semiconductor NS455A Terminal-Management Processor (TMP), 85, 92
RS-232C connectors for power-line carrier-current modem, 14
RTC-4 real-time controller, 1–11
Rule editor in text-to-speech algorithm, 116–117

Sags, electrical, 72–74
  from lightning, 75–76
Scanning communicators, 57
Scanning keyboards, 93–95
SCR relay devices, 213–215
Screen editor for Trump Card, 141, 153, 155
Scrolling alphanumeric LED display, 121–137
Serial interfaces for Micro D-Cam solid-state video camera, 32
SETRMDSK program for RAM disks, 150
Sieve of Eratosthenes (prime-number-generator) program as benchmark, 140, 155–156
Signal lines:
  in Intel 8748 microprocessor, 60
  in Trump Card, 148, 149
Silicon Systems Inc., SS1263 speech-synthesis chip by, 105, 107–114, 117, 119
Smart terminals, Term-Mite ST Smart Terminal, 80–91, 92–106
SN28827 sonar-ranging module (Texas Instruments), 181
Snubber networks, 216
Software:
  for AC power monitor, 176–179
  for AC/DC power interface, 219
  change-of-state detection by, 212–213
  for H-Com handicapped communicator, 64–70
  for Lis'ner 1000 speech-recognition system, 196, 199, 204–205
  for Micro D-Cam solid-state video camera, 31, 34–35, 50–55
  for scroller LED display, 128
  for speech synthesis, 110–119
  for Term-Mite ST Smart Terminal, 92–106
  for Trump Card, 140–141, 150, 153–160
Solid-state relays (SSRs), 213
SonarTape ultrasonic ranging system, 181–191
Sound:
  Lis'ner 1000 speech-recognition system for, 193–206
  musical telephone bell for, 161–167
  phonetic speech synthesizer for, 107–119
  SonarTape ultrasonic ranging system use of, 181–182
Source code, 154, 155

SP1000 voice-recognition chip (General Instrument), 195–199
Spark-gap devices, 77–78
Speaker-dependent speech-recognition systems, 193
Speaker-independent speech-recognition systems, 193
Speakers on H-Com handicapped communicator, 63
Speech-recognition system, Lis'ner 1000, 193–206
Speech synthesizers:
  phonetic, 107–119
  SP1000 voice-recognition chip used for, 197
Speed of TBASIC execution, 155–156
Spikes, voltage, 74
SS1263 speech-synthesis chip (Silicon Systems Inc.), 107–114, 119
  registers on, 117
Stacks:
  in Intel 8048 microprocessor, 84
  in Intel 8748 microprocessor, 64
Status signals in Trump Card, 148
Stress markers in text-to-speech algorithms, 117–119
Surges, electrical, 72, 74
Sweet Micro Systems, 114, 115
  Trump Card components available from, 151–152, 160
Sweet Talker voice synthesizer, 107
Sweet Talker II (phonetic speech synthesizer), 108–119
Switches:
  configuration, on Term-Mite ST Smart Terminal, 96
  controlled by RTC-4 real-time controller, 7–8
  DIP, on Term-Mite ST Smart Terminal, 95
  on H-Com handicapped communicator, 57
  on scanning keyboards, 94
  zero-voltage (ZVS), 216

Tape recording, 10–11
TBASIC (language), 140, 153, 156–157
  compiler for, 154–155
  programs in, 155–156
TCM1512A ring-detector chip, 167
Telephones:
  modems for use with, 13
  musical bell for, 161–167
Terminals:
  for programming of scrolling LED display, 136
  Term-Mite ST Smart Terminal, 80–91, 92–106
Term-Mite ST Smart Terminal:
  hardware for, 80–91
  programming and use of, 92–106
Texas Instruments:
  linear-predictive coding of speech in products by, 108
  ring-detector chips by, 164, 166

Texas Instruments (Cont.):
  SN28827 sonar-ranging module from, 181
  sonar-ranging circuit board from, 181
  TL851 and TL852 sonar-ranging controller/receiver chip set by, 182
  TMS 1000 series microcomputers-on-a-chip by, 1
  TMS1100 microprocessor chip by, 1
  TMS1121C Universal Timer Controller by, 1–4, 7, 11
Text-to-speech algorithms, 111–116
  customizing, 114–115
  intonation in, 117–119
  in Lis'ner 1000, 195
  rule editor for, 116–117
Throughput, 139–140
Time:
  alignment of, in speech recognition, 201–202
  in measurement of power consumption, 171
Timers:
  in Intel 8748 microprocessor, 60
  RTC-4 real-time controller, 1–11
Timing in SonarTape ultrasonic ranging system, 184, 188
TL851 and TL852 sonar-ranging controller/receiver chip set (Texas Instruments), 182
TMP (National Semiconductor NS455A Terminal-Management Processor), 82–87, 92–94, 105
TMS1000 series microcomputers-on-a-chip (Texas Instruments), 1
TMS1100 microprocessor chip (Texas Instruments), 1
TMS1121C Universal Timer Controller (Texas Instruments), 1–4, 7, 11
Tone synthesizer (General Instrument AY-3-1350), 167
Touch-Tone telephone dialing, 162
  decoders for, 206
Training:
  in Lis'ner 1000 speech-recognition system, 199, 204
  in speech-recognition units, 193
Transducers in SonarTape ultrasonic ranging system, 182
Transformers:
  in AC power monitor, 172
  for power-line carrier-current modem, 14
Transients, voltage (see Voltage transients)
Transistor optocouplers, 210
Transistor optoisolators, 211
Trump Card:
  hardware for, 138–152
  software for, 153–160
Tuned couplers, 14
TV Typewriter, 80

UART (universal asynchronous receiver/transmitter), 85, 93

Ultrasonic ranging system (SonarTape), 181–191
Unconnected speech recognition, 193–194
Uninterruptible power supply (UPS), 76

Vertical synchronization (sync) pulse, 87
Video camera, Micro D-Cam solid-state, 31–32
  hardware and RAM for, 20–30
  interfacing, 32–34
  software for, 34–35, 50–55
Video-controller (CRT-controller) chips, 81
Vidicon cameras, 20
Voice-recognition chip (General Instrument SP1000), 195–199
Voice-recognition system, Lis'ner 1000, 193–206
Voice synthesis, 195, 197
Voltage:
  in AC/DC power interface, 209
  fluctuations in, 72–74

Voltage (*Cont.*):
  in measuring electric power consumption, 147
  of telephones, 162–163
  as threshold for IS32 Optic RAM, 23
  transient suppressors for, 77–78
  zero-voltage switch (ZVS) for, 216
Voltage-clamping devices, 78
Voltage spikes, 74
Voltage transients, 58–61
  optocouplers for suppression of, 200
  snubber networks for suppression of, 206
  suppressors for, 63–64
Votrax SC-01A speech-synthesis circuit, 107, 108, 112

Watts, 171
Waveform digitization, 108
Western Electric, telephones by, 161
Whimsi-Bell (musical telephone bell), 167

Wiring, AC power, communicating through, 12

XR-2206 modulator (Exar), 15
XR-2211 demodulator (Exar), 15–18

Y multilevel language compiler for Trump Card, 153, 158–159

Z8 system-controller board, 186–187, 217
Z8-BASIC System Controller, 128
Zero-voltage switch (ZVS), 216
Zilog Z8 microprocessor, 128
Zilog Z80 microprocessor, 159
Zilog Z8000 microprocessor, 141–148
Zilog Z8001 microprocessor, 140–148, 153
Zilog Z8581 clock generator and controller (CGC), 148–149

DATE D

This book may be kept

FOURTEEN DAYS

A fine will be charged for each day the book is kept overtime.

AUG 21 1992			
AUG 24 1992			
~~MAR~~			
~~AUG 09~~			
GAYLORD 142			PRINTED IN U.S.A